Solid Dispersions for Drug Delivery: Applications and Preparation Methods

Solid Dispersions for Drug Delivery: Applications and Preparation Methods

Editors

Vitaliy Khutoryanskiy
Hisham Al-Obaidi

MDPI • Basel • Beijing • Wuhan • Barcelona • Belgrade • Manchester • Tokyo • Cluj • Tianjin

Editors

Vitaliy Khutoryanskiy
School of Pharmacy
University of Reading
Reading
United Kingdom

Hisham Al-Obaidi
School of Pharmacy
University of Reading
Reading
United Kingdom

Editorial Office
MDPI
St. Alban-Anlage 66
4052 Basel, Switzerland

This is a reprint of articles from the Special Issue published online in the open access journal *Pharmaceutics* (ISSN 1999-4923) (available at: www.mdpi.com/journal/pharmaceutics/special_issues/solid_dispersions).

For citation purposes, cite each article independently as indicated on the article page online and as indicated below:

LastName, A.A.; LastName, B.B.; LastName, C.C. Article Title. *Journal Name* **Year**, *Volume Number*, Page Range.

ISBN 978-3-0365-2639-3 (Hbk)
ISBN 978-3-0365-2638-6 (PDF)

Contents

About the Editors

Vitaliy Khutoryanskiy

Vitaliy Khutoryanskiy is the Professor of Formulation Science at the School of Pharmacy, University of Reading, United Kingdom. He is one of the leading researchers in the pharmaceutical and biomedical application of polymers. Prof. Khutoryanskiy's research interests include transmucosal drug delivery (ocular, nasal, oromucosal, gastrointestinal, vaginal, and intravesical), mucoadhesive materials, mucosa penetration, hydrogels, nanoparticles, pharmaceutical polymers, stimuli-responsive polymers, and encapsulation technologies. He has published over 172 original research papers and 26 reviews.

Hisham Al-Obaidi

Hisham Al-Obaidi is a lecturer in Pharmacy and Pharmaceutical Sciences at the School of Pharmacy, University of Reading, United Kingdom. He has an extensive track record in developing solid dispersions for applications in oral and pulmonary drug delivery. Dr. Al-Obaidi's research interests include the assessment of drug-polymer compatibility using thermal analysis methods and the development of particle engineering methods for novel applications such as lung infections and cancer.

Preface to "Solid Dispersions for Drug Delivery: Applications and Preparation Methods"

The use of solid dispersions has expanded in recent years to include different pharmaceutical applications such as oral drug delivery and drug delivery to the lungs. The success of this technology has allowed this expansion and requires an update of current understandings. Thus, the aim of this book is to summarize the achievements in the area of solid dispersions and suggest potential future applications. The formation of solid dispersions, where poorly soluble active pharmaceutical ingredients are dispersed in a solid carrier, is one of the strategies to improve drug dissolution profiles and to enhance bioavailability. These solid drug dispersions can be formulated with the use of water-soluble or -insoluble polymers as well as some materials with small molecular weight such as surfactants, sugars, and other drugs. Recent advances in this area of research have shown the potential of the application of co-amorphous and polymeric amorphous solid dispersions to enhance solubility as well as to deliver combination treatments. Likewise, pharmaceutical co-crystals have attracted significant interest to deliver combination treatments. The drug present in amorphous dispersions typically undergoes rapid dissolution, which may result in its better bioavailability. There are several methods used to prepare solid drug dispersions, which include solvent evaporation, melting, and supercritical fluid technology. Numerous solvent evaporation approaches may also involve the use of spray-drying, lyophilization, co-precipitation, electrospraying, electrospinning, etc. There are also several melting approaches such as hot-melt extrusion and melt agglomeration.

This book represents a collection of three reviews and seven original research papers contributed by different authors. Elshadeey and El-Dahmy [1] reported the formulation of oral fast-dissolving films containing nanosuspension of paroxetine. They characterized these films using in vitro techniques, studied ex vivo drug permeation, and demonstrated their applicability for buccal drug delivery in healthy human volunteers. Mudie and co-workers [2] described the streamlined in vitro and in silico approaches to formulate tablets based on solid dispersions of Calquence® (crystalline acalabrutinib). Shan et al. [3] reported the use of poly(N-vinyl pyrrolidone), poly(2-methyl-2-oxazoline), poly(2-ethyl-2-oxazoline), poly(2-n-propyl-2-oxazoline), and poly(2-isopropyl-2-oxazoline) to formulate solid dispersions with ibuprofen. They explored the effects of hydrogen bonding and polymer hydrophobicity on the inhibition of ibuprofen crystallization. In another study, Mudie and co-workers [4] formulated Calquence® as immediate-release tablets with spray-dried amorphous solid drug dispersions comprising 50% acalabrutinib and 50% hydroxypropyl methylcellulose acetate succinate, and they studied these in dogs. El-Helaly et al. [5] reported the development of solid dispersions of raloxifene using multilayered core-sheath nanofibers using an electrospinning technique and formulated it in the form of buccal films. Eedara et al. [6] prepared solid dispersions of fexofenadine hydrochloride using poly(ethylene glycol) 20 kDa and Poloxamer 188 and reported their improved dissolution properties. Oliveira et al. [7] reported the use of hydroxypropyl methylcellulose, poly(ethylene glycol) and poly(vinyl pyrrolidone) for formulating solid dispersions of naphthoquinone with physical mixing, kneading, and rotary evaporation methods. In addition to physicochemical characterization of these dispersions, they also evaluated antiparasitic activity in Trypanosoma cruzi Y strains in vitro.

Al-Obaidi et al. [8] reviewed recent advancements in the use of solid dispersions of anticancer and antimicrobial drugs for drug delivery to the lungs. The review covered the current state of the art for formulating drugs for inhalation with a focus on co-amorphous solid dispersions and cocrystals to deliver inhaled drugs. Qian et al. [9] reviewed the dissolution of amorphous solid dispersions and the theory of the formation of metastable drug-rich phases. The authors discussed the additional benefits of amorphous solid dispersions that induced drug-rich phases for bioavailability enhancements. Oliveira et al. [10] reviewed recent advancements in therapeutical applications of solid dispersions. The authors showed the breadth of applications of solid dispersions such as anticancer, anti-inflammatory, and antimicrobial applications that were reported in previous in vitro and in vivo studies.

This book can be particularly useful for researchers as well as postgraduate students in formulation sciences and drug delivery. Undergraduate students will also find elements of this book very relevant to scientific fundamentals such as solubility and crystallization of amorphous materials as well as drug delivery challenges.

References

1. Elshafeey AH, El-Dahmy RM. Formulation and Development of Oral Fast-Dissolving Films Loaded with Nanosuspension to Augment Paroxetine Bioavailability: In Vitro Characterization, Ex Vivo Permeation, and Pharmacokinetic Evaluation in Healthy Human Volunteers. Pharmaceutics. 2021;13(11):1869.

2. Mudie DM, Stewart AM, Rosales JA, Adam MS, Morgen MM, Vodak DT. In Vitro-In Silico Tools for Streamlined Development of Acalabrutinib Amorphous Solid Dispersion Tablets. Pharmaceutics. 2021;13(8):1257.

3. Shan X, Moghul MA, Williams AC, Khutoryanskiy VV. Mutual Effects of Hydrogen Bonding and Polymer Hydrophobicity on Ibuprofen Crystal Inhibition in Solid Dispersions with Poly(N-vinyl pyrrolidone) and Poly(2-oxazolines). Pharmaceutics. 2021;13(5):659.

4. Mudie DM, Stewart AM, Rosales JA, Biswas N, Adam MS, Smith A, Craig CD, Morgen MM, Vodak DT. Amorphous Solid Dispersion Tablets Overcome Acalabrutinib pH Effect in Dogs. Pharmaceutics. 2021;13(4):557.

5. Nageeb El-Helaly S, Abd-Elrasheed E, Salim SA, Fahmy RH, Salah S, EL-Ashmoony MM. Green Nanotechnology in the Formulation of a Novel Solid Dispersed Multilayered Core-Sheath Raloxifene-Loaded Nanofibrous Buccal Film; In Vitro and In Vivo Characterization. Pharmaceutics. 2021;13(4):474.

6. Eedara BB, Nyavanandi D, Narala S, Veerareddy PR, Bandari S. Improved Dissolution Rate and Intestinal Absorption of Fexofenadine Hydrochloride by the Preparation of Solid Dispersions: In Vitro and In Situ Evaluation. Pharmaceutics. 2021;13(3):310.

7. Oliveira VdS, Dantas ED, Queiroz ATdS, Oliveira JWdF, Silva MdSd, Ferreira PG, Siva FdCd, Ferreira VF, Lima ÁANd. Novel Solid Dispersions of Naphthoquinone Using Different Polymers for Improvement of Antichagasic Activity. Pharmaceutics. 2020;12(12):1136.

8. Al-Obaidi H, Granger A, Hibbard T, Opesanwo S. Pulmonary Drug Delivery of Antimicrobials and Anticancer Drugs Using Solid Dispersions. Pharmaceutics. 2021;13(7):1056.

9. Qian K, Stella L, Jones DS, Andrews GP, Du H, Tian Y. Drug-Rich Phases Induced by Amorphous Solid Dispersion: Arbitrary or Intentional Goal in Oral Drug Delivery? Pharmaceutics. 2021;13(6):889.

10. Oliveira VdS, de Almeida AS, Albuquerque IdS, Duarte FÍC, Queiroz BCSH, Converti A, Lima ÁANd. Therapeutic Applications of Solid Dispersions for Drugs and New Molecules: In Vitro and In Vivo Activities. Pharmaceutics. 2020;12(10):933.

Vitaliy Khutoryanskiy, Hisham Al-Obaidi
Editors

pharmaceutics

Article

Formulation and Development of Oral Fast-Dissolving Films Loaded with Nanosuspension to Augment Paroxetine Bioavailability: In Vitro Characterization, Ex Vivo Permeation, and Pharmacokinetic Evaluation in Healthy Human Volunteers

Ahmed Hassen Elshafeey [1,*] and Rania Moataz El-Dahmy [2]

1 Department of Pharmaceutics and Industrial Pharmacy, Faculty of Pharmacy, Cairo University, Cairo 11562, Egypt
2 Department of Pharmaceutics and Industrial Pharmacy, Faculty of Pharmacy, October 6 University, Central Axis, Cairo 12585, Egypt; rania-moataz@o6u.edu.eg
* Correspondence: ahmed.elshafeey@pharma.cu.edu.eg; Tel.: +20-100-584-0261

check for updates

Citation: Elshafeey, A.H.; El-Dahmy, R.M. Formulation and Development of Oral Fast-Dissolving Films Loaded with Nanosuspension to Augment Paroxetine Bioavailability: In Vitro Characterization, Ex Vivo Permeation, and Pharmacokinetic Evaluation in Healthy Human Volunteers. *Pharmaceutics* **2021**, *13*, 1869. https://doi.org/10.3390/pharmaceutics13111869

Academic Editors: Vitaliy Khutoryanskiy and Hisham Al-Obaidi

Received: 18 September 2021
Accepted: 1 November 2021
Published: 5 November 2021

Abstract: Paroxetine (PX) is the most potent serotonin reuptake inhibitor utilized in depression and anxiety treatment. It has drawbacks, such as having a very bitter taste, low water solubility, and undergoing extensive first pass metabolism, leading to poor oral bioavailability (<50%). This work aimed to develop and optimize palatable oral fast-dissolving films (OFDFs) loaded with a paroxetine nanosuspension. A PX nanosuspension was prepared to increase the PX solubility and permeability via the buccal mucosa. The OFDFs could increase PX bioavailability due to their rapid dissolution in saliva, without needing water, and the rapid absorption of the loaded drug through the buccal mucosa, thus decreasing the PX metabolism in the liver. OFDFs also offer better convenience to patients with mental illness, as well as pediatric, elderly, and developmentally disabled patients. The PX nanosuspension was characterized by particle size, poly dispersity index, and zeta potential. Twelve OFDFs were formulated using a solvent casting technique. A $2^2 \times 3^1$ full factorial design was applied to choose the optimized OFDF, utilizing Design-Expert® software (Stat-Ease Inc., Minneapolis, MN, USA). The optimized OFDF (F1) had a 3.89 ± 0.19 Mpa tensile strength, 53.08 ± 1.28% elongation%, 8.12 ± 0.13 MPa Young's modulus, 17.09 ± 1.30 s disintegration time, and 96.02 ± 3.46% PX dissolved after 10 min. This optimized OFDF was subjected to in vitro dissolution, ex vivo permeation, stability, and palatability studies. The permeation study, using chicken buccal pouch, revealed increased drug permeation from the optimized OFDF; with a more than three-fold increase in permeation over the pure drug. The relative bioavailability of the optimized OFDF in comparison with the market tablet was estimated clinically in healthy human volunteers and was found to be 178.43%. These findings confirmed the success of the OFDFs loaded with PX nanosuspension for increasing PX bioavailability.

Keywords: paroxetine; nanosuspension; oral fast dissolving film; factorial design; rapid release; ex vivo permeation; bioavailability

1. Introduction

Paroxetine (PX) is a phenylpiperidine compound and is the most selective and potent serotonin reuptake inhibitor. Additionally, it blocks muscarinic acetylcholine receptors and is even more effective than maprotiline or desipramine. PX is used to treat anxiety, depression, panic, obsessive-compulsive disorders, and various psychiatric disorders [1]. Although PX is absorbed well from the gastrointestinal tract, it has drawbacks such as a very bitter taste, low water solubility, and undergoing extensive metabolism in the liver, resulting in low oral bioavailability (about 30–50%) [2]. Some efforts have been made to increase PX bioavailability by developing PX in the form of intranasal nanoemulsions [2], transdermal liposomes [3], and as buccoadhesive bi-layer tablets [4]. Nanosuspension systems have many advantages over other drug delivery systems because of their efficient solubility and bioavailability enhancement of almost all low-soluble drugs and because of not utilizing the excipients utilized in other solubility enhancement systems [5]. Successes with nanosuspension formulations in enhancing the bioavailability of many drugs with poor water solubility have been previously reported [6]. However, nanosuspension solidification is always preferred, to improve their physical stability. Different techniques of solidification were utilized, such as vacuum drying, spray-drying, and freeze-drying [7]. However, the drug nanoparticles can irreversibly aggregate in case of an improper drying process, resulting in reducing the dissolution rate. Therefore, loading drug nanosuspensions on films could be the method of choice, because of its process efficiency and simplicity.

This work aimed to develop and optimize novel palatable oral fast dissolving films (OFDFs) loaded with a paroxetine nanosuspension, to improve PX oral bioavailability by decreasing its broad metabolism in the liver and increasing PX solubility and permeability through the buccal mucosa. OFDFs are palatable, robust, and unique delivery systems that offer several advantages over the other conventional dosage forms available [8]. They offer increased compliance by patients with mental illness, as well as by elderly, pediatric, and developmentally disabled patients. They also offer accurate dosing in a portable and convenient format, without the need for measuring devices or water [9]. OFDFs are very thin oral strips that rapidly hydrate and adhere onto the application site. They are dissolved instantly, within 1 min, by the saliva when placed on any oral mucosal tissue, allowing rapid drug absorption through the oral mucosa (pre-gastric absorption) and an early action onset [10]. Furthermore, they offer significantly greater drug bioavailability than the conventional dosage methods because of the pre-gastric absorption of the drug in the buccal cavity, which allows avoiding the first-pass effect, and because even if part of the drug released from the OFDF is swallowed, it will still enter the gastrointestinal tract either dissolved or suspended in the saliva. Thus, it will be present in a readily bioavailability form [11]. Moreover, it is worth mentioning that the oral mucosa is highly vascularized, which helps in rapidly achieving the drug therapeutic serum concentrations [12]. Considering these facts, it is very attractive to use the dual advantages of both, OFDFs loaded with drug nanosuspension, and buccal administration, to support PX indication for the treatment of depression and anxiety.

A PX nanosuspension, prepared using a solvent–antisolvent precipitation method, were loaded into OFDFs prepared using a solvent casting method. In preparing the OFDFs, pectin and carboxy methyl cellulose polymers were used to obtain the desired film toughness, to avoid any damages during transportation or handling [13]. Three types of plasticizers (glycerol, propylene glycol, and polyethylene glycol 400) were incorporated, to impart a certain degree of flexibility to the OFDFs; where these plasticizers convert the used polymer from a hard stiffy glassy state to soft rubbery state by decreasing the polymer's glass transition temperature [14]. Tween 80 was incorporated as a wetting agent to accelerate the disintegration of the prepared OFDFs within seconds, releasing the loaded drug rapidly [15,16]. Moreover, it was used as a permeation enhancer to increase the PX penetration through the buccal mucosa, because of its non-ionic nature, high HLB value, and low critical micelle concentration [17]. A full factorial design ($2^2 \times 3^1$) was implemented, to investigate the formulation variables affecting PX OFDFs properties, utilizing Design-Expert® software. All the prepared OFDFs were investigated by examining their physical characteristics, mechanical characteristics, and dissolution profiles. Comparative dissolution, ex vivo permeation, and a stability study were performed for the optimized PX OFDF. The bioavailability of the optimized PX OFDF was clinically investigated in healthy human volunteers and compared with the market tablet.

2. Materials and Method

2.1. Materials

Paroxetine (PX) was provided by Glaxosmithkline Beecham, England. Pectin was purchased from Alpha Chemika (Mumbai, India). Carboxymethyl cellulose (CMC) was provided by Dow Wolff Cellulosics (Bomlitz, Germany). Poloxamer 188, citric acid, menthol, and Tween 80 were provided by Sigma Aldrich Co. (St. Louis, MO, USA). Polyethylene glycol 400 (PEG 400), glycerol, and Propylene glycol (PG) were obtained from Prolabo Co. (Paris, France). Sucralose was from El- Gomhouria Chemical Co. (Cairo, Egypt). Paroxetine® 25 mg oral market tablets were from EVA Pharma Co., (Cairo, Egypt). The rest of the solvents and chemicals were of analytical grade.

2.2. Preparation of Paroxetine Nanosuspension

Paroxetine nanosuspension was formulated using a solvent–antisolvent precipitation technique [18]. In brief, 250 mg of PX was dissolved in a suitable volume of organic solvent (ethanol) using a magnetic stirrer (SCHOTT, Germany). The anti-solvent phase was prepared by dissolving poloxamer 188 (2% w/v), as a stabilizer, in 5 mL distilled water using a magnetic stirrer. Then, the organic solvent was injected using a syringe into the aqueous phase under continuous stirring. After precipitation of PX nanoparticles, stirring was continued for a further 2 h to assure complete evaporation of the organic solvent. Then, the resulting nanosuspension was subjected to ultrasonication (CPX 130, Cole Parmar) for 10 min to further reduce the PX particle size.

2.3. Assessment of Particle Size, Poly Dispersity Index, and Zeta Potential

Particle size (PS), poly dispersity index (PDI), and zeta potential (ZP) were detected utilizing the dynamic light scattering technique (Nano ZS-90, Malvern Zetasizer, Worcestershire, UK) [15]. Prior to starting the analysis, the PX nanosuspension was first diluted with an adequate amount of distilled water with sonication for 1 min. Measurements were performed three times and the mean was calculated.

2.4. Differential Scanning Calorimetry (DSC)

Paroxetine compatibility with both polymers (pectin and CMC), used to prepare the OFDFs, was investigated using a DSC (DSC-60; Shimadzu Corp., Kyoto, Japan) calibrated with indium. PX and physical mixtures of PX with each of the used polymers (in a ratio 1:1) were assessed for any possible interactions. Then, 3 mg from each sample was placed in aluminum pans. Each pan was heated from 20 to 200 °C at a rate of 10 °C/min under a nitrogen atmosphere [19].

2.5. Oral Fast Dissolving Films (OFDFs) Loaded with PX Nanosuspension Preparations

OFDFs containing 25 mg PX were prepared using a solvent casting technique [20,21]. Briefly, an precisely weighed quantity of the polymers (pectin or CMC) was added to 5 mL double distilled water and mixed using a magnetic stirrer (Wheaton, Rc-2, Kyoto, Japan) for 1 h. Then, tween 80 was added to that solution, with continuous stirring for 10 min. After, sucralose, citric acid, and menthol were added and stirred till the complete dissolving of all ingredients. After that, a plasticizer (Glycerol, PG, or PEG 400) was added and continually stirred for 1 h till the solution became clear. Then, the previously prepared PX nanosuspension was added to the mixture with continuous mixing for 1 h, till the formation of a homogenous mixture. The obtained mixture was left undisturbed and, after becoming totally free from any bubbles, was poured into a glass Petri dish and dried in a hot air oven at 60 °C for the first 2 h and at 40 °C for the following 24 h. The resulting OFDF was peeled from the glass Petri dish and cut into films with a size of 2×2 cm^2, each contained 25 mg PX. Then, they were wrapped, utilizing airtight aluminum foil. Any film with imperfections, cuts or air bubbles was excluded.

2.6. Full Factorial Statistical Design

A $2^2 \times 3^1$ full factorial design was implemented to detect the impact of the investigated factors on the PX OFDF's properties, utilizing Design-Expert® software (Stat-Ease Inc., Minneapolis, MN, USA). The investigated factors were polymer type (X_1), polymer concentration (X_2), and plasticizer type (X_3). The studied responses were tensile strength (Y_1), elongation % (Y_2), Young's modulus (Y_3), disintegration time (Y_4), and % PX dissolved after 10 min (Y_5).

Twelve experimental runs were prepared, where each run was performed three times, to determine the repeatability of the results. The detailed composition of the prepared OFDFs is displayed in Table 1. The optimized OFDF was selected based on those showing the highest desirability. The chosen optimized OFDF had the highest tensile strength, elongation %, and % PX dissolved after 10 min, and the lowest Young's modulus and disintegration time. Finally, the chosen optimized OFDF was characterized by comparing its observed and predicted responses to ensure the model performance accuracy.

Table 1. Experimental runs and independent variables of the full factorial experimental design for the prepared OFDFs loaded with PX nanosuspension.

Formulations	Factors (Independent Variables)		
	Polymer Type	**Polymer Concentration (%*w*/*v*)**	**Plasticizer Type**
F1	Pectin	1%	Glycerol
F2	Pectin	1%	PG
F3	Pectin	1%	PEG 400
F4	Pectin	2%	Glycerol
F5	Pectin	2%	PG
F6	Pectin	2%	PEG 400
F7	CMC	1%	Glycerol
F8	CMC	1%	PG
F9	CMC	1%	PEG 400
F10	CMC	2%	Glycerol
F11	CMC	2%	PG
F12	CMC	2%	PEG 400

Each prepared OFDF contained 250 mg plasticizer, 10 mg tween 80, 10 mg citric acid, 15 mg menthol, 50 mg sucralose, and distilled water to 10 mL.

2.7. *Characterization of Paroxetine OFDFs*

2.7.1. Average Weight

Each film was weighed individually, utilizing an electronic balance (type AX200, Shimadzu corp., Kyoto, Japan) [22]. Each test was repeated three times.

2.7.2. Film Thickness

The thickness of each OFDF was measured at five different positions, the center and the four corners, utilizing digital Vernier calipers (Mitutoyo, Kawasaki, Japan). This test was repeated three times, and the average ± SD was calculated [8].

2.7.3. Folding Endurance

This test was performed to check the flexibility of each film, needed to provide facile handling and administration. The film was folded or rolled repetitively until it broke. The number of folds that occurred before the breakage was counted and considered the folding endurance value. The folding times (folding endurance value) and the film flexibility were directly proportional [23].

2.7.4. Content Uniformity

This test was implemented to ensure a uniform distribution of PX in each OFDF. Each film was dissolved in a 100-mL volumetric flask containing 10 mL ethanol using a vortex (VM-300, Gemmy Industrial Corp., Taiwan) for five minutes [8]. The sample from the resulted liquid was diluted with an appropriate ethanol volume. PX content was detected utilizing a UV spectrophotometer (UV-1700; Shimadzu, Kyoto, Japan) at 294.3 nm [3]. PX average content ± SD was determined using the following equation [22]:

$$\text{Content uniformity} = \frac{\text{Actual PX amount in OFDFs}}{\text{Theoretical PX amount in OFDFs}} \times 100 \tag{1}$$

This test was repeated three times.

2.7.5. Surface pH

Each film was moistened in a closed Petri dish using distilled water (1 mL) and kept for 5 min at 25 °C. The pH of each OFDF was assessed utilizing a pH meter (Jenway 3510, Swedesboro, USA) by touching an electrode with a moistened surface to each OFDF. The test was performed in triplicate, and the mean value was calculated [24].

2.7.6. Moisture Content %

Percentage moisture content is an essential parameter to detect the integrity and the physical stability of formed films [21]. Moisture content %, included the two important parameters of moisture loss % and moisture absorption %.

Moisture Loss %

Moisture loss % was assessed by recording the initial weight of each OFDF, then putting these films in a desiccator that contained anhydrous calcium carbonate at room temperature for 72 h [25]. After that, OFDFs were taken out from the desiccator and reweighed. This parameter was determined according to the following equation [21]:

$$\text{Moisture loss \%} = \frac{\text{Initial weight} - \text{Final weight}}{\text{Initial weight}} \times 100 \tag{2}$$

Moisture Absorption %

This parameter was determined by putting pre-weighed films in a desiccator containing potassium chloride saturated solution, to achieve a 80% relative humidity for 24 h, at room temperature. Then, each film was weighed again and the moisture absorption % of each OFDF was assessed using the following equation [14]:

$$\text{Moisture absorption \%} = \frac{\text{Final weight} - \text{Initial weight}}{\text{Initial weight}} \times 100 \tag{3}$$

2.7.7. Mechanical Characteristics of PX OFDFs

The mechanical characteristics of the OFDFs were measured utilizing a digital tensile testing machine (Lloyd Instruments Ltd., LR 10K, UK). The thickness of each OFDF had previously been detected utilizing a digital Vernier caliper. Each OFDF was precisely cut into a rectangular shape using a metal scalpel blade [26]. The tensile tester was set with an initial grip separation of 25 mm and crosshead speed of 35 mm/minute. Each OFDF was placed vertically between the tensile tester's two clamps. Each OFDF was pulled apart by the two clamps until breakage [27]. For each OFDF, measurements were performed three times, and the mean value ± SD was determined.

Tensile Strength

Tensile strength is the maximum force required to break the films. This was performed to detect the strength of the OFDFs. Tensile strength was expressed as force/unit area or mega Pascal (MPa), which was estimated as follows [28]:

$$\text{Tensile strength} = \frac{\text{Force at breakage (kg)}}{\text{Film thickness (mm)} \times \text{film wedth (mm)}} \tag{4}$$

Percent Elongation

The % elongation (strain) was assessed to detect the toughness and stretching of the prepared OFDFs. The stretching and elongation of the film prior to breakage are referred to as the %Elongation or strain. Percent elongation was determined as follows [16]:

$$\%\ \text{Elongation} = \frac{\text{Increase in the OFDF length}}{\text{Initial length of the OFDF}} \times 100 \quad (5)$$

Young's Modulus

Young's modulus, or elastic modulus, was utilized to estimate the elasticity or stiffness of each film. It was calculated from the slope obtained from the linear portion of the stress–strain curve, as shown in this equation [26]:

$$\text{Young's modulus} = \frac{\text{Slope}}{\text{Thickness of the Film X Crosshead speed}} \times 100 \quad (6)$$

2.7.8. In Vitro Disintegration Time

Each OFDF was placed in a glass Petri dish, then, 10 mL of Sorensen's phosphate buffer (pH 6.8) was added to the petri dish at 25 °C. The time required to disintegrate or break each OFDF was recorded [27]. For each OFDF, measurements were performed three times, and the mean value was calculated.

2.7.9. In Vitro Dissolution Study

The PX in vitro release from all the prepared OFDFs was evaluated in a USP dissolution apparatus (type II) (Pharm Test, Hainburg, Germany). The dissolution medium was composed of 250 mL Sorensen's phosphate buffer with pH of 6.8 [29]. Throughout the study, the temperature was adjusted to 37 ± 0.5 °C, and the rotation speed was fixed at 50 rpm. Each PX OFDF was put at the bottom of the dissolution apparatus vessel. Then, 4 mL of sample was withdrawn at 2, 4, 6, 8, 10, 15, 30, 45, and 60 min time intervals. Then, 4 mL of fresh phosphate buffer was replenished to keep a constant medium volume [30,31]. Finally, PX concentration was detected spectrophotometrically in each sample at λ_{max} 294.3 nm [3]. The test was performed three times and the mean % PX dissolved was estimated [32].

2.8. Characterization of the Optimized OFDF Loaded with PX Nanosuspension

2.8.1. Re-Dispersion of PX Nanoparticles from the Optimized OFDF

The optimized OFDF (F1) was dispersed in 10 mL deionized water. After 10 min of magnetic stirring, the size of nanoparticles was measured, as explained earlier [15].

2.8.2. Comparative Dissolution Study of Optimized PX OFDF, Drug Powder, and the Market Tablet

A comparative dissolution study between the optimized PX OFDF (F1), drug powder, and the market tablet (Paroxetine® 25 mg) was performed to investigate the success of the optimized OFDF in increasing the dissolution rate [16,27]. The study was implemented using the same conditions as mentioned previously for the in vitro dissolution study. Dissolution $T_{50\%}$ and similarity factor (f_2) were estimated to compare PX release profiles of the optimized PX OFDF (F1), drug powder, and the market tablet.

2.8.3. Ex Vivo Permeation

Tissue Preparation

A permeation study was carried out to evaluate the PX permeability across a buccal chicken pouch membrane, to simulate buccal mucosa permeability. After obtaining the membrane from a slaughtered chicken, the underlying connective tissues was separated utilizing forceps. Then, the excised membrane was carefully washed and stored in normal saline at −20 °C. Prior to starting the permeation studies, the buccal mucosa was hydrated with a simulated saliva solution (pH 6.8), prepared by dissolving 0.2 g KH_2PO_4, 2.4 g Na_2HPO_4, and 8.0 g NaCl in distilled water (1000 mL) [33]. Then the buccal mucosa was thawed till it reached room temperature, prior to use.

Ex Vivo Permeation Testing

A USP dissolution apparatus (type I) (Pharm Test, Hainburg, Germany) was used to compare the PX permeation of both the optimized OFDF (F1) and the equivalent drug powder [15,16]. The buccal mucosa was firmly fixed to a glass cylinder (internal diameter 1 cm), with one side as the donor chamber. The optimized film (F1) was put in close contact with the surface of the membrane mucosa at one end of the glass tube. Then 1 mL of phosphate buffer with a pH of 6.8 was added to the glass tube, which was then attached to the dissolution apparatus shaft [16]. The same study was carried out using an equivalent amount of the pure drug, which was put inside the donor chamber and hydrated using 1 mL of pH 6.8 phosphate buffer. An ex vivo permeation study was performed utilizing the same method used in the in vitro release study. Whereby, 4 mL samples were withdrawn at time intervals of 5, 10, 15, 20, 25, 30, 45, and 60 min [15,16]. PX concentration in the gathered samples was analyzed utilizing HPLC (Shimadzu, Tokyo, Japan). Where, a C18 column (5 μm, 150 × 4.6 mm) (Thermo Fisher Scientific, NJ, USA) was the utilized column, and the used mobile phase was composed of 0.04 M phosphate buffer/acetonitrile (65:35), with a 1 mL/min flow rate for all separations. Permeated PX was detected at 294.3 nm [33]. The lower limit of quantification (LLOQ), accuracy, and linearity were estimated. The permeation study was implemented in triplicate and the mean value was recorded.

Permeation Parameter Calculation

The cumulative amount of PX permeated per cm^2 was plotted versus time.

The flux (J) was assessed utilizing the following equation [34]:

$$J = \frac{\text{Amount of PX permeated from the optimized film}}{\text{Time} \times \text{memberane area}} \quad (7)$$

Moreover, the permeation enhancement ratio (ER) was estimated utilizing the following equation [35]:

$$\text{ER} = \frac{J \text{ of the optimized film}}{J \text{ of drug reference}} \times 100 \quad (8)$$

Statistical Analysis

One-way ANOVA was utilized to statistically detect any significant differences in the obtained results, where the difference was considered significant when the p-value < 0.05 [36].

2.8.4. Stability Study

A stability study was implemented with the optimized PX OFDF (F1), according to the guidelines of ICH Q1A (R2) [37]. In this study, the temperature was set at 40 ± 2 °C and the relative humidity was adjusted to 75 ± 5%. Each film was covered with aluminum foil and then put in a desiccator for six months. After half of, and the full, storage period, each film was investigated for any changes in content uniformity, tensile strength, % elongation, Young's modulus, disintegration time, and % PX dissolved after 10 min. The resulting values were subjected to statistical analysis using Student's t-test and utilizing SPSS 17.0 software (SPSS Inc., Chicago, IL, USA). Differences were considered significant when the p-value was less than 0.05.

2.9. In Vivo Clinical Studies

2.9.1. In Vivo Disintegration Time and Palatability Studies

The taste masking of bitter drugs represents a fundamental challenge during the preparation of oral fast dissolving films, as most drugs are unpalatable; influencing patient's compliance and acceptance. The study was executed according to the Helsinki Declaration for bio-medical research, including human and good clinical practices (GCP) rules [38,39]. The protocol of this study was revised and accepted by the institutional review board of the Genuine Research Center, (Cairo, Egypt). Eight healthy human volunteers participated in this double-blind crossover study. All the enrolled volunteers signed a written informed consent, after telling them about the study purpose, nature, duration, and risks. Moreover, all the volunteers were instructed to maintain their standard dietary condition and normal physical activity for five days prior to intaking the dosage form [40]. The volunteers were administrated the optimized PX OFDF (F1) by putting it in the buccal cavity till its complete disintegration. All the enrolled volunteers were asked to record the texture, disintegration time, acceptance (bitterness taste masking), and aftertaste of both the optimized OFDF (F1) and PX pure powder. The scale of bitterness intensity was as follows: 1 = very bitter, 2 = bitter, 3 = acceptable (no specific feeling), 4 = pleasant, 5 = very pleasant [41]. Both the tested optimized OFDF (F1) and the drug pure powder contained the same quantity (25 mg of) PX.

2.9.2. Pharmacokinetic Evaluation in Healthy Human Volunteers

Study Design and Subjects

In this comparative study, six healthy male adult volunteers were enrolled. Whereby, a randomized crossover, two period, single dose design was implemented to assess the pharmacokinetic parameters of the optimized PX OFDF (F1) compared to the market tablet (Paroxetine® 25 mg). Their average body weight and age was 72.4 $\pm$ 6.7 kg and 40.6 $\pm$ 5.0 years, respectively. All the enrolled volunteers signed a written informed consent after receiving the required information about the nature, purpose, and risks of the study. In addition, all volunteers were asked to stop smoking or administrating any medicines for 2 weeks before starting the study, till its end. This study was executed according to the Helsinki Declaration for bio-medical research, including human and the good clinical practice (GCP) rules [38,39]. The protocol of this study was revised and accepted by the institutional review board of Genuine Research Center (Cairo, Egypt). The approval number is GRC/1/21/R4. A specialized physician was employed to observe all volunteers during the study.

Drug Administration and Sample Collection

The participating volunteers were randomly distributed into two equal groups. The first treatment was the optimized PX OFDF (F1), and the second treatment was the market tablet. Both the optimized PX OFDF and the market tablet contained 25 mg paroxetine. A seven day washout period was implemented to separate the two phases from each other. After that, the study was repeated using the same administrated conditions, to assure that each of the six volunteers received both treatments. Prior to the study, all the volunteers fasted overnight (ten hours). The participating volunteers placed the optimized PX OFDF (F1) in their buccal cavity till it completely disintegrated and disappeared in their mouths, followed by receiving 200 mL of water. An equivalent volume of water was utilized to swallow the market tablet. Two hours after intaking the dose, volunteers were allowed to drink water, then after an additional 2 h, they were allowed to eat. After that, the volunteers were given standard breakfast, lunch, and dinner meals, in accordance with a planned time schedule.

For PX analysis, an indwelling cannula was utilized to transfer the withdrawn blood samples (5 mL) to heparinized glass tubes at different time intervals of 0.0, 0.33, 0.66, 1, 1.5, 2, 2.5, 3, 3.5, 4, 4.5, 5, 6, 8, 12, 24, and 48 h after dose intake. The gathered blood samples were centrifuged (Hettich EBA 21 Centrifuge, Germany) at 3500 rpm at 4 °C for 10 min. Finally, the obtained plasma was frozen at −20 °C (Ultra-Low Freezer, Cincinnati, OH, USA) until PX analysis.

Sample Preparation

All the frozen human plasma samples were left to reach room temperature. Then, a 0.5 mL aliquot from each sample was mixed with 100 μL paroxetine-D6 maleate (internal standard) (IS). Then, the aliquots were vortexed for 1 min. Next, after the addition of 4 mL of methyl-t-butyl-ether (MTBE), the aliquots were vortexed again for two minutes. Thereafter, the resulting samples underwent centrifugation for ten minutes at room temperature, at a speed of 4000 rpm. A centrifugal vacuum concentrator (Vacufuge® 5301, Germany) was utilized to evaporate the obtained clear supernatant till dryness at 40 °C. Then, 100 μL of mobile phase was utilized to reconstitute the resulting dry residues and an autosampler was used to inject 10 μL of the resulting samples [42].

Chromatographic Conditions

The LC-MS/MS method is an accurate, selective, and sensitive method, utilized to analyze Paroxetine human plasma samples. Before the study, this method was developed and validated according to international guidelines [43]. To prepare the stock solution of Paroxetine-D6 (the internal standard), 10 mg was dissolved in methanol, then serial dilutions were prepared using the mobile phase. The utilized pump was an LC-20AD (Shimadzu, Japan). The degasser was a DGU-20AS (Shimadzu, Japan), and the utilized auto-sampler for injection of 10 μL samples was an SIL-20A (Shimadzu, Japan). The used mobile phase was composed of formic acid 0.1% in acetonitrile:water (6:4; *v/v*) and delivered at a flow rate of 0.15 mL/minute using a Luna C18 (Phenomenex, Torrance, CA, USA) (50 × 2 mm, 5μm) column [44]. An API-4000 mass spectrometer (Applied Biosystems, Canada), equipped with a nitrogen generator, was used for the quantitation, with a positive ion polarity mode for both paroxetine and Paroxetine-D6 and adjusted with an ion spray voltage of 5500 V. The ion source parameters of entrance potential, collision exit potential, collision energy, and declustering potential were 10, 10, 29, and 100, respectively, for Paroxetine, and 10, 16, 31, and 131 for Paroxetine-D6. Ion detection was implemented in the multiple reaction monitoring mode (MRM), monitoring the transition of m/z 329.6 precursor ion to m/z 192.3 for Paroxetine and the m/z 336.2 precursor ion to m/z 198.1 for Paroxetine-D6. Analytical data processing was implemented utilizing version 1.6.3 of Analyst® software.

Pharmacokinetic and Statistical Analysis

A paroxetine pharmacokinetic analysis was made for each subject using a noncompartmental approach, applying Kinetica® 4.4.1 SPSS 14 software (Thermo Fisher Scientific Inc., Waltham, MA, USA). Paroxetine concentration–time curves were used to obtain the PX peak concentration (C_{max}, ng/mL) and the time of PX peak concentration (T_{max}, hours). The area under the concentration–time curve from zero to the last analyzed point (AUC_{0-48}, ng.h/mL), and from zero to infinity ($AUC_{0-\infty}$, ng.h/mL), was estimated using the linear trapezoidal rule. $T_{1/2}$ (hours) was assessed as 0.693/K. All the pharmacokinetic parameters of the two treatments were compared utilizing an ANOVA test. Differences between any two related parameters were considered statistically significant if the p-value was ≤ 0.05.

3. Results and Discussion

3.1. Particle size, Poly Dispersity Index, and Zeta Potential

PS and PDI are important characterization parameters for nanosuspension evaluation. Both of these parameters have a significant effect on an active ingredient's stability, solubility, dissolution rate, and bioavailability. The prepared PX nanosuspension had PS and PDI values of 217.09 ± 4.18 nm and 0.46 ± 0.27, respectively. The low value of the PDI indicates the small distribution of the particle sizes, and, hence, the homogeneity of the diameter of particles. This low PDI value decreased the PX concentration gradients in the medium, which aids in inhibiting the occurrence of Ostwald ripening [45]. ZP is an essential parameter for indicating the physical stability of prepared nanosuspensions. The ZP value of the formulated PX nanosuspension was −33.49 ± 2.08 mV. A higher ZP value indicates the presence of a large number of particles with the same charges, leading to electrostatic repulsion between them, which prevents their aggregation and maintains their physical stability [46].

3.2. *Differential Scanning Calorimetry (DSC)*

DSC was utilized to test the PX compatibility with the studied polymers (pectin and CMC). Figure 1 shows the DSC thermograms of PX, pectin, CMC, and the physical mixtures of PX with each polymer, with a ratio (1:1), as this ratio maximizes the possibility of monitoring any interactions [47]. DSC thermograms showed that pure PX had an endothermic peak at 123.75 °C [5]. Pectin showed an endothermic peak at 154 °C, while CMC showed a broad endothermic peak at 78.21 °C [48,49]. It is worth mentioning that there was no shift in PX peak in the case of both physical mixtures, revealing chemical and physical compatibility.

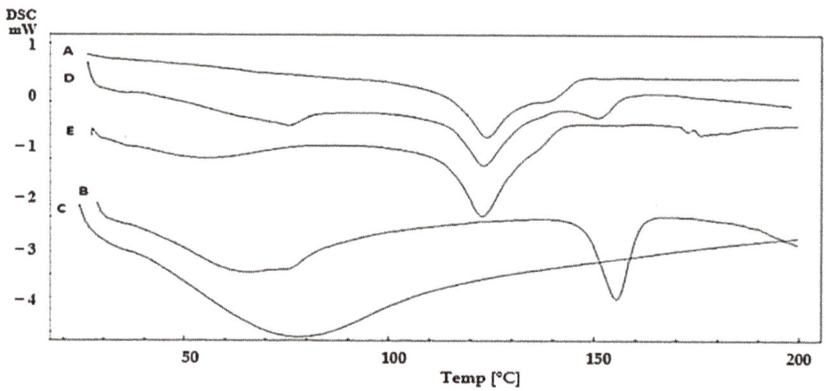

Figure 1. Differential scanning calorimetric (DSC) thermograms of (A) Paroxetine, (B) pectin, (C) CMC, physical mixture of Paroxetine with (D) pectin and (E) CMC.

3.3. *Preparation of OFDFs Loaded with PX Nanosuspension*

A solvent casting method was successfully utilized to prepare the PX OFDFs. The prepared OFDFs were durable and were able to withstand normal handling without any imperfections, cuts, cracks, or air bubbles. Pectin and CMC polymers were successful in obtaining the desired film toughness, to avoid any damages during transportation or handling [13]. Glycerol, PG, and PEG 400, which were used as plasticizers, imparted a certain degree of flexibility to the OFDFs [50]. Citric acid was used as a salivary stimulant to stimulate the saliva secretion in the buccal cavity, thus improving the disintegration of the OFDFs [51]. Sucralose was utilized as a sweetening agent to mask paroxetine's bitter taste. Whereby, sucralose has 200–600 times more sweetness than sucrose [52]. Menthol was utilized as a flavoring agent to provide a refreshing sensation in the mouth upon administration to the buccal cavity [53].

3.4. *Full Factorial Design Statistical Analysis*

A full factorial ($2^2 \times 3^1$) design was implemented using ANOVA because it is beneficial for evaluating the effect of different formulation factors, with their different levels, on the investigated responses. Sum-of-squares (Type II model) was the chosen model. Adequate precision was utilized, to assure that the chosen model was appropriate for navigation of the design space. Table 2 shows that the adequate precision value was >4 in all the responses, which was desirable. For all dependent variables, there was a good harmony between both the predicted and adjusted R^2 values. Table 2 displays all the investigated factors and the detected responses of the statistical design. The *p*-value of all responses was considered significant (<0.05). The effects of each studied factor on the responses and the graphical response surface plots were obtained utilizing Design expert 8.0.7 software (Stat-Ease Inc., Minneapolis, MN, USA).

Table 2. Output data of the factorial design and predicted and observed values for the optimized OFDF (F1).

Responses	Tensile Strength (Mpa)	% Elongation	Young's Modulus (Mpa)	Disintegration Time (s)	% PX Dissolved after 10 Minutes
Minimum	1.04 ± 0.11	6.03 ± 0.45	8.09 ± 0.15	17.09 ± 1.30	12.14 ± 0.08
Maximum	15.5 ± 0.68	53.08 ± 1.28	383.66 ± 11.06	160.06 ± 4.20	96.02 ± 3.46
F value	17.21	20.73	15.61	79.52	11.31
p-value	0.0010	0.0006	0.0013	< 0.0001	0.0036
Adequate precision	13.39	9.83	18.01	22.99	10.88
Adjusted R^2	0.855	0.877	0.842	0.966	0.789
Predicted R^2	0.729	0.772	0.704	0.934	0.686
R^2	0.908	0.922	0.899	0.978	0.896
Significant factors	X_1, X_2 and X_3	X_3	X_3	X_1, X_2 and X_3	X_1, X_2 and X_3
Observed values of optimum OFDF (F1)	3.89	53.08	8.12	17.09	96.02
Predicted values of optimum OFDF (F1)	3.46	50.07	9.90	20.28	97.14

Data represented as mean ± SD (n = 3).

3.5. *Characterization of the Prepared PX OFDFs*

3.5.1. Average Weight

Table 3 displays that weight of the different prepared OFDFs varied from 36.86 ± 3.96 mg to 52.61 ± 2.67 mg. These results show the lack of any significant weight variations, which reveals the efficiency of the implemented method and indicates the uniformity of PX distribution.

3.5.2. Films Thickness

Measuring the thickness of all the prepared OFDFs was important, to assure the uniformity of their thickness, and because the dose accuracy is directly related to the film thickness. Generally, an ideal OFDF should exhibit a thickness between 0.05 and 1 mm [8]. Table 3 shows that the thickness of all the prepared OFDFs ranged from (0.11 ± 0.02 mm to 0.23 ± 0.03 mm). The presence of slight variations in films thickness could be attributed to their containing different amounts of polymers. It is worth noting that the thickness of the prepared OFDFs was increased by increasing the amount of the used polymers.

3.5.3. Folding Endurance

The elasticity and flexibility of the prepared OFDFs were important physical characteristics, required for easy handling and the application of the OFDF to the administration site [29]. The folding endurance of all OFDFs ranged from 187 ± 5.71 to >300, as displayed in Table 3. All the prepared OFDFs exhibited good folding endurance values, indicating their good flexibility, possibly due to incorporation of the plasticizers.

3.5.4. Content Uniformity

The PX content was from 89.48 ± 1.09 to 96.68 ± 3.62% of the labeled claim, as shown in Table 3. These results indicate a good drug uniformity in all the prepared OFDFs, assuring that PX was uniformly dispersed in each prepared film.

Table 3. Pharmaceutical evaluation and measured responses of the prepared oral fast-dissolving films loaded with PX nanosuspension.

	Average Weight (mg)	Film Thickness (mm)	Folding Endurance	Content Uniformity (%)	pH	Moisture Loss %	Moisture Absorption %	Tensile Strength (Mpa)	Percentage Elongation	Young's Modulus (Mpa)	Disintegration Time (s)	% PX Dissolved after 10 Minutes
F1	38.02 ± 1.45	0.11 ± 0.02	>300	96.68 ± 3.62	6.80 ± 0.17	1.08 ± 0.02	1.39 ± 0.11	3.89 ± 0.19	53.08 ± 1.28	8.12 ± 0.13	17.09 ± 1.30	96.02 ± 3.46
F2	40.17 ± 0.98	0.13 ± 0.04	>300	93.14 ± 4.24	6.92 ± 0.11	1.24 ± 0.03	1.06 ± 0.08	9.68 ± 0.12	11.31 ± 1.06	306.41 ± 12.73	26.14 ± 3.06	41.74 ± 3.08
F3	39.29 ± 2.37	0.13 ± 0.01	>300	92.35 ± 1.28	6.64 ± 0.25	0.93 ± 0.02	1.21 ± 0.10	6.24 ± 0.26	11.06 ± 1.00	383.66 ± 9.06	20.27 ± 2.00	49.36 ± 2.63
F4	52.03 ± 1.88	0.15 ± 0.03	>300	91.50 ± 4.68	7.00 ± 0.16	1.16 ± 0.04	3.04 ± 0.27	6.63 ± 0.34	46.07 ± 2.46	13.98 ± 1.41	38.04 ± 1.38	35.15 ± 1.70
F5	52.61 ± 2.67	0.20 ± 0.01	203 ± 8.40	91.27 ± 2.89	6.62 ± 0.32	1.04 ± 0.02	3.62 ± 0.30	15.5 ± 0.68	18.1 ± 1.88	378.95 ± 16.86	34.25 ± 2.01	18.5 ± 1.55
F6	50.12 ± 1.48	0.18 ± 0.03	187 ± 5.71	94.04 ± 5.94	6.83 ± 0.24	1.17 ± 0.05	2.98 ± 0.12	6.71 ± 0.15	9.69 ± 2.08	337.75 ± 11.28	50.18 ± 1.33	21.62 ± 2.91
F7	41.07 ± 0.17	0.16 ± 0.06	>300	90.66 ± 3.36	7.01 ± 0.40	1.20 ± 0.04	5.66 ± 0.20	1.04 ± 0.11	49.82 ± 1.36	9.44 ± 0.27	97.23 ± 5.40	52.14 ± 4.07
F8	36.86 ± 3.96	0.15 ± 0.02	237 ± 10.00	93.78 ± 2.06	6.57 ±0.24	1.14 ± 0.03	6.35 ± 0.41	6.33 ± 0.32	6.03 ± 1.05	497.75 ± 22.47	110.79 ± 4.51	32.03 ± 1.08
F9	40.33 ± 2.07	0.13 ± 0.06	>300	92.04 ± 2.31	6.90 ± 0.14	0.96 ± 0.01	6.09 ± 0.26	3.68 ± 0.18	23.84 ± 2.04	211.5 ± 13.20	127.04 ± 4.38	40.44 ± 2.59
F10	49.20 ± 1.43	0.22 ± 0.01	>300	90.13 ± 4.63	6.84 ± 0.36	1.00 ± 0.04	8.73 ± 0.38	1.83 ± 0.10	46.93 ± 3.02	8.09 ± 0.15	107.23 ± 2.46	23.87 ± 1.16
F11	52.20 ± 3.75	0.22 ± 0.04	198 ± 6.70	89.48 ± 1.09	6.78 ± 0.12	1.18 ± 0.06	7.22 ± 0.19	8.95 ± 0.28	19.46 ± 1.80	361.87 ± 20.89	123.51 ± 3.87	12.19 ± 1.64
F12	50.49 ± 2.03	0.23 ± 0.03	240 ± 12.00	90.29 ± 2.43	6.61 ± 0.38	0.99 ± 0.05	8.64 ± 0.42	6.19 ± 0.27	20.13 ± 1.07	296.52 ± 14.04	160.06 ± 4.20	12.14 ± 0.08

3.5.5. Surface pH

The pH of each OFDF should be close to the buccal cavity pH (neutral pH) [8]. The pH of all the prepared OFDFs was varied from 6.61 ± 0.38 to 7.01 ± 0.40, as displayed in Table 3. These results assured the absence of any possible irritation to the buccal mucosal lining that could happen due to alkalinity or acidity.

3.5.6. Moisture content %

Moisture Loss %

Studying the moisture loss % was undertaken to evaluate the integrity and the physical stability of each OFDF [21]. The moisture loss % of all the prepared OFDFs ranged from 0.93 ± 0.02 to 1.24 ± 0.03%, as shown in Table 3. These results were in an acceptable range, indicating little moisture loss and the good physical integrity and stability of the prepared OFDFs [27]

Moisture Absorption %

The moisture absorption % of the films is important, because it influences the friability, mechanical strength, adhesive properties, disintegration, and dissolution behaviors of each film [16]. Table 3 shows that the moisture absorption % of all the prepared OFDFs were from 1.06 ± 0.08% to 8.73 ± 0.38%. It was noticed that the moisture absorption % of the OFDFDs was slightly increased when increasing the polymer concentration. This could have been because of the hydrophilic nature of the used polymers [54]. It was also noticed that the OFDFs made of CMC polymer showed a high water sorption; therefore, they seemed to be not suitable for use at high humidity, which makes the films sticky and unsuitable for this application [13].

3.5.7. Mechanical Characteristics of the OFDFs

Tensile Strength

An ideal OFDF should exhibit an adequately high tensile strength value to be able to withstand normal handling. In spite of this, a very high value (very high rigidity) is not desired, because it could retard the drug release from the polymer matrix [29]. The prepared OFDFs had tensile strength values from 1.04 ± 0.11 Mpa to 15.5 ± 0.68 Mpa, as displayed in Table 3. Tensile strength values were analyzed using the following equation:

$$\text{Tensile strength} = 6.39 - 1.72 \times X_1 + 1.25 \times X_2 - 3.04 \times X_3\,[1] + 3.73 \times X_3\,[2] \quad (9)$$

where X_1 is the polymer type, X_2 is the polymer concentration, and X_3 is the plasticizer type. X_3 [1] represents the first plasticizer type (glycerol) and X_3 [2] represents the second plasticizer type (propylene glycol).

Table 3 and Figure 2a show that all the independent variables (X_1, X_2, and X_3) had a significant impact on the tensile strength (rigidity) of all the OFDFs loaded with PX. It was obvious that the tensile strength values were significantly changed when changing the type of polymer used (X_1). Where OFDFs prepared using pectin polymer showed significantly higher tensile strength values than those prepared with CMC polymer. This could be accredited to the differences in the nature and the molecular weight between the two polymers. These findings are in good agreement with that detected by Maher et al., who found that the type of polymer used (hydroxypropyl methyl cellulose and carboxymethyl cellulose) had a significant effect on the tensile strength of the prepared films [55].

Moreover, it was noticed that increasing the polymer concentration (X_2) significantly increased the tensile strength values of the prepared OFDFs. This could have been because of the formation of a densely packed network of the used polymer chains at higher concentration, leading to formation of a stronger matrix. Bharti et al. stated similar findings, where they found that the tensile strength of the prepared films was increased by increasing the film's former concentration (hydroxypropyl methyl cellulose polymer) [33].

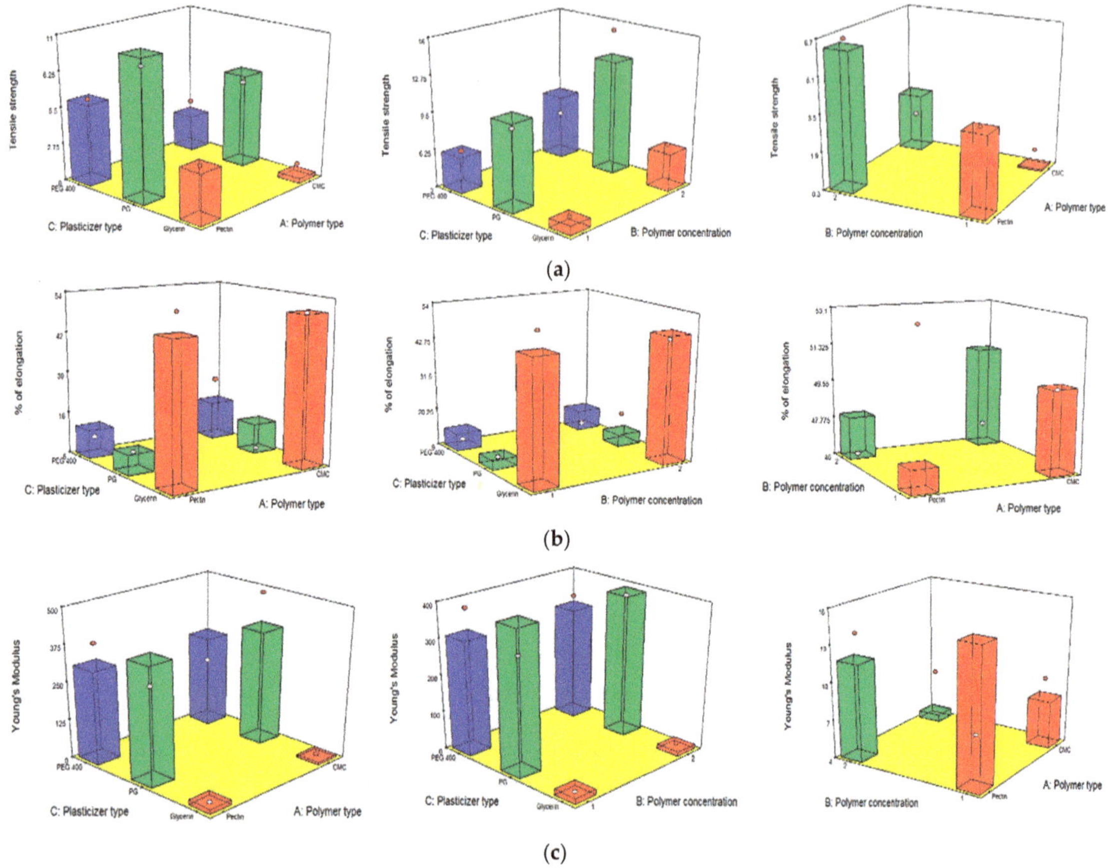

Figure 2. Response 3-D plots for the effect of polymer type (X_1), polymer concentration (X_2), and plasticizer type (X_3) on (**a**) tensile strength, (**b**) elongation %, and (**c**) Young's modulus of the OFDFs loaded with PX nanosuspension.

On the other hand, the tensile strength of PX OFDFs prepared with different plasticizer types (X_3) was increased in this order: PG > PEG 400 > glycerol. It can be noticed that OFDFs prepared with glycerol as a plasticizer showed the lowest tensile strength values. This could be because of the high hygroscopicity of glycerol, which causes humidity absorption and consequently gives softness to the prepared films and decreases the tensile strength values [56].

Percentage Elongation

OFDFs should possess a large elongation percentage, in order to exhibit the desired flexibility and stretchability, which is important for facile handling and application of the film to the buccal cavity [15]. The percentage elongation values for all the prepared OFDFs were found to be from 6.03 ± 0.45% to 53.08 ± 1.28%, as displayed in Table 3. The following equation was used to analyze the percentage elongation values:

$$\text{Percent Elongation} = 26.29 + 1.41 \times X_1 + 0.44 \times X_2 + 22.68 \times X_3\,[1] - 12.57 \times X_3\,[2] \quad (10)$$

where X_1 is the polymer type, X_2 is the polymer concentration, and X_3 is the plasticizer type. X_3 [1] represents the first plasticizer type (glycerol) and X_3 [2] represents the second plasticizer type (propylene glycol).

Table 3 and Figure 2b show that the percentage elongation of the prepared OFDFs was significantly affected only by the plasticizer type (X_3). Where, the percentage elongation of the OFDFs prepared with glycerol had the highest percentage elongation values.

The increase in the OFDFs elongation percentage can be attributed to the fact that glycerol replaces the intermolecular bonds present between polymer matrixes with hydrogen bonds created between polymer and glycerol molecules. This disruption and reconstruction of polymer molecular chains allows greater chain mobility, resulting in decreasing the rigidity and providing flexibility and stretching to the films [57]. These findings are in accordance with those stated by Junmahasathien et al., who reported that glycerol was the best plasticizer for increasing the elongation percentage of the prepared films, in comparison with other plasticizers [58].

Young's Modulus

OFDFs should have low Young's modulus values, to exhibit the desired elasticity; whereby, high values of Young's modulus lead to the formation of stiff and brittle films [59]. All the prepared OFDFs showed Young's modulus values from 8.09 ± 0.15 to 383.66 ± 11.06 Mpa, as displayed in Table 3. The following equation was used to analyze the Young's modulus values:

$$\text{Young's modulus} = 234.50 - 3.64 \times X_1 - 1.64 \times X_2 - 224.60 \times X_3\,[1] + 151.74 \times X_3\,[2] \quad (11)$$

where X_1 is the polymer type, X_2 is the polymer concentration, and X_3 is the plasticizer type. X_3 [1] represents the first plasticizer type (glycerol) and X_3 [2] represents the second plasticizer type (propylene glycol).

Table 3 and Figure 2c show that the plasticizer type (X_3) was the only factor significantly impacting the Young's modulus of the prepared OFDFs. Where, the OFDFs prepared using glycerol had the lowest Young's modulus values.

This could be attributed to glycerol being capable of dispersing between the spaces of the polymer chains, reducing their intermolecular attraction and, hence, providing flexibility to the films [60].

3.5.8. In Vitro Disintegration Time

Disintegration time is a very important parameter for OFDFs, to indicate the onset of drug action. A low value of disintegration time, allows a faster release and absorption of the loaded drug through the buccal mucosa. The mean disintegration time of all the prepared OFDFs ranged from 17.09 ± 1.30 to 160.06 ± 4.20 s, as displayed in Table 3. The following equation was used to analyze disintegration time values:

$$\text{Disintegration time} = 75.57 + 45.41 \times X_1 + 9.98 \times X_2 - 11.92 \times X_3\,[1] - 3.36 \times X_3\,[2] \quad (12)$$

where X_1 is the polymer type, X_2 is the polymer concentration, and X_3 is the plasticizer type. X_3 [1] represents the first plasticizer type (glycerol) and X_3 [2] represents the second plasticizer type (propylene glycol).

Table 3 and Figure 3a show that all the independent variables (X_1, X_2 and X_3) had a significant effect on the disintegration time of the prepared OFDFs. Disintegration time values were significantly changed when changing the type of polymer used (X_1). Where, OFDFs prepared using pectin polymer, showed significantly lower disintegration time values than those prepared with CMC polymer. This could be because of the difference in the nature of the used polymers; wherein, pectin is more hydrophilic than CMC [61]. Thus, OFDFs prepared with pectin had a faster hydration, and hence faster disintegration, than those prepared with CMC.

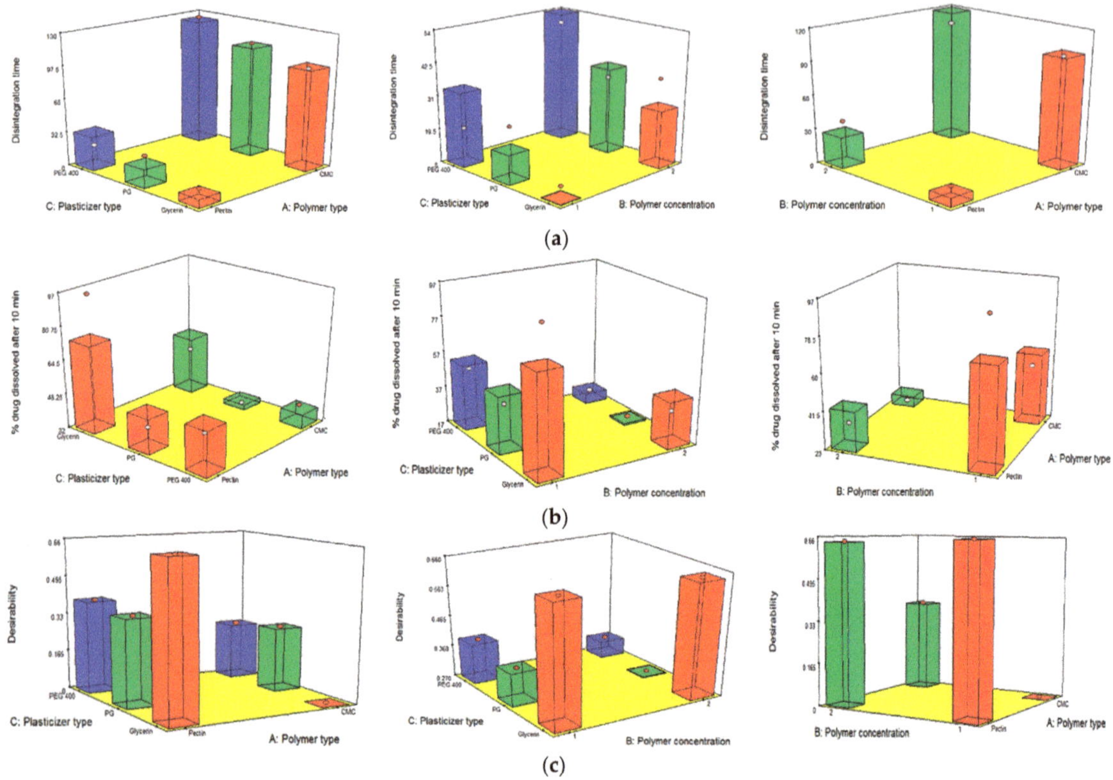

Figure 3. Response 3-D plots for the effect of polymer type (X_1), polymer concentration (X_2), and plasticizer type (X_3) on (**a**) disintegration time, (**b**) % PX dissolved after 10 min, and (**c**) desirability of OFDFs loaded with PX nanosuspension.

Additionally, it is worth noting that increasing the polymer concentration (X_2) resulted in a significant increase in the disintegration time of the prepared OFDFs. This could have been because increasing polymer concentration leads a need for more fluids to wet the films and increasing the film thickness, which retarded the penetration of water. Moreover, Shen et al., who formulated fast-dissolving films loaded with herpetrione nanoparticles, stated that the disintegration time of the formulated films increased with increasing HPMC concentration. He attributed this to the fact that increasing the HPMC concentration increases water viscosity when the film comes into contact with it, which inhibited its intake and retarded the film disintegration [20].

On the other hand, changing the plasticizer type (X_3) significantly influenced the disintegration time of the prepared OFDFs. Where, OFDFs containing glycerol as a plasticizer were found to have a lower disintegration time than OFDFs prepared with PG and PEG 400. These results were in harmony with those stated by Singh et al., who prepared oral films loaded with desloratadine [50].

3.5.9. In Vitro Dissolution Studies

For OFDFs, time is an important factor, because the loaded drug should be dissolved within a minute. Figure 4 illustrates the dissolution profiles of all the prepared OFDFs; where, the % PX dissolved after 10 min from all the prepared OFDFs was found to be from $12.14 \pm 0.08\%$ to $96.02 \pm 3.46\%$, as displayed in Table 3. The following equation was used to analyze the % PX dissolved after 10 min:

$$\%\ \text{PX dissolved after 10 min} = 36.27 - 7.47 \times X_1 - 15.69 \times X_2 + 15.53 \times X_3\,[1] - 10.15 \times X_3\,[2] \quad (13)$$

where X_1 is the polymer type, X_2 is the polymer concentration, and X_3 is the plasticizer type. X_3 [1] represents the first plasticizer type (glycerol) and X_3 [2] represents the second plasticizer type (propylene glycol).

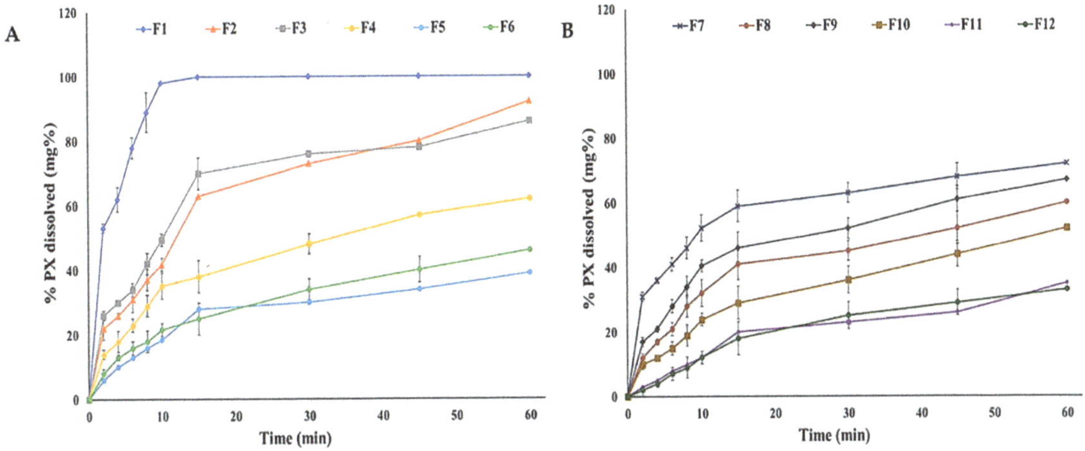

Figure 4. Dissolution profiles of the prepared (**A**) pectin and (**B**) CMC OFDFs loaded with PX nanosuspension.

Table 3 and Figure 3b show that all the independent variables (X_1, X_2, and X_3) had a significant influence on the % PX dissolved after 10 min from the prepared OFDFs.

The polymer type (X_1) had a significant influence on the % PX dissolved after 10 min. Where, OFDFs prepared by pectin polymer showed higher % PX dissolved after 10 min upon comparison with CMC-based OFDFs. This could have been because pectin has a more hydrophilic nature than CMC, resulting in faster hydration of pectin OFDFs, as explained previously [61]. Moreover, this could be because of the ionization of pectin at pH 6.8 (pH of the utilized dissolution medium), which is >pKa value of pectin (3.5) [62]. The ionization of pectin resulted in the presence of negative charges on the pectin backbone. Thus, the pectin polymer was uncoiled in the form of an extended structure, because of the negative charge repulsion, and the diffusion of positive charges within the pectin matrix generated an extra difference in the osmotic pressure across the matrix, which caused a higher water uptake. Hence, the pectin polymer swelled, resulting in drug diffusion from the films at a higher rate [63].

In addition, the statistical analyses clarified that polymer concentration (X_2) had a significant impact on % PX dissolved after 10 min. Where, higher % PX dissolved after 10 min was shown in OFDFs with lower polymer concentration. Shaikh et al. reported similar findings, where they found that using a low polymer concentration led to needing a lower amount of water to dissolve the film and leading to faster drug release [64].

On the other hand, plasticizer type (X_3) had a significant impact on %PX dissolved after 10 min. Where, OFDFs contained glycerol as a plasticizer had a higher %PX dissolved than those films contained PG or PEG 400. As explained earlier, this could be attributed to the high glycerol hygroscopicity, which led to more humidity absorption. This resulted in increasing the film hydrophilic character and increasing the %PX dissolved [56].

3.6. Selection of the Optimized OFDF Loaded with PX Nanosuspension

To identify the optimal OFDF, it was nearly impossible to fulfill all the desired responses at the same time, because the optimum condition fulfilled for one response could adversely affect other responses [22]. However, the desirability function combined all the desired responses in one variable, in order to determine the optimum levels of the examined factors [31]. Figure 3c shows that the highest desirability value was 0.659 for the optimized PX OFDF (F1) containing pectin polymer with a concentration of 1% *w/v* and glycerol as a plasticizer. This optimized OFDF (F1) collectively showed the maximal tensile strength, elongation %, % PX dissolved after 10 min, and minimal Young's modulus and disintegration time. Where, OFDF (F1) showed a tensile strength of 3.89 ± 0.19 Mpa, elongation % of 53.08 ± 1.28, Young's modulus of 8.12 ± 0.13 Mpa, disintegration time of 17.09 ± 1.30 s, and 96.02 ± 3.46% PX dissolved after 10 min. Upon comparing the observed and predicted values, they were found to be very similar; as shown in Table 2. Consequently, the optimized OFDF (F1) was chosen for further investigation.

3.7. Characterization of the Optimized OFDF Loaded with PX Nanosuspension

3.7.1. Re-Dispersion of PX Nanoparticles from the Optimized OFDF

There was no significant difference between the PS of PX nanosuspension (217.09 ± 4.18 nm) and PS measures after re-dispersion of the optimized OFDF (231.88 ± 3.50 nm) ($p < 0.05$). This slight increase in PS could be attributed to coating the embedded nanoparticles with the polymeric matrix (pectin) and the plasticizer (glycerol) used in preparing the film. These results indicate the stability of the PX nanoparticles within the polymeric matrix [15].

3.7.2. Comparative Dissolution Study of the Optimized OFDF (F1), Pure Drug, and the Market Tablet

The optimized OFDF (F1) showed a significant increase in PX dissolution rate and extent in comparison with the PX dissolution from pure drug powder and the market tablet, with f_2 values of 6 and 11, respectively. Figure 5 illustrates that 96.02% (more than 75%) PX was dissolved within just 10 min from the optimized OFDF (F1), compared with 10.04 and 26.37% from pure PX powder and the market tablet, respectively. The extent of dissolution of the optimized OFDF (F1) after 10 min was increased by more than 9.5 and 3.6 fold compared to the drug released from the pure drug and the market tablet, respectively. Additionally, the release $T_{50\%}$ of the optimized OFDF (F1) was 3.96 min, while the release $T_{50\%}$ of PX pure powder and the market tablet were 112.24 min and 54.16 min, respectively.

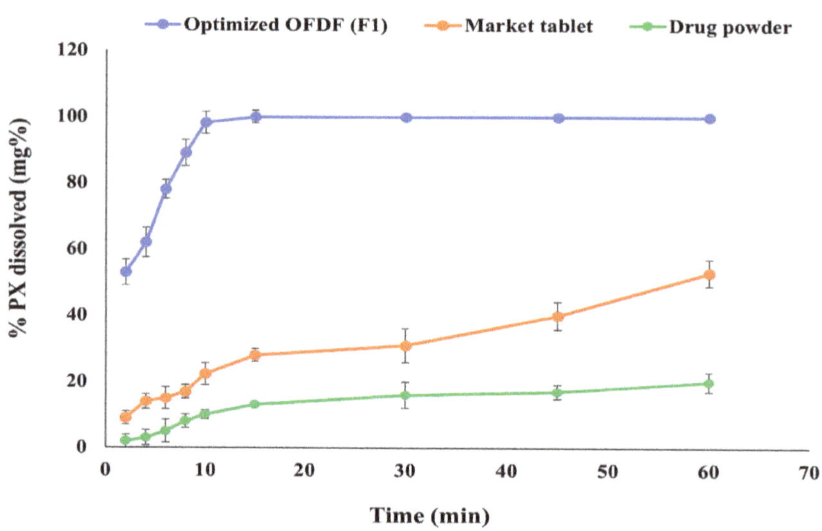

Figure 5. Dissolution profiles of the optimized OFDF loaded with PX nanosuspension (F1), PX pure powder, and the market tablet.

This could be attributed to the addition of tween 80, which enhanced the drug release from the optimized OFDF (F1), due to aiding in the acceleration of the disintegration of the film and releasing the incorporated drug more rapidly [15].

Another important reason for enhancement of the PX dissolution from the optimized OFDF was the formulation of the loaded drug as nanosized drug particles, which increased the drug solubility due to embedding the nanosized PX particles in the hydrophilic matrix. Moreover, the presence of the drug at nano-size could also have decreased the diffusion layer thickness, increasing the concentration gradient, which consequently increased the drug dissolution rate from the optimized film [65].

The low PX release from the pure PX powder could be attributed to the poor dissolution rate of paroxetine [6].

3.7.3. Ex Vivo Permeation Studies

Many permeability studies have revealed the importance of choosing an appropriate mucosal membrane; where, the buccal mucosa of many experimental animals, such as rabbits and rats, is entirely covered with keratin [66]. On the other hand, chicken buccal mucosa (pouch) is considered the best alternative, because it resembles the human non-keratinized and thin oral lining mucosa [67].

The PX permeation from the optimized OFDF (F1) and PX pure powder through the freshly excised chicken buccal mucosa (pouch) was examined. HPLC analysis showed a linearity coefficient, LLOQ, and accuracy range of 0.9994, 10 µg/mL, and 100% ± 10, respectively. The optimized OFDF (F1) showed a significant increase in PX permeation rate and extent when compared to PX pure powder, as illustrated in Figure 6. The flux (J) values were detected and there was a significant difference (p value < 0.001) between the optimized OFDF (F1) (85.00 µg/h/cm^2) when compared with PX pure powder (26.66 µg/h/cm^2). The enhancement ratio (ER) value of the optimized OFDF (F1) was also estimated and was found to be 3.18, reflecting a more than three-fold increase in PX permeation through the buccal mucosa.

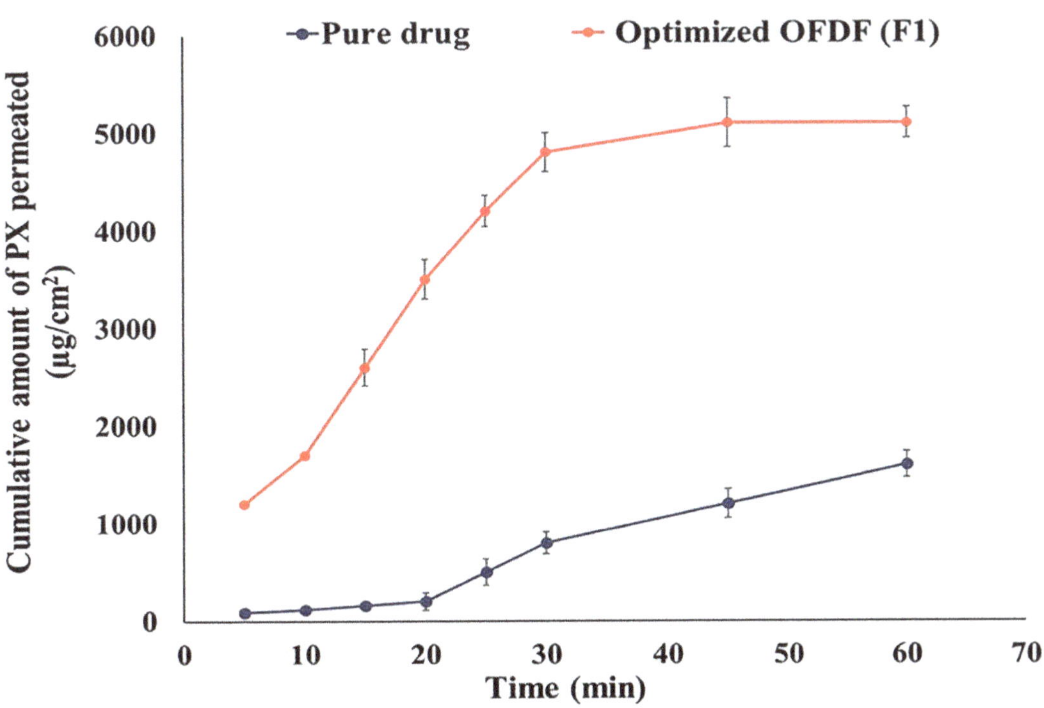

Figure 6. Permeation profiles of PX from the optimized OFDF (F1) and the pure drug through the chicken buccal mucosa.

These results could be attributed to coating the loaded PX NPs with hydrophilic polymer, which enhanced the drug solubility and increased the surface area in contact with the mucosal membrane surface, resulting in increased permeability [15]. These results could also be attributed to the incorporation of tween 80 into the optimized film. Whereby, this surfactant provided an elastic effect that loosened or fluidized the mucosal membrane lipid bilayer and made the drug capable of squeezing into deeper layers of the biological membrane, leading to the enhancement of drug permeability through the buccal mucosa [31]. Regarding the mechanism of PX transport across the buccal mucosal membrane, the uptake process was concentration-dependent, via simple diffusion [67].

3.7.4. Stability Study

Table 4 illustrates that there were no significant changes ($p > 0.05$) concerning all the investigated parameters of the optimized OFDF (F1) upon storage under the applied stability conditions.

Table 4. Stability test parameters for the optimized OFDF (F1).

Optimized OFDF (F1)	Content Uniformity (%)	Tensile Strength (Mpa)	% Elongation	Young's Modulus (Mpa)	Disintegration Time (s)	% PX Dissolved after 10 min
Freshly prepared	96.68 ± 3.62	3.89 ± 0.19	53.08 ±1.28	8.12 ± 0.13	17.09 ± 1.30	96.02 ± 3.46
After 3 months	95.70 ± 3.14	4.02 ± 0.25	48.34 ± 0.03	8.06 ± 0.32	15.24 ± 0.87	96.50 ± 1.78
After 6 months	93.89 ± 4.08	3.93 ± 0.12	48.29 ± 0.16	7.99 ± 0.30	20.33 ± 1.01	95.63 ± 2.44

3.8. *In Vivo Clinical Studies*

3.8.1. In Situ Disintegration Time and Palatability Studies

The recorded results for texture evaluation showed that 87.5% of the enrolled volunteers stated that the optimized OFDF (F1) was flexible, easy to handle, and non-sticky.

The enrolled volunteers declared that the mean in situ disintegration time of the optimized OFDF (F1) was less than 1 min (14.84 ± 2.15 s), which is similar to the previously recorded in vitro disintegration time of the optimized OFDF (17.09 ± 1.30 s). This fast-disintegration could have been due to citric acid's incorporation in the OFDF, which stimulated the saliva secretion in the buccal cavity, promoting rapid disintegration of the OFDF [51].

Regarding the results of the taste evaluation, 100% of the volunteers stated that PX pure powder was very bitter. While, in the case of the optimized OFDF (F1), 12.5% of the volunteers stated an acceptable taste, 25% stated pleasant taste, and 62.5% stated a very pleasant taste. These findings represent a success in masking PX's bitter taste by incorporating sucralose as a sweetening agent [52]. Moreover, approximately 87.5% of the volunteers reported a mouth refreshment feeling. This feeling can be attributed to the incorporation of menthol in the OFDF [53].

All the enrolled volunteers also stated that the aftertaste of the optimized OFDF (F1) was significantly improved when compared to PX pure powder. Overall, these findings indicate that the optimized OFDF has the desired properties to be an easily handled and palatable fast-dissolving film.

3.8.2. Pharmacokinetic Parameters of PX in Healthy Human Volunteer s

LC-MS/MS Method for Detection of Paroxetine in Human Plasma

No significant interference with paroxetine or Paroxetine-6D maleate (IS) was noticed in the chromatographed human plasma utilized in the preparation of quality control samples and calibration standards. The retention times of paroxetine and Paroxetine-6D maleate were 1.6 and 1.7 min, respectively. The linear relationship between PX concentrations and peak area ratio of PX/Paroxetine-6D maleate exhibited a linearity coefficient equal to 0.995, and the LLOQ was 1 ng/mL.

Estimation of Bioequivalence

All the volunteers tolerated the procedures executed in this study and the investigated drug well. Figure 7 illustrates PX mean plasma concentration–time profiles following oral administration of both the treatments. The PX pharmacokinetic parameters determined for the two treatments are shown in Table 5.

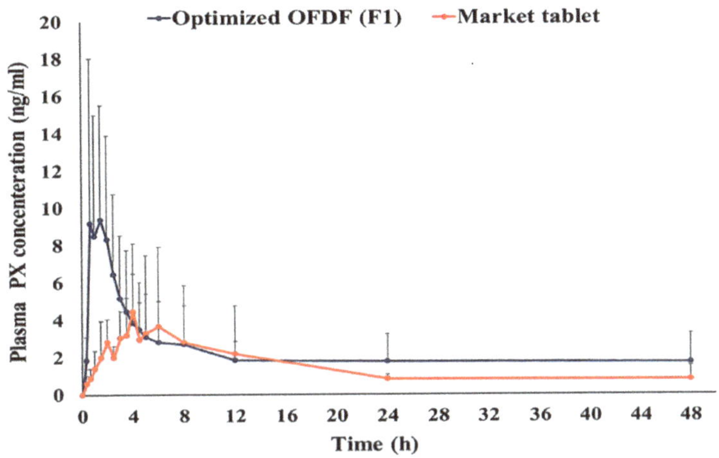

Figure 7. Mean paroxetine plasma concentration time curve after administration of the optimized OFDF (F1) and the market tablet to six healthy human volunteers.

Table 5. Drug pharmacokinetic parameters after the administration of the optimized OFDF (F1) compared to the market tablet.

Pharmacokinetics Parameter	Treatment (Mean ± SD)	
	Optimized OFDF (F1)	Market Tablet
C_{max} (ng/mL) [a]	11.18 ± 7.86	6.44 ± 3.77
$AUC_{0–48}$ (ng.h/mL) [a]	108.92 ± 81.31	69.79 ± 52.92
$AUC_{0-\infty}$ (ng.h/mL) [a]	165.07 ± 135.10	92.51 ± 67.35
t_{max} (h) [a]	0.94 ± 0.54	3.08 ± 1.88
$t_{1/2}$ (h) [a]	22.54 ± 4.11	22.32 ± 4.81
K (l/h) [a]	0.030 ± 0.01	0.030 ± 0.01
MRT [a]	37.90 ± 7.29	34.06 ± 8.24
% Relative bioavailability (%RB)	178.43	-

[a] Data are the mean values (n = 6) ± SD.

Statistical Analysis of Paroxetine Pharmacokinetic Parameters

It was noticed that the optimized OFDF (F1) had a significant increase in C_{max} (1.74 folds), $AUC_{0–48}$ (1.56 folds), and $AUC_{0-\infty}$ (1.78 folds), when compared with the market tablet (p-value < 0.05). These results clarified that the extent of PX absorption from the optimized OFDF (F1) was significantly higher than the absorption from the market tablet. The optimized OFDF (F1) had a significantly lower T_{max} value when compared with the market tablet (p-value less than 0.05). This is might have been because of the rapid disintegration of the optimized OFDF and rapid dissolution of the PX in saliva, leading to fast absorption of PX through the buccal mucosa, and reaching higher plasma concentrations more rapidly [55]. The relative bioavailability of the optimized OFDF (F1) upon comparison with the market tablet was 178.43%. These findings fulfill the goal of this study, for enhancement of paroxetine bioavailability.

This increase in PX bioavailability can be credited to many reasons. First, the presence of the drug at nano-size within the OFDF, which led to increased drug solubility and dissolution rate, in comparison with the market product [16]. Second, the rapid disintegration of the optimized OFDF resulted in rapid absorption through the buccal mucosa and the prevention of large amounts of PX being metabolized in liver, achieving a higher bioavailability [53]. Third, the incorporation of tween 80 in the film formulation could have led to increased PX particle permeability at the absorption sites, which boosted the absorbed fraction of PX [17,21,27].

4. Conclusions

A PX nanosuspension, prepared using a solvent–antisolvent precipitation method, was successfully loaded into OFDFs prepared using a solvent casting method. The OFDFs loaded with PX nanosuspension represent a palatable and stable dosage method, which can be easily taken by pediatric, geriatric, and psychiatric patients. More than 90% of PX was dissolved within 10 min from the optimized OFDF, compared with 10.04 and 26.37% from the pure drug and the market tablet, respectively. A permeation study utilizing chicken buccal pouch revealed increasing drug permeation with the optimized OFDF, with a more than three-fold increase in permeation over the pure drug. Moreover, an in vivo bioavailability estimation in healthy human volunteers clarified that the optimized OFDF (F1) increased the PX bioavailability significantly more than the market tablet. Hence, the prepared OFDF can be considered a promising, convenient, and economical approach to boosting paroxetine bioavailability.

Author Contributions: Conceptualization, R.M.E.-D.; Methodology, A.H.E. and R.M.E.-D.; Software, R.M.E.-D.; Formal Analysis, A.H.E. and R.M.E.-D.; Investigation, R.M.E.-D.; Resources, R.M.E.-D.; Data Curation, A.H.E.; Writing-Original Draft Preparation, R.M.E.-D.; Writing-Review & Editing, A.H.E.; Visualization, R.M.E.-D.; Supervision, A.H.E. All authors have read and agreed to the published version of the manuscript.

Funding: This research received no external funding.

Institutional Review Board Statement: The study was conducted according to the guidelines of the Declaration of Helsinki, and approved by the Institutional Review Board of Genuine Research Center (GRC/1/21/R4). The date of approval was (14/03/2021).

Informed Consent Statement: Written informed consent has been obtained from the patients to publish this paper.

Conflicts of Interest: The authors declare no conflict of interest.

References

1. Jin, S.J.; Yoo, Y.H.; Kim, M.S.; Kim, J.S.; Park, J.S.; Hwang, S.J. Paroxetine hydrochloride-controlled release POLYOX® matrix tablets: Screening of formulation variables using Plackett–Burman screening design. *Arch. Pharm. Res.* **2008**, *31*, 399–405. [CrossRef] [PubMed]
2. Pandey, Y.R.; Kumar, S.; Gupta, B.K.; Ali, J.; Baboota, S. Intranasal delivery of paroxetine nanoemulsion via the olfactory region for the management of depression: Formulation, behavioural and biochemical estimation. *Nanotechnology* **2016**, *27*, 25102. [CrossRef]
3. El-Nabarawi, M.A.; Bendas, E.R.; Tag, R.; El Rehem, A.; Abary, M.Y.S. Transdermal drug delivery of paroxetine through lipid-vesicular formulation to augment its bioavailability. *Int. J. Pharm.* **2013**, *443*, 307–317. [CrossRef]
4. Gudas, G.K.; Battacharjee, C. Design and evaluation of buccoadhesive bi-layer tablet of paroxetine hydrochloride. *Indian J. Pathol. Res. Pract.* **2011**, *2*, 167–169.
5. Rabinow, B.E. Nanosuspensions in drug delivery. *Nat. Rev. Drug Discov.* **2004**, *3*, 785–796. [CrossRef]
6. Xu, Y.; Liu, X.; Lian, R.; Zheng, S.; Yin, Z.; Lu, Y.; Wu, W. Enhanced dissolution and oral bioavailability of aripiprazole nanosuspensions prepared by nanoprecipitation/homogenization based on acid-base neutralization. *Int. J. Pharm.* **2012**, *438*, 287–295. [CrossRef] [PubMed]
7. Keck, C.M.; Muller, R.H. Drug nanocrystals of poorly soluble drugs produced by high pressure homogenisation. *Eur. J. Pharm. Biopharm.* **2006**, *62*, 3–16. [CrossRef]
8. El-Feky, Y.A.; Mostafa, D.A.; Al-Sawahli, M.M.; El-Telbany, R.F.; Zakaria, S.; Fayez, A.M.; Ahmed, K.A.; Alolayan, E.M.; El-Telbany, D.F. Reduction of intraocular pressure using timolol orally dissolving strips in the treatment of induced primary open-angle glaucoma in rabbits. *J. Pharm. Pharmacol.* **2020**, *72*, 682–698. [CrossRef] [PubMed]
9. Siddiqui, M.N.; Garg, G.; Sharma, P.K. A short review on-a novel approach in oral fast dissolving drug delivery system and their patents. *Adv. Biol. Res.* **2011**, *5*, 291–303.

10. Liu, C.; Chang, D.; Zhang, X.; Sui, H.; Kong, Y.; Zhu, R.; Wang, W. Oral fast-dissolving films containing lutein nanocrystals for improved bioavailability: Formulation development, in vitro and in vivo evaluation. *AAPS PharmSciTech* **2017**, *18*, 2957–2964. [CrossRef]
11. Lai, K.L.; Fang, Y.; Han, H.; Li, Q.; Zhang, S.; Li, H.W.; Chow, S.F.; Lam, T.N.; Lee, W.Y.T. Orally-dissolving film for sublingual and buccal delivery of ropinirole. *Colloids Surf. B* **2018**, *163*, 9–18. [CrossRef]
12. Mohapatra, S.S.; Ranjan, S.; Dasgupta, N.; Mishra, R.K.; Tomas, S. (Eds.) *Nanocarriers for Drug Delivery: Nanoscience and Nanotechnology in Drug Delivery*; Elsevier: Amsterdam, The Netherlands, 2018.
13. Garsuch, V.; Breitkreutz, J. Comparative investigations on different polymers for the preparation of fast-dissolving oral films. *J. Pharm. Pharmacol.* **2010**, *62*, 539–545. [CrossRef] [PubMed]
14. Irfan, M.; Rabel, S.; Bukhtar, Q.; Qadir, M.I.; Jabeen, F.; Khan, A. Orally disintegrating films: A modern expansion in drug delivery system. *Saudi Pharm. J.* **2015**, *24*, 537–546. [CrossRef] [PubMed]
15. Nakarani, M.; Misra, A.K.; Patel, J.K.; Vaghani, S.S. Itraconazole nanosuspension for oral delivery: Formulation, characterization and in vitro comparison with marketed formulation. *Daru* **2010**, *18*, 84–90. [PubMed]
16. Londhe, V.; Shirsat, R. Formulation and characterization of fast-dissolving sublingual film of iloperidone using Box–Behnken design for enhancement of oral bioavailability. *AAPS PharmSciTech* **2018**, *19*, 1392–1400. [CrossRef]
17. Rai, V.; Tan, H.S.; Michniak-Kohn, B. Effect of surfactants and pH on naltrexone (NTX) permeation across buccal mucosa. *Inter. J. Pharm.* **2011**, *411*, 92–97. [CrossRef]
18. Shariare, M.; Sharmin, S.; Jahan, I.; Reza, H.; Mohsin, K. The impact of process parameters on carrier free paracetamol nanosuspension prepared using different stabilizers by antisolvent precipitation method. *J. Drug Deliv. Sci. Technol.* **2018**, *43*, 122–128. [CrossRef]
19. Chandra, A.; Chondkar, A.D.; Shirodkar, R.; Lewis, S.A. Rapidly dissolving lacidipine nanoparticle strips for transbuccal administration. *J. Drug Deliv. Sci. Technol.* **2018**, *47*, 259–267. [CrossRef]
20. Shen, C.; Shen, B.; Xu, H.; Bai, J.; Dai, L.; Lv, Q.; Han, J.; Yuan, H. Formulation and optimization of a novel oral fast dissolving film containing drug nanoparticles by Box–Behnken design response surface methodology. *Drug Dev. Ind. Pharm.* **2014**, *40*, 649–656. [CrossRef]
21. Chavan, D.U.; Marques, S.M.; Bhide, P.J.; Kumar, L.; Shirodkar, R.K. Rapidly dissolving Felodipine nanoparticle strips—Formulation using design of experiment and characterisation. *J. Drug Deliv. Sci. Technol.* **2020**, *60*, 102053. [CrossRef]
22. Elsayed, I.; El-Dahmy, R.M.; Elshafeey, A.H.; Abd El Gawad, N.A.; El Gazayerly, O.N. Tripling the bioavailability of rosuvastatin calcium through development and optimization of an in situ forming nanovesicular system. *Pharmaceutics* **2019**, *11*, 275. [CrossRef]
23. Shivhare, U.D.; Bodkhe, P.D.; Bhusari, K.P.; Mathur, V.B. Formulation and evaluation of buccoadhesive films of losartan potassium. *Pharm. Lett.* **2010**, *2*, 251–260.
24. Smriti, T. Mouth dissolving films: A review. *Int. J. Pharma Bio. Sci.* **2013**, *4*, 899–908.
25. Satyanarayana, D.A.; Keshavarao, K.P. Fast disintegrating films containing anastrazole as a dosage form for dysphagia patients. *Arch. Pharm. Res.* **2012**, *35*, 2171–2182. [CrossRef] [PubMed]
26. Miles, K.B.; Ball, R.L.; Matthew, H.W.T. Chitosan films with improved tensile strength and toughness from *N*-acetyl-cysteine mediated disulfide bonds. *Carbohydr Polym.* **2016**, *139*, 1–9. [CrossRef]
27. Bharti, K.; Mittal, P.; Mishra, B. Formulation and characterization of fast dissolving oral films containing buspirone hydrochloride nanoparticles using design of experiment. *J. Drug Deliv. Sci. Technol.* **2019**, *49*, 420–432. [CrossRef]
28. Ali, M.; Vijendar, C.; Kumar, S.; Krishnaveni, J. Formulation and evaluation of fast dissolving oral films of diazepam. *Aust. J. Pharm.* **2016**, *4*, 1–5.
29. Nair, A.B.; Kumria, R.; Harsha, S.; Attimarad, M.; Al-Dhubiab, B.E.; Alhaider, I.A. In vitro techniques to evaluate buccal films. *J. Control. Release* **2013**, *166*, 10–21. [CrossRef] [PubMed]
30. Al-Nemrawi, N.K.; Dave, R.H. Formulation and characterization of acetaminophen nanoparticles in orally disintegrating films. *Drug Deliv.* **2016**, *23*, 540–549. [CrossRef]
31. Elsayed, I.; El-Dahmy, R.M.; El-Emam, S.Z.; Elshafeey, A.H.; Abd El Gawad, N.A.; El-Gazayerly, O.N. Response surface optimization of biocompatible elastic nanovesicles loaded with rosuvastatin calcium: Enhanced bioavailability and anticancer efficacy. *Drug Deliv. Transl. Res.* **2020**, *10*, 1459–1475. [CrossRef]
32. El-Dahmy, R.M.; Elsayed, I.; Elshafeey, A.H.; Abd El Gawad, N.A.; El-Gazayerly, O.N. Optimization of long circulating mixed polymeric micelles containing vinpocetine using simple lattice mixture design, in vitro and in vivo characterization. *Int. J. Pharm.* **2014**, *477*, 39–46. [CrossRef]
33. Wong, C.F.; Yuen, K.H.; Peh, K.K. An in vitro method for buccal adhesion studies: Importance of instrument variables. *Int. J. Pharm.* **1999**, *180*, 47–57. [CrossRef]
34. Giovino, C.; Ayensu, I.; Tetteh, J.; Boateng, J.S. An integrated buccal delivery system combining chitosan films impregnated with peptide loaded PEG-b-PLA nanoparticles. *Colloids Surf. B. Biointerfaces* **2013**, *112*, 9–15. [CrossRef] [PubMed]
35. Elkomy, M.H.; El Menshawe, S.F.; Abou-Taleb, H.A.; Elkarmalawy, M.H. Loratadine bioavailability via buccal transferosomal gel: Formulation, statistical optimization, in vitro/in vivo characterization, and pharmacokinetics in human volunteers. *Drug Deliv.* **2017**, *24*, 781–791. [CrossRef] [PubMed]

36. Ali, A.A.; Hassan, A.H.; Eissa, E.M.; Aboud, H.M. Response surface optimization of ultra-elastic nanovesicles loaded with deflazacort tailored for transdermal delivery: Accentuated bioavailability and anti-inflammatory efficacy. *Int. J. Nanomed.* **2021**, *2021*, 591–607. [CrossRef] [PubMed]
37. International Conference on Harmonisation of Technical Requirements for Registration of Pharmaceuticals for Human Use (ICH). Stability Testing of New Drug Substances and Products Q1A (R2). ICH Harmonized Tripartite Guidelines. 2003. Available online: http://www.alz.org/what-is-dementia.asp (accessed on 6 February 2021).
38. World Medical Association Declaration of Helsinki. Ethical principles for medical research involving human subjects. *JAMA* **2013**, *310*, 2191–2194. [CrossRef]
39. Smith, P.E. Third international conference on harmonization of technical requirements for registration of pharmaceuticals for human use—A toxicologist's perspective. *Toxicol. Pathol.* **1996**, *24*, 519–528. [CrossRef]
40. Auda, S.H.; Elbadry, M.; Ibrahim, M.A. Design, formulation and characterization of fast dissolving films containing dextromethorphan. *Dig. J. Nanomat. Biostruc.* **2014**, *9*, 133–141.
41. Khan, S.; Kataria, P.; Nakhat, P.; Yeole, P. Taste masking of ondansetron hydrochloride by polymer carrier system and formulation of rapid-disintegrating tablets. *AAPS PharmSciTech* **2007**, *8*, 46. [CrossRef] [PubMed]
42. Elshafeey, A.H.; Kamel, A.O.; Awad, G.A.S. Ammonium methacrylate units polymer content and their effect on acyclovir colloidal nanoparticles properties and bioavailability in human volunteers. *Colloids Surf. B Biointerfaces* **2010**, *75*, 398–404. [CrossRef]
43. Shah, V.P.; Midha, K.K.; Dighe, S.; McGilveray, I.J.; Skelly, J.P.; Yacobi, A.; Layloff, T.; Viswanathan, C.T.; Cook, C.E.; McDowall, R.D.; et al. Analytical methods validation: Bioavailability, bioequivalence and pharmacokinetic studies. *Eur. J. Drug Metab. Pharmacokinet.* **1991**, *16*, 249. [PubMed]
44. Massaroti, P.; Cassiano, N.M.; Duarte, L.F.; Campos, D.R.; Marchioretto, M.A.M.; Bernasconi, G.; Calafatti, S.; Barros, F.A.P.; Meurer, E.C.; Pedrazzoli, J. Validation of a selective method for determination of paroxetine in human plasma by LC-MS/MS. *J. Pharm. Pharm. Sci.* **2005**, *8*, 340–347. [PubMed]
45. Wu, L.; Zhang, J.; Watanabe, W. Physical and chemical stability of drug nanoparticles. *Adv. Drug Deliv. Rev.* **2011**, *63*, 456–469. [CrossRef] [PubMed]
46. Muller, R.H.; Jacobs, C.; Kayser, O. Nanosuspensions as particulate drug formulations in therapy. Rationale for development and what we can expect for the future. *Adv. Drug Deliv. Rev.* **2001**, *47*, 3–19. [CrossRef]
47. Vueba, M.L.; Batista, D.; Carvalho, L.A.E.; Veiga, F.; Sousa, J.J.; Pina, M.E. Influence of cellulose ether polymers on ketoprofen release from hydrophilic matrix tablets. *Eur. J. Pharm. Biopharm.* **2004**, *58*, 51–59. [CrossRef] [PubMed]
48. Iijima, M.; Nakamura, K.; Hatakeyama, T.; Hatakeyama, H. Phase transition of pectin with sorbed water. *Carbohydr. Polym.* **2000**, *41*, 101–106. [CrossRef]
49. Akhtar, H.M.S.; Riaz, A.; Hamed, Y.S.; Abdin, M.; Chen, G.; Wan, P.; Zeng, X. Production and characterization of CMC-based antioxidant and antimicrobial films enriched with chickpea hull polysaccharides. *Int. J. Biol. Macromol.* **2018**, *118*, 469–477. [CrossRef]
50. Singh, H.; Kaur, M.; Verma, H. Optimization and evaluation of desloratadine oral strip: An innovation in paediatric medication. *Sci. World J.* **2013**, *2013*, 395681. [CrossRef]
51. Panda, B.; Dey, N.; Rao, M. Development of innovative orally fast disintegrating film dosage forms: A review. *Int. J. Pharm. Sci. Nanotechnol.* **2012**, *5*, 1666–1674. [CrossRef]
52. Cilurzo, F.; Cupone, I.E.; Minghetti, P.; Buratti, S.; Gennari, C.G.; Montanari, L. Diclofenac fast-dissolving film: Suppression of bitterness by a taste-sensing system. *Drug Dev. Ind. Pharm.* **2011**, *37*, 252–259. [CrossRef]
53. Allam, A.; Fetih, G. Sublingual fast dissolving niosomal films for enhanced bioavailability and prolonged effect of metoprolol tartrate. *Drug Des. Dev. Ther.* **2016**, *10*, 2421–2433.
54. Dinge, A.; Nagarsenker, M. Formulation and evaluation of fast dissolving films for delivery of triclosan to the oral cavity. *AAPS PharmSciTech.* **2008**, *9*, 349–356. [CrossRef]
55. Maher, E.M.; Mahmoud, A.; Ali, A.; Farouk, H.; Abdelrahman, A.A.; Magdy, E.; Mahmoud, A.; Ali, A.; Salem, H.F.; Maher, E.M.; et al. In vitro/in vivo evaluation of an optimized fast dissolving oral film containing olanzapine co-amorphous dispersion with selected carboxylic acids. *Drug Deliv.* **2016**, *23*, 3088–3100. [CrossRef] [PubMed]
56. Vieira, M.G.A.; da Silva, M.A.; dos Santos, L.O.; Beppu, M.M. Natural-based plasticizers and biopolymer films: A review. *Eur. Polym. J.* **2011**, *47*, 254–263. [CrossRef]
57. Fishman, M.L.; Coffin, D.R.; Konstance, R.P.; Onwulata, C.I. Extrusion of pectin/starch blends plasticized with glycerol. *Carbohydr. Polym.* **2000**, *41*, 317–325. [CrossRef]
58. Junmahasathien, T.; Panraksa, P.; Protiarn, P.; Hormdee, D.; Noisombut, R.; Kantrong, N.; Jantrawut, P. Preparation and evaluation of metronidazole-loaded pectin films for potentially targeting a microbial infection associated with periodontal disease. *Polymers* **2018**, *10*, 1021. [CrossRef]
59. Roberts, R.J.; Rowe, R.C. The Young's modulus of pharmaceutical materials. *Int. J. Pharm.* **1987**, *37*, 15–18. [CrossRef]
60. Bermudez-Oria, A.; Rodriguez- Guiterrez, G.; Voque, B.; Rubio-Senet, F.; Demandez-Bolanos, J. Physical and functional properties of pectin-fish gelatin films containing the olive phenols hydroxytyrosol and 3,4-dihydroxyphenylglycol. *Carbohyd. Polym.* **2017**, *178*, 368–377. [CrossRef]
61. Pathare, Y.S.; Hastak, V.S.; Bajaj, A.N. Polymers used for fast disintegrating oral films: A review. *Int. J. Pharm. Sci. Rev. Res.* **2013**, *21*, 169–178.

62. Opanasopit, P.; Apirakaramwong, A.; Ngawhirunpat, T.; Rojanarata, T.; Ruktanonchai, U. Development and characterization of pectinate micro/nanoparticles for gene delivery. *AAPS PharmSciTech* **2008**, *9*, 67–74. [CrossRef] [PubMed]
63. Sujja-Areevath, J.; Munday, D.L.; Cox, P.J.; Khan, K.A. Relationship between swelling, erosion and drug release in hydrophillic natural gum mini-matrix formulations. *Eur. J. Pharm. Sci.* **1998**, *6*, 207–217. [CrossRef]
64. Shaikh, M.T.M.; Gore, A.A.; Salunkhe, K.S.; Chaudhari, S.R. Formulation development and evaluation of fast dissolving oral film of amlodipine besilate by solvent casting technique. *Int. J. Pharm. Biol. Sci.* **2013**, *2*, 534–544.
65. Abdelbary, A.A.; Li, X.; El-Nabarawi, M.; Elassasy, A.; Jasti, B. Comparison of nanomilling and coprecipitation on the enhancement of in vitro dissolution rate of poorly water-soluble model drug aripiprazole. *Pharm. Dev. Technol.* **2014**, *19*, 491–500. [CrossRef] [PubMed]
66. Shojaei, A.H. Buccal mucosa as a route for systemic drug delivery: A review. *J. Pharm. Pharm. Sci.* **1998**, *1*, 15–30.
67. Abd El Azim, H.; Nafee, N.; Ramadan, A.; Khalafallah, N. Liposomal buccal mucoadhesive film for improved delivery and permeation of water-soluble vitamins. *Inter. J. Pharm.* **2015**, *488*, 78–85. [CrossRef]

pharmaceutics

Article

In Vitro-In Silico Tools for Streamlined Development of Acalabrutinib Amorphous Solid Dispersion Tablets

Deanna M. Mudie *, Aaron M. Stewart , Jesus A. Rosales †, Molly S. Adam, Michael M. Morgen and David T. Vodak

Global Research & Development, Lonza, Bend, OR 97703, USA; aaron.stewart@lonza.com (A.M.S.); rosaleja@uw.edu (J.A.R.); molly.adam@lonza.com (M.S.A.); michael.morgen@lonza.com (M.M.M.); david.vodak@lonza.com (D.T.V.)
* Correspondence: deanna.mudie@lonza.com
† Current address: Pharmaceutics Department, University of Washington, Seattle, WA 98195, USA.

Abstract: Amorphous solid dispersion (ASD) dosage forms can improve the oral bioavailability of poorly water-soluble drugs, enabling the commercialization of new chemical entities and improving the efficacy and patient compliance of existing drugs. However, the development of robust, high-performing ASD dosage forms can be challenging, often requiring multiple formulation iterations, long timelines, and high cost. In a previous study, acalabrutinib/hydroxypropyl methylcellulose acetate succinate (HPMCAS)-H grade ASD tablets were shown to overcome the pH effect of commercially marketed Calquence in beagle dogs. This study describes the streamlined in vitro and in silico approach used to develop those ASD tablets. HPMCAS-H and -M grade polymers provided the longest acalabrutinib supersaturation sustainment in an initial screening study, and HPMCAS-H grade ASDs provided the highest in vitro area under the curve (AUC) in gastric to intestinal transfer dissolution tests at elevated gastric pH. In silico simulations of the HPMCAS-H ASD tablet and Calquence capsule provided good in vivo study prediction accuracy using a bottom–up approach (absolute average fold error of $AUC_{0\text{-inf}} < 2$). This streamlined approach combined an understanding of key drug, polymer, and gastrointestinal properties with in vitro and in silico tools to overcome the acalabrutinib pH effect without the need for reformulation or multiple studies, showing promise for reducing time and costs to develop ASD drug products.

Keywords: acalabrutinib; amorphous solid dispersion; bioavailability enhancement; acid reducing agent; proton pump inhibitor; kinase inhibitor; in silico prediction; absorption modeling; spray drying; GastroPlus

Citation: Mudie, D.M.; Stewart, A.M.; Rosales, J.A.; Adam, M.S.; Morgen, M.M.; Vodak, D.T. In Vitro-In Silico Tools for Streamlined Development of Acalabrutinib Amorphous Solid Dispersion Tablets. *Pharmaceutics* **2021**, *13*, 1257. https://doi.org/10.3390/pharmaceutics13081257

Academic Editors: Vitaliy Khutoryanskiy and Hisham Al-Obaidi

Received: 18 June 2021
Accepted: 9 August 2021
Published: 13 August 2021
Corrected: 2 December 2021

1. Introduction

Oncology is the top therapeutic area in pharmaceutical drug development, accounting for 25% of drugs approved by the FDA over the last decade [1]. However, many oncology drugs are poorly water soluble across at least part of the gastrointestinal (GI) pH range, often manifesting as drug–drug or food–drug interactions that can limit oral bioavailability [2]. Calquence® (crystalline acalabrutinib) is an example of an oral oncology drug that exhibits pH-dependent absorption, whereby the area under the plasma drug concentration–time curve (AUC) is reduced by 43% when taken with a proton pump inhibitor (PPI), which elevates gastric pH [3,4]. This drug–drug interaction (DDI) can result in decreased patient compliance and efficacy, since many cancer patients are prescribed PPIs and other gastric acid reducing agents (ARAs) [5].

The mechanism of reduced absorption of Calquence when taken with ARAs is decreased solubility at elevated pH, which is a common mechanism for Biopharmaceutics Classification System (BCS) 2 weak base drugs such as acalabrutinib [6,7]. A previous publication from our laboratory demonstrated that amorphous solid dispersion (ASD) tablets overcame the ARA effect at the human-prescribed 100 mg dose in a fasted beagle dog model [8]. Results demonstrated that ASD tablets achieved similar AUC values at low and high gastric pH conditions and outperformed Calquence capsules 2.4-fold at high gastric pH.

The improved performance of the ASD tablet compared to Calquence was driven by the increased aqueous solubility of the amorphous compared to the crystalline form, which is enough to provide rapid and high extent of GI dissolution at high and low gastric pH values (i.e., in the presence and absence of ARAs). In addition, ASD tablets were 60% smaller than Calquence capsules and showed good physical stability, chemical stability when stored desiccated or refrigerated, and manufacturability as described by Mudie et al. [8]. This outcome highlighted the utility of ASD dosage forms for improving the performance of the numerous weak base and oral oncology drugs that show decreased absorption when taking ARAs [3,9,10].

While the previous publication focused on the in vivo study outcome, the current publication describes a streamlined approach to develop ASD tablets, including formulation selection, in vitro dissolution testing, and in silico simulations used to build confidence in the ASD tablet, overcoming the ARA effect observed with Calquence. A priori in silico plasma concentration–time profiles are compared with in vivo plasma profiles, and in silico prediction accuracy is calculated. This paper describes:

1. Dispersion polymer screening
2. Spray drying and characterization of ASD intermediates
3. In vitro dissolution testing of ASD intermediates versus crystalline acalabrutinib
4. In vitro dissolution testing of lead 50/50 acalabrutinib/HPMCAS-H ASD tablet versus commercially available Calquence
5. Physicochemical property measurements of lead ASD intermediate
6. In silico predictions of in vivo performance

2. Materials and Methods

2.1. Material Sourcing

Acalabrutinib (CAS 1420477-60-6, >98% purity) was purchased from LC Laboratories (Woburn, MA, USA). Form I was prepared by recrystallizing the purchased acalabrutinib according to WO 2017/002095 Al, Example 1 [11]. See Appendix A.1 for Form I verification. Hydroxypropyl methylcellulose acetate succinate (HPMCAS) (Aqoat, HF grade, MF grade and LF grade) was purchased from Shin-Etsu Chemical Co., Ltd. (Tokyo, Japan). Poly(methyl methacrylate-co-methacrylic acid (Eudragit L100®) was purchased from Evonik (Evonik Industries AG, Essen, Germany). Vinylpyrrolidone-vinyl acetate copolymer (PVPVA) (Kollidon® VA64) was purchased from BASF (Ludwigshafen, Germany). Polyvinylpyrrolidone (PVP) (Kollidon® 30) was purchased from BASF. Hydroxypropyl methylcellulose (HPMC) (Methocel™ E3 LV) was purchased from DuPont de Nemours, Inc. (Wilmington, DE, USA). Methocel™ A4M was purchased from ThermoFisher Scientific (Waltham, MA, USA). Sodium acetate, sodium phosphate, potassium phosphate, hydrochloric acid (HCl), and sodium chloride (NaCl) were purchased from Sigma Aldrich Chemical Company (St. Louis, MO, USA). Fasted-state simulated intestinal fluid (FaSSIF) powder was purchased from Biorelevant.com Ltd. (London, UK). Methanol (HPLC grade) was purchased from Honeywell (Morris Plains, NJ, USA). Tetrahydrofuran (THF) (Optima grade) was purchased from Thermo Fisher Scientific (Waltham, MA, USA). Calquence capsules were purchased from Drug World (Cold Spring, NY, USA). Avicel PH-101 (microcrystalline cellulose) was purchased from FMC Corporation (Philadelphia, PA, USA). Pearlitol 25 (mannitol) was purchased from Roquette America (Geneva, IL, USA). Ac-Di-Sol (croscarmellose sodium) was purchased from Dupont (Wilmington, DE, USA). Cab-O-Sil M5P (fumed silica) was purchased from Cabot Corporation (Alpharetta, GA, USA). Magnesium stearate was purchased from Macron Fine Chemicals/Avantor (Radnor, PA, USA).

2.2. Dispersion Polymer Screening

A polymer screening test of seven different polymers was performed using a solvent shift method to assess the polymers' abilities to sustain acalabrutinib supersaturated drug concentrations [12,13]. The polymers tested included HPMCAS-H, HPMCAS-M, HPMCAS-L, Eudragit L100, HPMC E3, PVP K30, and PVP-VA64. A 20 mg/mL acalabrutinib stock solution (95/5 (*w/w*) tetrahydrofuran (THF)/water) was delivered into an aqueous buffer containing 200 µg/mL of a dissolved polymer using a calibrated pipette. The aqueous buffer consisted of 67 mM phosphate containing 0.5% (*w/w*) (6.7 mM) FaSSIF powder and 82 mM NaCl at a pH of 6.5. A total of 200 µL stock solution was added to achieve a target final concentration of 400 µg/mL acalabrutinib.

During addition of the stock solution, light scattering was monitored by Pion Rainbow™ (Pion Inc., Billerica, MA, USA) ultraviolet (UV) probes (2 mm path length) at 500 to 600 nm (outside the UV absorbance range of acalabrutinib) to observe light scattering with increasing concentration. The scatter signal was corroborated by monitoring direct UV absorbance from 370 to 374 nm to verify that the light-scattering signal could be attributed to crystallization (loss of UV absorbance).

2.3. ASD and ASD Tablet Manufacturing and Characterization

Six different acalabrutinib ASDs were prepared using three different grades of HPMCAS (HPMCAS-H, HPMCAS-M, and HPMCAS-L) and two different drug loadings (25 and 50 wt %). HPMCAS-H and HPMCAS-M were chosen since they performed best in the polymer screening study. HPMCAS-L was also selected to determine the impact of HPMCAS grade on relative dissolution and precipitation rates of acalabrutinib. Specifically, in progressing from -L, to -M, to -H grades, HPMCAS becomes more hydrophobic, and the minimum pH above which it dissolves increases [14]. According to Friesen et al., HPMCAS grades are sparingly soluble above pH values of 4.8 (-L), 5.2 (-M), and 5.7 (-H), dispersing to form colloidal solutions [14]. When partially ionized, hydrophobic regions of HPMCAS can interact with hydrophobic drugs, and carboxylate groups can interact with the aqueous phase at the drug–water interface to inhibit crystal nucleation and growth [14–17].

Compositions, spray solvents, and total solids loadings are shown in Table 1. Solutions were spray dried with an outlet temperature of 45–50 °C and inlet temperature of 142–150 °C on a custom laboratory scale spray dryer with a 35 kg/h drying gas capacity and a 0.3 m chamber diameter. A Schlick 1.5 pressure-swirl nozzle was used for the 25% drug loading formulations, and a Schlick 2.0 pressure-swirl nozzle was used for the 50% formulations (model 121, 150 um and 200 um orifice, Schlick Americas, Bluffton, SC, USA). After material was collected in a cyclone, residual solvent was removed by secondary drying in a vacuum dryer (Model TVO-2, Cascade TEK, Cornelius, OR, USA) for >16 h at 40 °C with a nitrogen sweep gas (−60 cm Hg, 3 standard liters per minute). Solvent removal was below the International Council for Harmonization (ICH) thresholds for methanol (<3000 ppm) as confirmed using a gas chromatograph with a headspace sampler (GC).

Table 1. Compositions of ASDs and spray-drying solutions.

Drug Loading (wt %)	Dispersion Polymer	Spray Solvent	Total Dissolved Drug and Polymer in Spray Solvent (wt %)
25	HPMCAS-H	Methanol	6
25	HPMCAS-M	Methanol	6
25	HPMCAS-L	Methanol	6
50	HPMCAS-H	Methanol	5.4
50	HPMCAS-M	Methanol	5.4
50	HPMCAS-L	Methanol	5.4

Powder X-ray diffraction (PXRD) was used to confirm that ASDs were amorphous, and modulated differential scanning calorimetry (mDSC) was used to ensure a single T_g as an indication of drug–polymer homogeneity. Scanning electron microscopy (SEM) was performed to visualize morphology and to detect the presence of any surface crystals. Method details can be found in Appendix A.2.

ASD immediate release (IR) tablets were made using the 50/50 (*w/w*) acalabrutinib/ HPMCAS-H ASD. This ASD was chosen as the lead after dissolution testing, since it maximized both in vitro performance (i.e., AUC) and drug loading. The 50 wt % drug loading ASD resulted in ASD tablets that were 60% smaller than Calquence capsules (by volume) at an equivalent 100 mg unit dosage strength. ASD tablets had a 400 mg total tablet mass and a drug loading of 25 wt %. Refer to Mudie et al. [8] for tablet formulation, manufacturing methods, and characterization.

2.4. In Vitro Dissolution Testing of Intermediates

ASDs were evaluated for dissolution performance in a gastric to intestinal transfer dissolution test using a Pion µDiss™ Profiler with Rainbow™ fiber optic UV probe detection. Tests using all six ASDs were first conducted using simulated gastric media with elevated pH representative of dogs (or humans) taking an ARA. While gastric pH can vary from ≈4 to 7 with ARAs, gastric pH values of 5 and 6 were chosen, since the solubilities of different HPMCAS polymer grades (-L, -M, and -H) are most sensitive in this range [18–22].

Relative ASD performance in the elevated gastric pH tests was used to select lead ASDs (25 and 50 wt % drug loading HPMCAS-H ASDs and 50 wt % drug loading HPMCAS-M ASD). Then, these ASDs were tested at a gastric pH representative of dogs pre-treated with pentagastrin (i.e., pH 2). Simulated intestinal medium was representative of fasted dogs, with a pH of 6.5 and 0.5 wt % (6.7 mM) FaSSIF powder (see Appendix A.3 for detailed medium compositions) [23–25].

Tests were conducted at dose concentrations of 2 (gastric) and 1 (intestinal) mg/mL, which approximated dose/volume in the stomach of fasted beagle dogs taking a 100 mg dose of acalabrutinib and resulted in 'non-sink' conditions (dose/volume/solubility > 1) with respect to both the apparent amorphous and crystalline solubilites in intestinal medium [26]. Refer to Appendix A.3 for dose number calculations.

For the tests conducted using pH 5 and pH 6 simulated gastric media, crystalline acalabrutinib and ASDs were prepared as a suspension in 0.5% Methocel A4M at a concentration of 25 mg/mL acalabrutinib. To begin the experiment, 0.8 mL of suspension was added to 9.2 mL of gastric fluid to achieve a dose concentration of 2 mg/mL acalabrutinib. For the tests conducted using pH 2 simulated gastric medium, neat crystalline or ASD powder was added directly to the dissolution vessel to begin the experiment. Samples were stirred at 100 rpm and held at 37 ± 2 °C by circulating water through a heating block mounted to the Pion µDiss™ profiler. After 30 min, gastric medium was diluted 1:1 with a concentrated intestinal medium to a final volume of 20 mL and a concentration of 1 mg/mL acalabrutinib.

Data were collected for 30 min in gastric medium and for approximately 160 min in intestinal medium with Pion Rainbow™ UV probes. See Appendix A.3 for UV analysis parameters. The apparent concentrations measured consisted of (1) drug dissolved in aqueous medium and (2) drug partitioned into bile salt micelles (present in intestinal medium) as a micelle-bound drug. Each sample was measured in duplicate.

AUC in intestinal medium for each test was calculated in Microsoft Excel (Microsoft Corporation, Seattle, WA, USA) using the rectangular rule using 0.25 min time increments between times 30 and 176 min of the dissolution test (i.e., for a 146 min duration in intestinal medium). AUC enhancement was determined for each ASD sample by dividing AUC in intestinal medium for the ASD by AUC in intestinal medium for crystalline acalabrutinib tested using the same dissolution conditions.

2.5. In Vitro Dissolution Testing of Dosage Forms

ASD tablets and commercially available Calquence capsules were evaluated for dissolution performance in a gastric to intestinal transfer dissolution test using a Vankel VK7000 (now Agilent, Palo Alto, CA, USA) United States Pharmacopeia (USP) 2 dissolution apparatus equipped with 500 mL vessels. Tests were conducted using simulated gastric media (HCl) representative of dogs taking ARAs (pH 6) or dogs pretreated with pentagastrin (pH 2). Simulated intestinal medium was representative of fasted dogs, with a pH of 6.5 and 0.5 wt % (6.7 mM) FaSSIF powder (see Appendix A.3 for detailed medium compositions).

Tests were conducted at dose concentrations of 0.4 (gastric) and 0.2 (intestinal) mg/mL to allow enough of the dose to dissolve in pH 2 and pH 6 gastric media, and in intestinal medium for both the ASD tablet and Calquence capsule to facilitate the determination of z-factors (i.e., dissolution rates) for in silico predictions [27]. This dose concentration allowed for ≈20–100% dose dissolved for the Calquence capsule and ≈40–100% dose dissolved for the ASD tablet across the three media. Refer to Appendix A.3 for dose number calculations. In vito testing at a higher, non-sink dose concentration was conducted in a controlled transfer dissolution (CTD) test as described by Mudie et al. [8].

To begin the test, 250 mL of gastric medium was added to the dissolution vessel, which was followed by a single ASD tablet, or a Calquence capsule (100 mg dose) contained in a capsule sinker. After the dosage form was added, the paddles were started. Samples were stirred at 75 rpm and held at 37 ± 2 °C by circulating water through a heater attached to the USP 2 dissolution apparatus. After 30 min, gastric medium was diluted 1:1 with a concentrated intestinal medium to a final volume of 500 mL and a concentration of 0.2 mg/mL acalabrutinib.

Dissolution performance was monitored for 30 min in gastric medium and for 150 min in intestinal medium with Pion Rainbow™ UV probes. See Appendix A.3 for UV analysis parameters. The apparent concentrations measured consisted of (1) drug dissolved in aqueous medium and (2) drug partitioned into bile salt micelles (present in intestinal medium) as a micelle-bound drug. All samples were analyzed in duplicate.

AUC in intestinal medium for each test was calculated in Microsoft Excel using the rectangular rule using 0.25 min time increments between times 30 and 150 min of the dissolution test. AUC enhancement was determined for the ASD tablet by dividing AUC in intestinal medium for the ASD tablet by AUC in intestinal medium for the Calquence capsule tested using the same dissolution conditions.

2.6. Physicochemical Property Measurements of Lead ASD Intermediate

2.6.1. Amorphous Acalabrutinib Slurry pH

Slurry pH was measured as a means to estimate solid surface pH of the 50/50 (*w/w*) acalabrutinib/HPMCAS-H ASD in different HCl molarities (pH 1.6, 2.0, 3.0, 4.0, 5.0, and 6.0) containing 34 mM of NaCl [28,29]. Briefly, the 50/50 (*w/w*) acalabrutinib/HPMCAS-H ASD was added to each HCl medium and allowed to dissolve until a plateau in the pH was achieved with excess solids present. During the measurements, slurries were mixed and controlled to a temperature of 37 ± 1 °C. Measurements were performed with one sample each. Refer to Appendix A.4 for full method details.

Moderate intrinsic solubility of acalabrutinib coupled with its basic pK_as of 3.5 and 5.8 result in a solid surface pH higher than the bulk pH when solubility at the solid particle surface exceeds the buffering capacity at the solid particle surface [29,30]. Increased solid surface pH can decrease acalabrutinib solubility due to decreased acalabrutinib ionization at the solid particle surface, resulting in a decreased dissolution rate [30]. Therefore, knowledge of solid surface pH as a function of bulk pH and medium composition is critical for accurate dissolution rate predictions.

A similar measurement method was used by Pepin et al. using neat crystalline acalabrutinib [29]. For the ASD, acalabrutinib is present in the amorphous form with a higher intrinsic solubility than crystalline acalabrutinib. In addition, the ASD contains HPMCAS-H, which contains acidic groups ionizable in the GI pH range [14]. Therefore, measurements were repeated for ASDs, since amorphous acalabrutinib and HPMCAS-H in the formulation could result in a different solid surface pH than for neat crystalline acalabrutinib.

2.6.2. Amorphous Acalabrutinib pH-Solubility

Solubilities of amorphous acalabrutinib dosed as the 50/50 (*w/w*) acalabrutinib/HPMCAS-H ASD were determined in different HCl molarities (pH 4.0, 4.5, 5.0, and 6.0) containing 34 mM NaCl. This approach was used rather than measuring amorphous solubility in the absence of polymer, since polymer has been shown to impact the apparent amorphous solubility [13,31]. After dosing the 50/50 (*w/w*) acalabrutinib/HPMCAS-H ASD to each medium, HCl was automatically titrated to maintain the target starting pH during dissolution using a Metrohm Titrado apparatus (Metohm, Tampa, FL, USA). During the measurements, slurries were mixed and controlled to a temperature of 37 ± 1 °C. Once a plateau in pH was achieved, indicating the end of dissolution and saturation of acalabrutinib, two aliquots of each slurry were spun down using microcentrifuge. Drug concentration in the supernatant was analyzed by high-performance liquid chromotography (HPLC). See Appendix A.4 for HPLC method details.

2.7. *In Silico Predictions of In Vivo Performance*

In silico predictions for acalabrutinib blood plasma versus time concentration (C_p) profiles were performed using GastroPlus™ v9.8 simulation software (Simulations Plus Inc. Lancaster, CA, USA) with an "IR:tablet" (ASD tablet) or "IR:capsule" (Calquence capsule) dosage form setting (see detailed parameters in Table 2). All noncompartmental pharmacokinetic (PK) analyses were performed in Microsoft Excel. Values for AUC extrapolated to infinity ($AUC_{0\text{-}inf}$), maximum drug plasma concentration (C_{max}), and time to maximum plasma concentration (T_{max}) were calculated for both simulated and observed plasma profiles. In silico prediction accuracy was evaluated by calculating the absolute average fold error (*AAFE*) of the predicted values for $AUC_{0\text{-}inf}$, C_{max}, T_{max}, and C_p versus time. *AAFE* can be calculated using Equation (1), where n is the number of individuals and subscript i is the parameter of interest for determining *AAFE* (e.g., AUC) [32,33].

$$AAFE = 10^{\frac{1}{n} \times \Sigma \left| Log\left(\frac{predicted_i}{observed_i}\right) \right|} \quad (1)$$

For example, an *AAFE* value of 1.3 suggests that the spread of the observed value around the predicted value is 1.3-fold, with a value of 1 having zero spread and exact agreement. An *AAFE* of <2 from a "bottom–up" in silico prediction was considered acceptable for this study.

2.7.1. Physicochemical Properties and In Vitro Bioperformance Inputs

Physicochemical properties for acalabrutinib including molecular weight, pK_a, log D, and effective permeability (P_{eff}) were either referenced from prior literature studies, extracted from regulatory filing documents, or estimated from ADMET Predictor v9.5 (Simulations Plus Inc. Lancaster, CA, USA) (see Table 2) [4,6]. Solubility versus pH was referenced from Pepin et al. for the Calquence capsule and measured for the ASD tablet [6,29].

Dissolution rate in the in silico prediction was calculated using the z-factor model from the in vitro dissolution data in gastric media (see Section 3.4) for both the Calquence capsule and ASD tablet at low (starting pH 2) and high gastric pH (starting pH 6) [27]. In addition, the z-factor was also calculated from simulated intestinal medium in the presence of bile salt micelles at a pH of 6.5, using the experimental data from the pH 6 to 6.5 gastric to intestinal transfer test (see Section 3.4). Subsequently, a z-factor versus pH profile was constructed for each formulation and physiological condition (i.e., pentagastrin or famotidine pretreatment) and used for each simulation to account for changes in dissolution rate versus pH down the GI tract.

Pepin et al. reported that precipitation does not have a significant impact on acalabrutinib absorption via dissolution testing at the clinically prescribed 100 mg dose [6]. In support of this assumption, the measured in vitro dissolution data for the Calquence capsule and for the ASD tablet demonstrated no precipitation to a lower dissolved drug concentration during the duration of the experiment in the USP 2 gastric to intestinal transfer test (see Section 3.4) or in the CTD apparatus at a physiological dose concentration (see Mudie et al. [8]). The mean precipitation time in the in silico prediction was set to 100,000 seconds to reflect negligible precipitation.

2.7.2. Physiology

In silico predictions were performed in GastroPlus v9.8 using default physiology settings for mean transit time, fluid volume, bile salt concentration, and small/large intestinal pH for a fasted beagle dog. Stomach pH was first adjusted to account for pentagastrin (bulk pH 2) or famotidine (bulk pH 6) pretreatment [21,22]. Subsequently, the surface pH of the dissolving solid was taken into account using the method of Pepin et al., where the bulk pH in the gastric compartment was adjusted to the estimated surface pH (i.e., slurry pH values) [6]. Doing so more accurately captures the solubility and dissolution driving force of acalabrutinib at the solid surface, resulting in a more accurate dissolution rate prediction than would be obtained using the bulk stomach pH (see Appendix A.5) [30,32].

In the intestinal compartments, bulk pH was not modified, since neither amorphous nor crystalline acalabrutinib should have a solid-surface pH that differs significantly from the buffered media at pH > 6 according to the recommendations by Mudie et al. [30]. Bulk and surface pH values used in the in silico predictions and rationale for maintaining bulk pH in the intestinal compartments are included in Appendix A.5.

The theoretical GI tract "compartment" surface-area-to-volume (SA/V) was selected as the absorption scaling factor (ASF) model in GastroPlus to scale SA/V in the different segments of the GI tract.

2.7.3. PK

PK input parameters for in silico predictions are reported in Table 2. Clearance, volume of distribution, and elimination half-life for acalabrutinib in beagle dogs were calculated from the blood plasma concentration versus time profile for an acalabrutinib acidified oral solution published by Podoll et al. at a 30 mg/kg dose (see Appendix A.6 for calculation details) [34]. Oral rather than intravenous (IV) data were used for these calculations, since IV data in dogs were not available.

Lastly, first-pass extraction (FPE) was set to 25% based on a study by Pepin et al. where the average bioavailability (%F) was 75% when an acalabrutinib oral solution was dosed at 100 mg to beagle dogs (n = 12). This calculation (FPE = 100% − %F) assumes complete absorption of the oral solution, and therefore, %F less than 100 is due to metabolism in the gut and liver [6].

The assumption that acalabrutinib pharmacokinetics can be described via one central compartment was demonstrated by sufficient agreement between simulations using a one compartment model with observed data (see Appendix A.6).

Table 2. Summary of in silico prediction input parameters used for simulating blood plasma concentration versus time profiles of acalabrutinib in beagle dogs.

Drug Properties					
Parameter	**Value**				**Source**
Molecular weight (g/mol)	465.52				ADMET Predictor
pK_a	3.5, 5.8 (basic)				Pepin et al. [a], FDA Biopharm review
log D (pH 7.4)	1.14				ADMET Predictor
Effective permeability ($\times 10^{-4}$ cm/s)	5.4				Pepin et al. for humans [a] Assumed dog permeability is similar to human
Solubility vs. pH (mg/mL)	ASD Tablet		Calquence Capsule		ASD Tablet (measured) Calquence capsule (Pepin et al. [a])
	pH	Solubility	pH	Solubility	
	4	6.47	4	3.9	
	4.5	2.62	5	0.34	

Table 2. *Cont.*

Drug Properties					
Parameter	**Value**				**Source**
Solubility vs. pH (mg/mL)	5	0.81	6	0.08	ASD Tablet (measured)
	6	0.43	7	0.05	Calquence capsule (Pepin et al. [a])
Biorelevant solubility—6.7 mM SIF [b], pH 6.5 (mg/mL)	0.71 (ASD Tablet) 0.11 (Calquence capsule)				Measured
Dissolution model (z-factor, mL/mg/s)	ASD Tablet		Calquence capsule		Calculated from in vitro dissolution data for Calquence capsule and ASD tablet (Section 3.4)
	pH	z-factor	pH	z-factor	
	4.6	0.007	4	0.001	
	6.3	0.005	6	0.03	
	6.5 (SIF [b])	0.011	6.5 (SIF [b])	0.02	
Mean precipitation time (s)	100,000				Pepin et al. [a]
PK Parameters in Beagle Dog (Single Compartment)					
Parameter	**Value**				**Source**
Clearance (L/h/kg)	0.75				Calculated from Podoll et al. [c]
Volume of distribution (L/kg)	1.27				
Elimination half-life (h)	1.18				
First-pass extraction (%)	25				Pepin et al. [a]

[a] Reference [6]. [b] SIF—simulated intestinal fluid comprised of 6.7 mM FaSSIF powder (Biorelevant.com, London, UK) dissolved in 67 mM phosphate-buffered saline (pH 6.5). [c] Reference [34].

3. Results

3.1. Dispersion Polymer Screening

Of the seven polymers screened, HPMCAS-H and HPMCAS-M showed the longest sustainment of supersaturation, with no evidence of precipitation over the 16 h test duration (see Figure 1). Eudragit L100 showed the second longest sustainment, followed by HPMC, HPMCAS-L, and PVP VA. PVP resulted in the shortest level of supersaturation sustainment at 4.8 h. Even without polymer present, acalabrutinib remained supersaturated for a relatively long duration of 3 h.

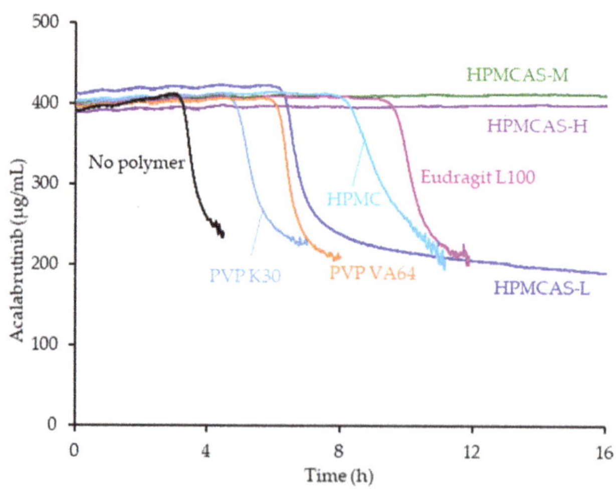

Figure 1. Acalabrutinib concentration versus time in pH 6.5 phosphate buffer containing 6.7 mM FaSSIF powder and 82 mM NaCl measured in the absence and presence of 200 µg/mL dissolved polymer.

3.2. ASD Manufacturing and Characterization

Each spray-dried ASD displayed a single T_g upon mDSC measurement and showed no evidence of crystallinity (i.e., absence of sharp diffraction peaks) upon PXRD analysis. See Appendix A.7 for PXRD diffractograms, mDSC thermograms, T_g values, and SEM images.

3.3. In Vitro Dissolution Testing of Intermediates

ASD intermediates and crystalline acalabrutinib were evaluated in gastric to intestinal transfer dissolution tests using a Pion μDiss Profiler. All six ASDs were evaluated in tests with pH 5 and pH 6 gastric medium representative of fasted dogs taking an ARA. All six ASDs achieved higher acalabrutinib concentrations in gastric and intestinal media and higher intestinal AUCs than crystalline acalabrutinib (See Figures 2–4). Whereas ASDs were supersaturated in gastric and intestinal media, low solubility of acalabrutinib at elevated gastric pH resulted in limited release in gastric medium, preventing supersaturation upon transfer to intestinal medium.

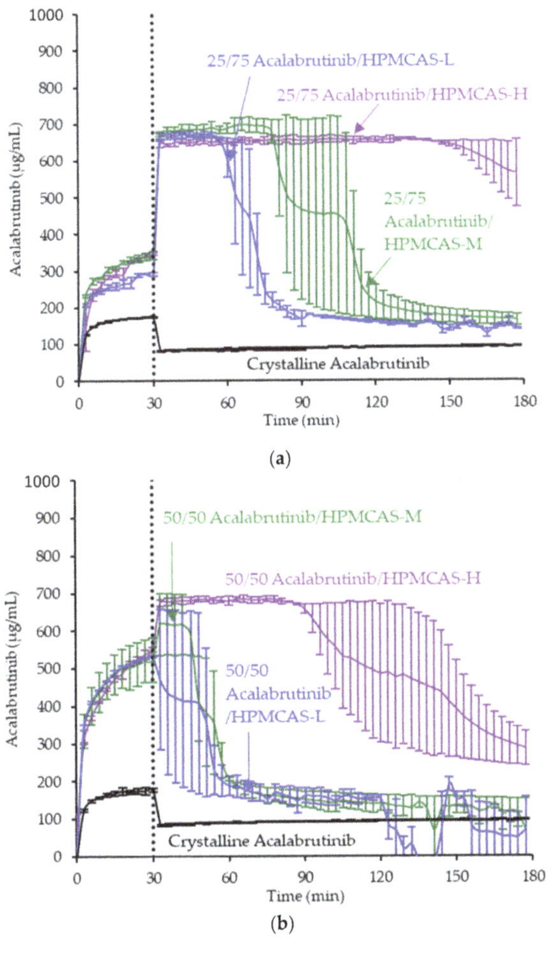

Figure 2. Dissolution profiles of ASD intermediates in a gastric to intestinal transfer dissolution test using a Pion μDiss Profiler under conditions simulating fasted dogs treated with an ARA (gastric pH 5) (curves represent average values whereas error bars represent range, $n = 2$). Panel (**a**) 25 wt % drug loading ASDs. Panel (**b**) 50 wt % drug loading ASDs.

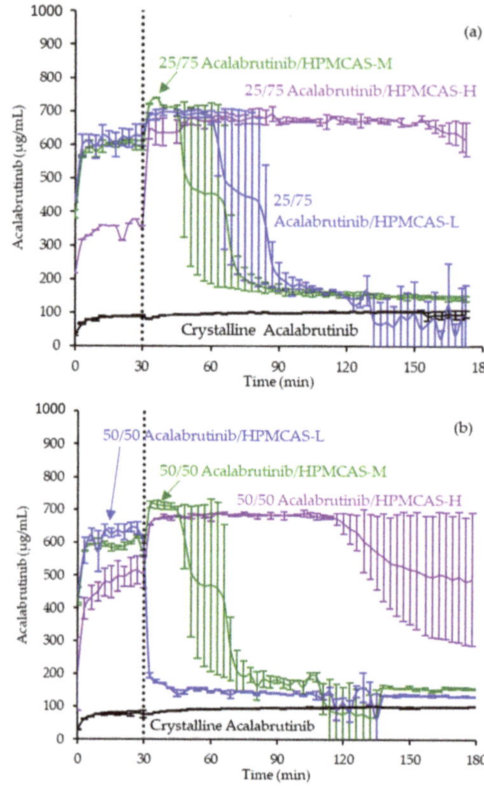

Figure 3. Dissolution profiles of ASD intermediates in a gastric to intestinal transfer dissolution test using a Pion μDiss Profiler under conditions simulating fasted dogs treated with an ARA (gastric pH 6) (curves represent average values whereas error bars represent range, n = 2). Panel (**a**) 25 wt % drug loading ASDs. Panel (**b**) 50 wt % drug loading ASDs.

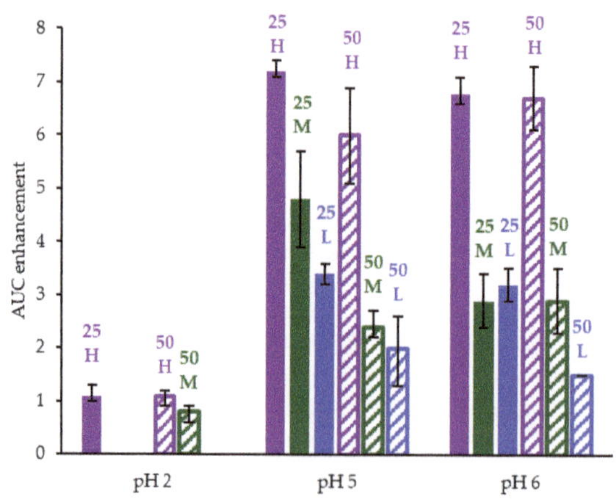

Figure 4. AUC enhancement for ASD intermediates tested in gastric to intestinal transfer dissolution tests using a Pion μDiss Profiler. *X*-axis labels represent starting gastric pH. AUC enhancement = AUC(ASD)/AUC(crystalline acalabrutinib), reported as average with error bars representing the range of two measurements.

HPMCAS-H ASDs achieved the longest supersaturation durations in intestinal medium and the highest AUC enhancements (see Figure 4 and Appendix A.8 for tabulated values). AUC enhancements for HPMCAS-H ASDs were independent of gastric pH and drug loading within the error of the measurements. With the exception of the 25 wt % drug loading HPMCAS-M ASD tested at pH 5, which showed the second best performance, differences between AUC enhancements for the -M and -L grade ASDs were within the error of the measurements.

Since the 25 and 50 wt % drug loading HPMCAS-H ASDs achieved the highest AUC enhancements in the tests representative of fasted dogs taking an ARA, these ASDs were selected as lead candidates and carried forward into pH 2 gastric tests representative of fasted dogs pretreated with pentagastrin. The 50 wt % drug loading HPMCAS-M ASD was included as a back-up ASD. All three ASDs and crystalline acalabrutinib reached much higher extents of gastric dissolution in simulated gastric medium at pH 2 than at elevated gastric pH, and they were supersaturated upon transfer to intestinal medium (see Figure 5). It should be noted that for the gastric portion of the test, measured concentrations in gastric media may be somewhat lower than actual values due to UV detector saturation. At pH 2, the ionization of weakly basic acalabrutinib results in high crystalline solubility, driving the rapid dissolution of both crystalline acalabrutinib and ASDs. Similar performance of ASDs and crystalline acalabrutinib in intestinal medium resulted in AUC enhancements near unity (see Figure 4). Overall, in vitro testing of ASD intermediates showed that HPMCAS-H ASDs maximized performance (i.e., AUC enhancement) at elevated pH and showed equivalent performance to Calquence at low gastric pH.

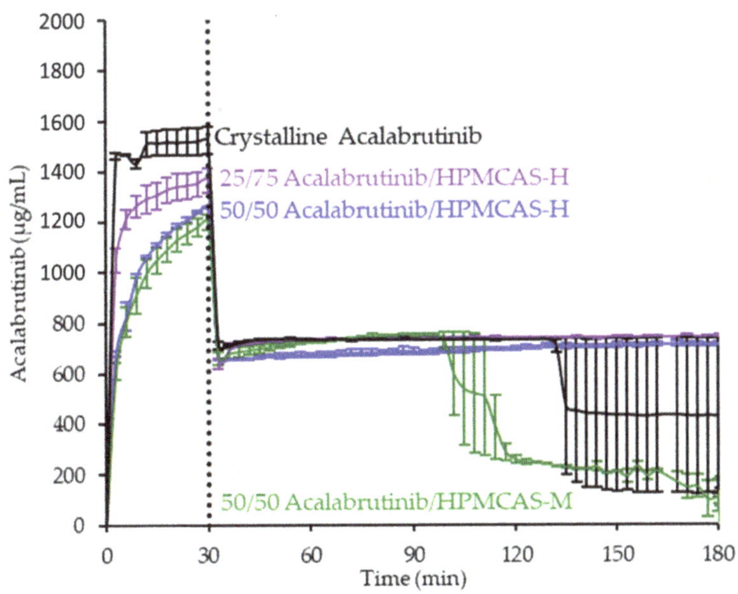

Figure 5. Dissolution profiles of ASD intermediates in a gastric to intestinal transfer dissolution test using a Pion µDiss Profiler under conditions simulating fasted dogs treated with pentagastrin (gastric pH 2) (curves represent average values whereas error bars represent range, $n = 2$).

3.4. *In Vitro Dissolution Testing of Dosage Forms*

ASD tablets using the 50/50 acalabrutinib/HPMCAS-H ASD and commercially available Calquence capsules were evaluated for dissolution performance in gastric to intestinal transfer dissolution tests in a USP 2 dissolution apparatus. Tests were conducted using gastric media representative of dogs taking ARAs (pH 6) or dogs pretreated with pentagastrin (pH 2) at a dose concentration lower than that used for the ASD intermediates to facilitate an adequate determination of dissolution rates for in silico predictions. The 50/50 acalabrutinib/HPMCAS-H ASD was chosen as the lead ASD, since in vitro dissolution testing of ASD intermediates demonstrated that HPMCAS-H ASDs achieved the highest AUC enhancements and because the higher drug loading minimizes tablet size.

In pH 2 gastric medium, the ASD tablet and Calquence capsule achieve similar and rapid release of acalabrutinib to a value approaching the dose concentration of 0.4 mg/mL (see Figure 6). Upon 1:1 dilution with concentrated intestinal medium, acalabrutinib concentration is halved, maintaining supersaturation at the 0.2 mg/mL intestinal dose concentration for the duration of the test.

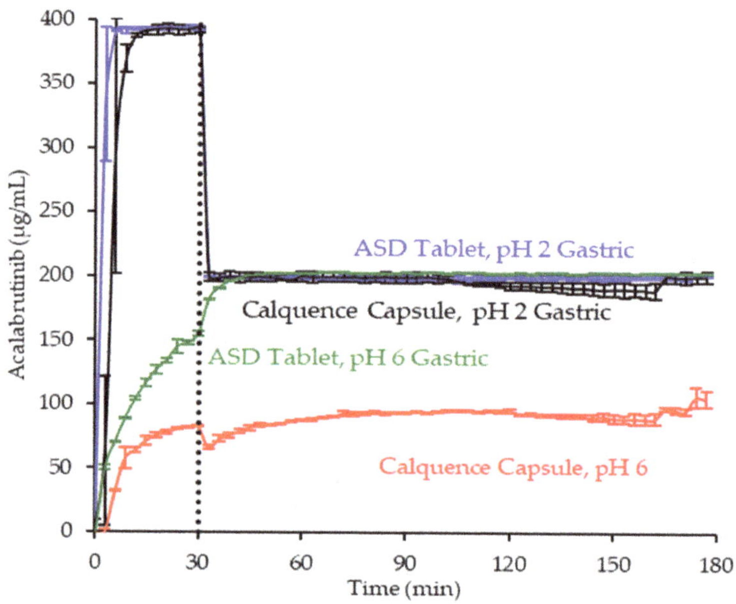

Figure 6. Dissolution profiles of ASD tablet and Calquence capsule in gastric to intestinal transfer dissolution tests in the USP 2 apparatus under conditions simulating fasted dogs treated with pentagastrin (gastric pH 2) or an ARA (gastric pH 6) (curves represent average values, whereas error bars represent range, $n = 2$).

In pH 6 gastric medium, neither the ASD tablet nor Calquence capsule show complete release of acalabrutinib in gastric medium (see Figure 6). The ASD tablet reaches a 1.8-fold higher acalabrutinib concentration in gastric medium compared to the Calquence capsule, achieving complete release upon dilution with concentrated intestinal medium and maintaining supersaturation at the 0.2 mg/mL intestinal dose concentration for the duration of the test. In contrast, the Calquence capsule is solubility-limited in both pH 6 gastric medium and intestinal medium. Acalabrutinib concentrations plateau near the crystalline solubility in both media, reaching 53% of the dose dissolved in intestinal medium by the end of the test.

Overall, in vitro testing of dosage forms showed that the ASD tablet matched the performance of the Calquence capsule at gastric pH 2 and exceeded the performance of the Calquence capsule at gastric pH 6 with an AUC enhancement of 2.2 (2.2–2.3). In vitro testing at a higher dose concentration in an in vitro CTD test demonstrated similar relative performance between dosage forms at a given initial gastric pH (see Mudie et al. [8]).

3.5. Physicochemical Property Measurements of Lead ASD Intermediate

Slurry pH and amorphous solubility values of the 50/50 (*w/w*) acalabrutinib/HPMCAS-H ASD were measured at different HCl concentrations (i.e., bulk pH values). Results are shown in Table 3. Molar concentrations of acalabrutinib in solution are high enough to increase slurry (ending) pH above the starting values, and they are higher than the values reported by Pepin et al. for crystalline acalabrutinib due to the higher solubility of the ASD [29]. The measured amorphous solubilities shown in Table 3 are higher than crystalline solubilities reported by Pepin et al.; for example, they are 5.6-fold higher at pH 6.0 where acalabrutinib is ≈63% non-ionized [29].

Table 3. Slurry pH and solubilities of amorphous acalabrutinib dosed as the 50/50 (*w/w*) acalabrutinib/HPMCAS-H ASD in different molarities of HCl containing 34 mM NaCl (n = 1).

Starting pH	Slurry (Ending) pH	Amorphous Acalabrutinib Solubility (mg/mL) [a]
1.6	4.4	-
2.0	4.6	-
3.0	5.4	-
4.0	6.2	6.47 (6.43–6.51)
4.5	-	2.62 (2.48–2.76)
5.0	6.3	0.81 (0.80–0.82)
6.0	6.3	0.43 (0.43–0.44)

[a] Average (range of two measurements per one sample).

3.6. In Silico Predictions Versus In Vivo Performance

The in silico PK predictions for acalabrutinib are shown in Figure 7 with tabulated results comparing simulated versus observed data in Table 4. The simulations provided good prediction accuracy of the observed data. The AAFE of the in silico predictions for $AUC_{0\text{-}inf}$, C_{max}, T_{max}, and C_p versus time for all formulation treatments were <2-fold (ideal value = 1) with the exception of C_p versus time for the Calquence capsule + famotidine treatment, indicating that the in silico prediction framework is sufficient for simulating acalabrutinib blood plasma concentrations within this in vivo study.

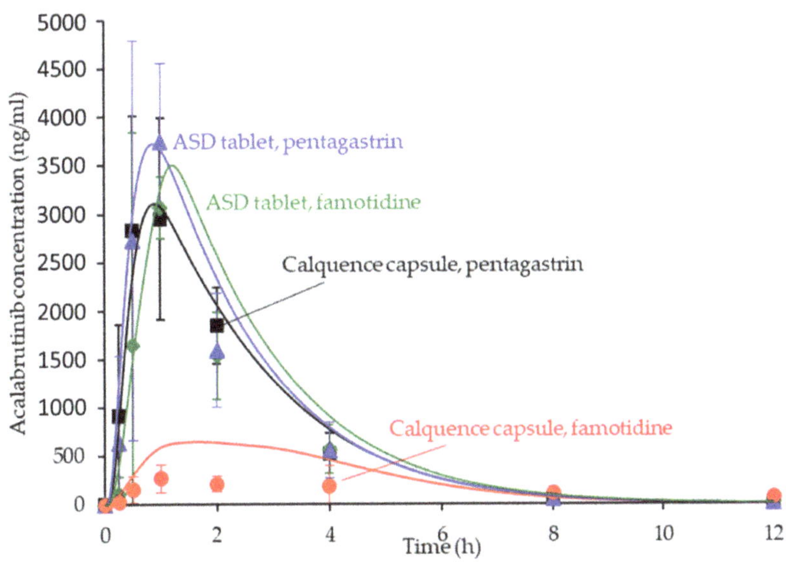

Figure 7. Simulated versus observed acalabrutinib blood plasma concentrations versus time at 100 mg dose in beagle dogs with pentagastrin or famotidine pretreatment. Markers are the observed average data points with error bars representing the standard deviation, and solid lines are the in silico predictions.

Table 4. Noncompartmental analysis comparing simulated (sim) versus observed (obs) data for all formulation treatments in the dog study. Absolute average fold error (AAFE) was calculated for $AUC_{0\text{-inf}}$, C_{max}, T_{max}, and C_p versus time to determine the accuracy of the in silico prediction exercise (ideal value = 1).

Formulation	$AUC_{0\text{-inf}}$ (ng h/mL)		C_{max} (ng/mL)		T_{max} (h)		AAFE			
	Obs	Sim	Obs	Sim	Obs	Sim	$AUC_{0\text{-inf}}$	C_{max}	T_{max}	C_p vs. Time
ASD tablet, pentagastrin	8161	9766	3332	3727	0.9	0.9	1.2	1.2	1.6	1.3
ASD tablet, famotidine	7579	9555	3443	3508	0.9	1.6	1.3	1.2	1.8	1.6
Calquence capsule, pentagastrin	8365	8607	4480	3110	0.8	0.9	1.1	1.4	1.3	1.3
Calquence capsule, famotidine	3112	3096	355	648	1.6	1.2	1.6	1.9	1.7	3.0

4. Discussion

4.1. Benefits of Streamlined, In Vitro–In Silico ASD Development Approach

ASD dosage forms are useful for increasing the bioavailability of poorly soluble drugs, thereby enabling effective oral delivery and improving patient safety and compliance, such as by removing DDIs, reducing plasma variability, reducing dose, and mitigating food effects [35]. As highlighted in this study, ASDs can be effective for formulating weakly basic drugs, which make up 78% of new molecular entities approved between 2003 and 2013 that showed a clinical DDI with ARAs [3].

To be robust, ASD dosage forms must achieve good in vivo performance, stability, and manufacturability, while minimizing dosage form size. Attaining all of these attributes can be challenging, often resulting in numerous formulation iterations that increase development timelines and costs. As described in this study, knowledge of key drug, polymer, and GI properties can be combined with in vitro and in silico tools to design ASD formulations that achieve good performance in preclinical studies, thereby helping to mitigate these issues.

This streamlined ASD development approach not only achieved the in vivo study goal on the first try with a single ASD tablet formulation but also minimized the time and amount of drug needed for formulation development. For example, this approach included the use of a high-throughput, material-sparing solvent-shift UV test, which identified the most effective dispersion polymers prior to spray-drying ASDs, reducing the number of ASD formulations. Next, Pion µDiss in vitro dissolution testing was used to screen performance and select the lead ASD intermediate, before making tablets, allowing for minimal use of drug and ASD material while achieving high throughput. Once the lead ASD was identified, a tablet dosage form was made of this ASD alone, which was evaluated against the Calquence capsule in a USP 2 dissolution apparatus, as well as a physiologically relevant CTD apparatus (as described by Mudie et al.) [8]. Finally, in vitro data together with drug physicochemical and physiological properties were input into in silico software to maximize the likelihood of the ASD tablet working as intended vivo.

Throughout this approach, knowledge of the problem statement (i.e., pH-dependent absorption of a BCS 2 weak base) along with key drug, polymer, and physiological properties facilitated effective design of in vitro testing methods and media compositions. For example, gastric to intestinal transfer and controlled transfer tests were used to evaluate the precipitation of a weakly basic drug, and gastric and intestinal media were selected to test performance over a pH range relevant to dogs treated with pentagastrin and ARAs. Further, nuances of acalabrutinib, such as surface pH effects on dissolution rate were accounted for in the in silico simulations. Taken together, this streamlined approach resulted in successful removal of the Calquence ARA effect in a beagle dog model without the need for multiple in vivo studies or reformulation. In addition, ASD tablets were 60% smaller than Calquence capsules, and they achieved good stability and manufacturability, as described by Mudie et al. [8].

4.2. Relative In Vitro Performance of HPMCAS ASDs

This study examined the effects of HPMCAS grade and drug loading on the in vitro performance of acalabrutinib ASDs. While all ASD intermediates achieved higher AUCs than crystalline acalabrutinib in intestinal medium (see Figures 2–4), HPMCAS-H ASDs achieved the highest intestinal AUC enhancements of the three grades. These highest AUC enhancements were due to the longest supersaturation durations in intestinal medium, which was potentially due to having the highest degree of hydrophobicity and therefore best crystallization inhibition of the three grades, as shown for some drugs [15,36,37].

While intestinal AUC enhancement was the dominant metric used to select the lead ASD, the extent of gastric dissolution varied with polymer grade, drug loading, and initial gastric pH. Notwithstanding these differences, all -M and -H ASDs dissolved nearly instantaneously to the apparent amorphous solubility in intestinal medium representative of dogs, with AUC enhancements for the lead, 50 wt % drug loading HPMCAS-H ASD $\geq$ 2 fold greater than for all other ASDs. The goal of the study was to develop an ASD tablet that could achieve equivalent AUC at low and high gastric pH and match the performance of Calquence as prescribed (i.e., at low gastric pH). Indeed, this metric was achieved in a dog model described herein and by Mudie et al. [8].

4.3. In Silico PK Simulation Tools in Early Development

In silico PK simulations are emerging tools for ASD formulation development, since they can simulate many dynamic physiological factors that impact ASD bioperformance [13,38–40]. Simulations using physiologically-based pharmacokinetic (PBPK) modeling software such as GastroPlus can help design and optimize in vivo studies with respect to dose, prandial state, or pH modification (i.e., selecting ARAs for pH effect study) to maximize the likelihood of achieving desired ASD in vivo performance.

Since ASD formulations are typically first evaluated in a preclinical setting, early in silico predictions in the absence of a validated biomodel are important for understanding formulation or physiological parameter sensitivity and rank ordering formulations. In fact, hypothesis testing via in silico predictions connects all aspects of ASD formulation selection from excipient screening to in vitro dissolution of the target dosage form—providing appropriate affirmation at each stage of development.

For acalabrutinib, in silico prediction was implemented early in formulation evaluation to aid in vivo study design and execution and to understand parameter sensitivity (solubility and dissolution rate as a function of pH) for lead formulations. Predictions provided confidence that the 100-mg dose of Calquence capsules in dogs would show reduced exposure (i.e., a pH effect) with the famotidine (high stomach pH) pretreatment compared to the pentagastrin (low stomach pH) pretreatment. Importantly, the predictions also forecasted that the ASD tablets could overcome the pH effect by showing equivalent AUC across both pentagastrin and famotidine pretreatments, highlighting their superior performance to the Calquence capsule. Therefore, the in silico predictions performed herein were an integral and necessary step to the successful in vivo result.

5. Conclusions

This study describes a streamlined in vitro and in silico development approach for designing an ASD tablet to overcome the pH-dependent absorption of the weak base drug, acalabrutinib. The methodologies described in this study can be applied directly to the development of ASDs of other weak base drugs for mitigating gastric pH effects. Furthermore, the general framework can be applied to developing ASDs of other poorly soluble drugs for bioavailability enhancement. When designed with key drug, polymer, and gastrointestinal properties in mind, in vitro and in silico tools can help identify the in vivo barriers to absorption and assess whether they can be overcome by an ASD drug product. As this approach achieved the in vivo study goal on the first try with a single ASD tablet formulation, it has the potential to reduce time and cost to introducing robust ASD drug products to the market.

Author Contributions: Conceptualization, D.M.M., A.M.S., M.M.M., and D.T.V.; methodology, D.M.M., A.M.S., J.A.R., M.S.A., and M.M.M.; formal analysis, D.M.M., A.M.S., and J.A.R.; investigation, J.A.R. and M.S.A.; data curation, D.M.M., A.M.S., and J.A.R.; writing—original draft preparation, D.M.M., A.M.S., and J.A.R.; writing—review and editing, D.M.M., A.M.S., M.S.A., and M.M.M.; visualization, D.M.M., A.M.S., and J.A.R.; supervision, D.M.M. and D.T.V.; project administration, D.M.M. All authors have read and agreed to the published version of the manuscript.

Funding: This research received no external funding.

Institutional Review Board Statement: Not applicable.

Informed Consent Statement: Not applicable.

Data Availability Statement: The data in this study in the form of Microsoft Excel worksheets are available from the corresponding author upon request.

Acknowledgments: We would like to thank Henny Zijlstra for identifying the Calquence ARA-effect problem statement and Ann Malkin for her review and editing of the manuscript. We would like to thank Lauren Switala for performing some analytical characterization. No financial support or funding was provided for this work.

Conflicts of Interest: D.M. Mudie and A.M. Stewart are co-inventors on a pending patent application filed by Bend Research, Inc., patent application 63/031685 filed 29 May 2020. This application discloses amorphous solid dispersions of acalabrutinib and a dispersion polymer (HPMCAS), pharmaceutical dosage forms thereof, and a method for preparing said amorphous solid dispersions. The authors declare no other conflicts of interest.

Appendix A.

The following are available in Appendix A: Appendix A.1 Acalabrutinib form I verification, Appendix A.2 Amorphous solid dispersion (ASD) and active pharmaceutical ingredient (API) characterization methods, Appendix A.3 In vitro dissolution methods, Appendix A.4 Amorphous acalabrutinib slurry pH values for in silico predictions, Appendix A.5 Bulk and surface pH values for in silico predictions, Appendix A.6 PK parameter calculations and in silico prediction accuracy of observed data from acalabrutinib acidified oral solution, Appendix A.7 ASD characterization results, Appendix A.8 In vitro dissolution testing of ASD intermediates tabulated results.

Appendix A.1. Acalabrutinib Form 1 Preparation

Acalabrutinib (CAS 1420477-60-6, >98% purity) was purchased from LC Laboratories (Woburn, MA, USA). Form I was prepared by recrystallizing the purchased acalabrutinib to obtain Form I, according to WO 2017/002095 Al, Example 1, Table 1, Sample 23 [11]. Verification of Form I was done by mDSC to determine the melting temperature (210 °C), which is close to the 214 °C reported in WO 2017/002095 Al, and by matching the PXRD diffractogram of the recrystallized material to Figure 1 of WO 2017/002095 Al, which matched as illustrated in Figure A1 herein. PXRD and mDSC measurement methods are described in Appendix A.2.

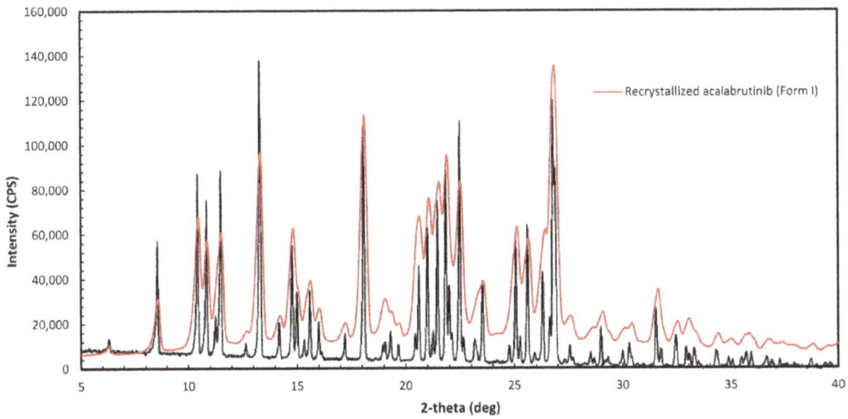

Figure A1. Overlay of recrystallized acalabrutinib and Form I as specified in Figure 1 of WO 2017/002095 Al.

Appendix A.2. Amorphous Solid Dispersion (ASD) and Active Pharmaceutical Ingredient (API) Characterization Methods

Appendix A.2.1. Powder X-ray Diffraction (PXRD)

ASD and API samples were analyzed using PXRD to confirm they were amorphous, as evidenced by the lack of sharp Bragg diffraction peaks in the X-ray pattern, using a Rigaku MiniFlex600 X-ray diffractometer (Rigaku, The Woodlands, TX, USA) equipped with a Cu-Kα source. The scan rate was set to 2.5°/min with a 0.02° step size from 3° to 40° on the 2Θ scale.

Appendix A.2.2. Modulated Differential Scanning Calorimetry (mDSC)

ASD samples were analyzed to confirm that they were homogeneous as evidenced by a single T_g using a TA Instruments Q2000 mDSC instrument (TA Instruments-Waters LLC, New Castle, DE, USA). Samples were prepared as loose powder, loaded into a Tzero, nonhermetically sealed pan (TA Instruments). The instrument was run in modulated mode at a scan rate of 2.5 °C/min, modulation of ±1.5 °C/min, and scan range of 0–160 °C (n = 3).

API sample was analyzed to determine the melting temperature of the crystallized acalabrutinib using a TA Instruments Q2000 mDSC instrument. The sample was prepared as loose powder and loaded into a Tzero nonhermetically sealed pan (TA Instruments). The instrument was run at a scan rate of 10 °C/min and a scan range of 100–230 °C. The sample was then immediately quenched with liquid nitrogen. The sample was then added into a Tzero nonhermetically sealed pan (TA Instruments). The instrument was run in modulated mode at a scan rate of 2.5 °C/min, modulation of ±1.5 °C/min, and scan range of 0–200 °C (n = 1).

Appendix A.2.3. Scanning Electron Micrography (SEM)

ASD samples were imaged using a Hitachi SU3500 scanning electron microscope (Hitachi High Technologies America Inc., Schaumburg, IL, USA) to assess morphology and confirm the absence of surface crystals. Samples were spread onto a post using double-sided tape and sputter-coated for 8 min at 8 V and 20 mA using an Anatech Hummer 6.2 sputter coater (Anatech USA, Hayward, CA, USA).

Appendix A.3. In Vitro Dissolution Methods

Appendix A.3.1. In Vitro Dissolution Testing of Intermediates

Three types of gastric medium were used in the gastric to intestinal transfer dissolution test using a Pion µDiss™ Profiler. pH 5 and pH 6 gastric media represented fasted dogs taking an ARA, whereas pH 2 gastric medium representated fasted dogs pre-treated with pentagastrin. After 30 min, gastric medium was diluted 1:1 with a concentrated intestinal medium. After dilution intestinal medium contained 67 mM phosphate, 0.5 wt % (6.7 mM) FaSSIF powder and either 82 mM NaCl (pH 5 and 6 gastric pH tests) or 99 mM NaCl (pH 2 gastric pH test). Media composition are shown in Table A1.

Table A1. Simulated gastric and concentrated intestinal media compositions used in Pion μDiss dissolution testing of ASD intermediates.

Ingredient	Mass Concentration (g/L)	Molar Concentration (mM)
	Gastric medium (pH 5)	
NaOAc	0.83	10
Adjust to pH 5.0 dropwise using 12.1 N HCl		
	Gastric medium (pH 6)	
$Na_2HPO_4 \cdot 7H_2O$	5.36	20
KH_2PO_4	6.36	47
NaCl	4.8	82
Adjust to pH 6.0 dropwise using 12.1 N HCl		
	Gastric medium (pH 2)	
NaCl	1.98	34
Adjust to pH 2.0 by diluting 12.1 N HCl to 0.01 N HCl		
	Concentrated intestinal medium (pH 5 and pH 2 gastric tests)	
$Na_2HPO_4 \cdot 7H_2O$	10.72	40
KH_2PO_4	12.72	94
NaCl	9.6	164
FaSSIF/FeSSIF/FaSSGF	10	—
Adjust to pH 6.55 dropwise using 10 N NaOH		
	Concentrated intestinal medium (pH 6 gastric tests)	
$Na_2HPO_4 \cdot 7H_2O$	5.36	20
KH_2PO_4	6.36	47
NaCl	4.8	82
FaSSIF/FeSSIF/FaSSGF	10	—
Adjust to pH 6.9 dropwise using 10 N NaOH		

Dose number (Do) calculations for the gastric to intestinal transfer dissolution tests of ASD intermediates in the Pion μDiss^{TM} Profiler are shown in Table A2. In fasted beagle dogs, Wang and coworkers measured average gastric volume to be 24.0 ± 4.2 mL [26] and since dogs in this study ingested 10 mL of water with the dosage form, average gastric volume in beagle dogs for this study is estimated to be 34 ± 4 mL. Dose number (dose/volume/solubility) in the stomach of fasted beagle dogs is therefore estimated to be 2.9 ± 0.3.

Table A2. Do (Dose number) calculations for gastric to intestinal transfer dissolution test using a Pion μDiss^{TM} Profiler.

Dose/Volume in Gastric Medium (mg/mL)	Do pH 2 Gastric Medium (ASD/Crystalline Acalabrutinib) [a]	Do pH 5 Gastric Medium (ASD/Crystalline Acalabrutinib) [a]	Do pH 6 Gastric Medium (ASD/Crystalline Acalabrutinib) [a]	Dose/Volume in Intestinal Medium (mg/mL)	Do intestinal Medium (ASD/Crystalline Acalabrutinib) [a]
2	0.03/0.04 [b]	2.6/5.9 [c]	4.6/26.0 [d]	1	1.4 [b]/9.1 [c]

[a] Do (Dose/volume/solubility) calculations use apparent amorphous solubility for ASD or apparent crystalline solubility for crystalline acalabrutinib. Calculations use the in vitro volume in the gastric or intestinal compartment. [b] Calculation used ASD or crystalline solubility from GastroPlus solubility-pH profile at pH 2.7 (calculated pH assuming 2 mg/mL acalabrutinib in solution). [c] Calculation used ASD or crystalline solubility from GastroPlus solubility-pH profile at pH 5.2 (ASD) or pH 5.0 (crystalline) (calculated pH based upon average acalabrutinib concentration at end of gastric portion of test). [d] Calculation used ASD or crystalline solubility from GastroPlus solubility-pH profile at pH 6.0.

Prior to each experiment, unique calibration curves were established for each UV probe (1 mm path length) by delivering aliquots of a known amount of stock acalabrutinib solution (10–20 mg/ml acalabrutinib in 95/5 THF/H_2O) to gastric or intestinal medium held at 37 ± 2 °C. UV analysis wavelengths and calibration ranges are shown in Table A3.

Table A3. UV analysis wavelengths and calibration ranges for gastric to intestinal transfer dissolution test using a Pion µDiss^{TM} Profiler.

UV Probe Path Length (mm)	pH 5 Gastric Medium UV Analysis Wavelength (nm)/Calibration Range (mg/mL)	pH 6 Gastric Medium UV Analysis Wavelength (nm)/Calibration Range (mg/mL)	pH 2 Gastric Medium UV Analysis Wavelength (nm)/Calibration Range (mg/mL)	Intestinal Medium UV Analysis Wavelength (nm)/Calibration Range (mg/mL)
1	298–308/0–0.6	340–344/0–0.6	370–376/0–0.6	340–344/0–0.6

Appendix A.3.2. In Vitro Dissolution Testing of Dosage Forms

Two types of gastric medium were used in the gastric to intestinal transfer dissolution test using a USP 2 apparatus. pH 6 gastric medium represented fasted dogs taking an ARA, whereas pH 2 gastric medium representated fasted dogs pre-treated with pentagastrin. After 30 min, gastric medium was diluted 1:1 with a concentrated intestinal medium. After dilution intestinal medium contained 67 mM phosphate, 0.5 wt % (6.7 mM) FaSSIF powder and 82 mM NaCl. Media compositions are shown in Table A4.

Table A4. Simulated gastric and concentrated intestinal media compositions used in USP 2 transfer dissolution testing of dosage forms.

Ingredient	Mass Concentration (g/L)	Molar Concentration (mM)
	Gastric medium (pH 6)	
Adjust to pH 6.0 dropwise using 0.1 N HCl		
	Gastric medium (pH 2)	
Adjust to pH 2.0 by diluting 12.1 N HCl to 0.01 N HCl		
	Concentrated intestinal medium	
$Na_2HPO_4{\cdot}7H_2O$	10.72	40
KH_2PO_4	12.72	94
NaCl	9.6	164
FaSSIF/FeSSIF/FaSSGF	10	—
Adjust to pH 6.55 dropwise using 10 N NaOH		

Do calculations for USP 2 transfer testing of ASD tablet and Calquence capsule are provided in Table A5. Tests were conducted in a USP 2 vessel with a total capacity of 500 mL.

Table A5. Do (Dose number) calculations for USP 2 dissolution testing of ASD tablet and Calquence capsule.

Dose/Volume in Gastric Medium (mg/mL)	Do pH 2 Gastric Medium (ASD Tablet/Calquence) [a]	Do pH 6 Gastric Medium (ASD Tablet/Calquence) [a]	Dose/Volume in Intestinal Medium (mg/mL)	Do intestinal Medium (ASD Tablet/Calquence) [a]
0.4	0.002 [b]/0.003 [b]	1.0 [c]/5.2 [c]	0.2	0.3 [b]/1.8 [c]

[a] Do (Dose/volume/solubility) calculations use apparent amorphous solubility for ASD or apparent crystalline solubility for Calquence. Calculations use the in vitro volume in the gastric or intestinal compartment. [b] Calculation used ASD or crystalline solubility from GastroPlus solubility-pH profile at pH 2.0 (calculated pH assuming 0.4 mg/mL acalabrutinib in solution). [c] Calculations used ASD or crystalline solubility from GastroPlus solubility-pH profile at bulk pH = 6.3 (ASD) from experiments herein or bulk pH 6.0 (crystalline acalabrutinib) from Pepin et al. [29].

Prior to each experiment, unique calibration curves were established for each UV probe (1 mm path length) by delivering aliquots of a known amount of stock acalabrutinib solution (10–20 mg/ml acalabrutinib in 95/5 THF/H_2O) to gastric or intestinal medium held at 37 ± 2 °C. UV analysis wavelengths and calibration ranges are shown in Table A6.

Table A6. UV analysis wavelengths and calibration ranges for gastric to intestinal transfer dissolution test using a USP 2 transfer test.

Dose/Volume in Gastric Medium (mg/mL)	UV Probe Path Length (mm)	pH 2 Gastric Medium UV Analysis Wavelength (nm)/Calibration Range (mg/mL)	pH 6 Gastric Medium UV Analysis Wavelength (nm)/Calibration Range (mg/mL)	Intestinal Medium UV Analysis Wavelength (nm)/Calibration Range (mg/mL)
0.4	1	330–340/0–0.7	330–340/0–0.7	330–340/0–0.7

Appendix A.4. Amorphous Acalabrutinb Slurry pH and pH-Solubility Method Details

Slurry pH was measured as a means to estimate solid surface pH of the 50/50 (*w/w*) acalabrutinib/HPMCAS-H ASD in different HCl molarities (pH 1.6, 2.0, 3.0, 4.0, 5.0, and 6.0) containing 34 mM NaCl. The 50/50 (*w/w*) acalabrutinib/HPMCAS-H ASD was dosed into 20 mL of medium controlled to 37 ± 1 °C using a beaker with a magnetic stirrer held in a custom warm box for pH 1.6 and 2.0 media, or using a Lauda Eco Silver E4 water circulated conical vessel (Lauda Scientific, Vancouver, WA, USA) with a Metrohm overhead stirrer (Metrohm, Tampa, FL, USA) for pH ≥ 3.0 media. The ASD was dosed at 65 mg/mL acalabrutinib (1.3 g total ASD) into pH 1.6 and 2.0 media, or at 3.3 mg/mL acalabrutinib (6.6 mg/mL total ASD) for pH ≥ 3.0 media. pH during dissolution was monitored using a Vernier Labquest 2 LQ2-LE (Vernier Software and Technology, Beaverton, OR, USA) data logger equipped with a Vernier PH-BTA pH probe for pH 1.6 and 2.0 media, or a Metrohm Titrado apparatus (Metrohm 804 Ti Stand, 802 Stirrer, 902 Titrando, 800 Dosino) for pH ≥ 3.0 media. Both pH meters were calibrated between pH 1.68 and pH 10 using known pH standards. Excess solids were present at the end of each measurement. Measurements were performed with one sample each.

Saturated solubilities of amorphous acalabrutinib dosed as the 50/50 (*w/w*) acalabrutinib/ HPMCAS-H ASD were determined at starting pH values of 4.0, 4.5, 5.0 and 6.0. the 50/50 (*w/w*) acalabrutinib/HPMCAS-H ASD was dosed to each medium. As the ASD dissolved HCl was automatically titrated to maintain the target starting pH using a Metrohm Titrado apparatus (Metohm, Tampa, FL, USA) calibrated between pH 1.68 and pH 10 using known pH standards. See Table A7 for concentrations of acalabrutinib added and molarities of titration media.

During the measurements slurries were mixed and controlled to a temperature of 37 ± 1 °C using a Lauda Eco Silver E4 water circulated conical vessel with a Metrohm overhead stirrer. Once pH was titrated back to the starting pH in the presence of excess solids, indicating the end of dissolution and saturation of acalabrutinib, two aliquots of each suspension were transferred to microcentrifuge tubes and spun down at 10,000 relative centrifugal force for 3 min. A sample of the supernatant was diluted in 8/2 MeOH/H_2O and analyzed for concentration by high performance liquid chromotography (HPLC). HPLC measurements were performed on a Shimadzu LC2010 (Shimadzu Corporation, Kyoto, Japan) system using an Eclipse Plus C8 4.6 × 50 mm, 1.8 μm particles column (Agilent Technologies) at a temperature of 30 °C. Injection volume was 5 μL. Mobile phase was pumped into the column at a flow rate of 1 mL/min using a gradient method. 10 mM sodium phosphate monobasic was used as mobile phase A and methanol was used as mobile phase B supplied according to the following sequence: 0 min: 30% B, 4 min: 90% B, 4.6 min: 30% B. The sample tray was held at uncontrolled room temperature and the analysis wavelength was 230 nm.

Table A7. Concentrations of acalabrutinib added and molarities of titration media for pH-solubility measurements.

pH	Concentration of Acalabrutinib Added (mg/mL) [a]	Molarity of HCl Titration Medium (M)
4.0	25	0.1
4.5	10	0.01
5.0	3.5	0.001
6.0	2.5	0.001

[a] Total concentration of ASD added = 2 × concentration of acalabrutinib.

Appendix A.5. Bulk and Surface pH Values for In Silico Predictions

Stomach pH was first adjusted to account for pentagastrin (bulk pH 2) or famotidine (bulk pH 6 pretreatment [21,22]. Subsequently, surface pH of the dissolving solid was taken into account using the method of Pepin and coworkers for the Calquence capsule, where the bulk pH in the gastric compartment was adjusted to the estimated surface pH (i.e. slurry pH values). See Table A8.

Table A8. GastroPlus fasted dog physiology bulk pH values and adjusted surface pH values used in the in silico simulations for ASD tablet and Calquence capsule.

Compartment	GastroPlus Fasted Dog Physiology Bulk pH	Adjusted (i.e. Surface) pH ASD Tablet	Adjusted (i.e. Surface) pH Calquence Capsule
Stomach (fasted)	2.0 [a]	4.6	4.0
Stomach (ARA)	6.0 [a]	6.3	6.0
Duodenum	6.2	6.2	6.2
Jejunum 1	6.2	6.2	6.2
Jejunum 2	6.2	6.2	6.2
Ileum 1	6.4	6.4	6.4
Ileum 2	6.6	6.6	6.6
Ileum 3	6.68	6.68	6.68
Caecum	6.75	6.75	6.75
Asc Colon	6.45	6.45	6.45

[a] Default pH in GastroPlus is 3.0. Bulk pH was assumed to be 2.0 for pentagastrin and 6.0 for famotidine treatments.

According to Mudie and coworkers, no appreciable change in solid surface pH nor dissolution rate is expected in buffered media representative of intestinal fluids for weak bases when the highest pK_a is ≤ 7 and $pK_w - pK_a - logS_o$ is ≥ 11 [30]. The highest acalabrutinib pK_a is 5.8. For the 50/50 acalabrutinib/HPMCAS-H ASD $pK_w - pK_a - logS_o = 11$ and for crystalline acalabrutinib $pK_w - pK_a - logS_o = 12$. Therefore no adjustment of bulk pH is needed for the intestinal compartments (i.e. surface pH = bulk pH) in the in silico predictions.

Appendix A.6. PK Parameter Calculations and In Silico Prediction Accuracy of Observed Data from Acalabrutinib Acidified Oral Solution

Clearance, volume of distribution and elimination half-life for acalabrutinib in beagle dogs were calculated from the observed blood plasma concentration versus time profile for an acalabrutinib acidified oral solution published by Podoll and coworkers at a 30 mg/kg dose in dogs weighing an average of 10 kg (See Figure A2) [34]. The terminal slope of the observed blood plasma concentration versus time profile (where no, or negligible absorption is occurring relative to elimination) (k_el) plotted on a logarithmic scale was determined using Microsoft Excel, and elimination half-life ($t_{1/2}$) was calculated by dividing the natural log of 2 (0.693) by k_{el}. AUC was calculated using the rectangular rule and extrapolated to infinity using the ratio of the last measured concentration and k_{el}. This value together with bioavailability (%F) (reported by Podoll and coworkers as 54%), and average dose (300 mg) was used to calculate clearance as shown in Equation (A1). Volume of distribution was calculated according to Equation (A2).

$$\text{Clearance} = \frac{\text{Dose} \times \text{F}}{\text{AUC}_{\text{oral solution}}} \tag{A1}$$

$$Volume\ of\ Distribution = \frac{Clearance \times t_{1/2}}{0.693} \tag{A2}$$

To determine if a one compartment model is appropriate to describe acalabrutinib pharmacokinetics in beagle dogs, a simulation was run at a 30 mg/kg (assuming 10 kg dog) dose in fasted beagle dogs and compared to observed data from an acidified oral solution published by Podoll et al [34]. IR:oral solution was selected as the dosage form treatment in the simulation to reflect what was dosed in the referenced study. Given that Pepin et al. reports an absolute bioavailability for acalabrutinib in beagle dogs of ca. 75% at a 100 mg (10 mg/kg) dose [6] and the reported bioavailability in the Podoll et al. study was ca. 54%, it is assumed the oral solution achieves approximately 72% dose absorbed at 30 mg/kg. Operating under this assumption and evaluating single compartment PK in dogs, the simulation provides a good prediction of the oral solution data (See Figure A2). Lack of multiple distinct slopes in the terminal phase when plotted in log scale suggests the use of a single compartment model is sufficient for subsequent simulation and prediction of the in vivo results generated for this study.

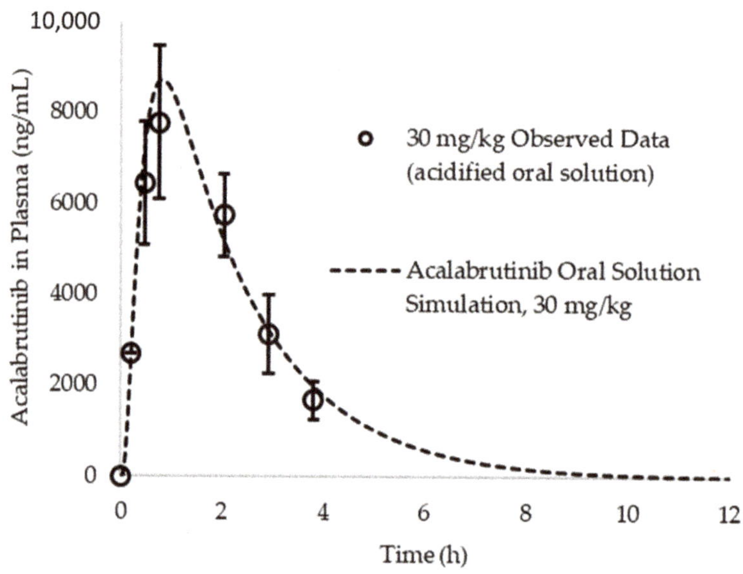

Figure A2. Acalabrutinib simulated vs. observed blood plasma concentration vs. time profile for a 30 mg/kg dose of an acidified oral solution. Observed data originally reported by Podoll et al. Simulations assumed a one compartment model for acalabrutinib in dogs.

Appendix A.7. ASD Characterization Results

SEM images, PXRD diffractograms, and mDSC thermograms of all six ASDs after spray drying and secondary drying are shown in Figures A3–A5, respectively. Tabulated ASD T_g values are shown in Table A9. ASDs showed collapsed hollow sphere morphology with no evidence of surface crystals. PXRD diffractograms showed no signs of crystallinity. All ASDs had high T_g values, above 114 °C.

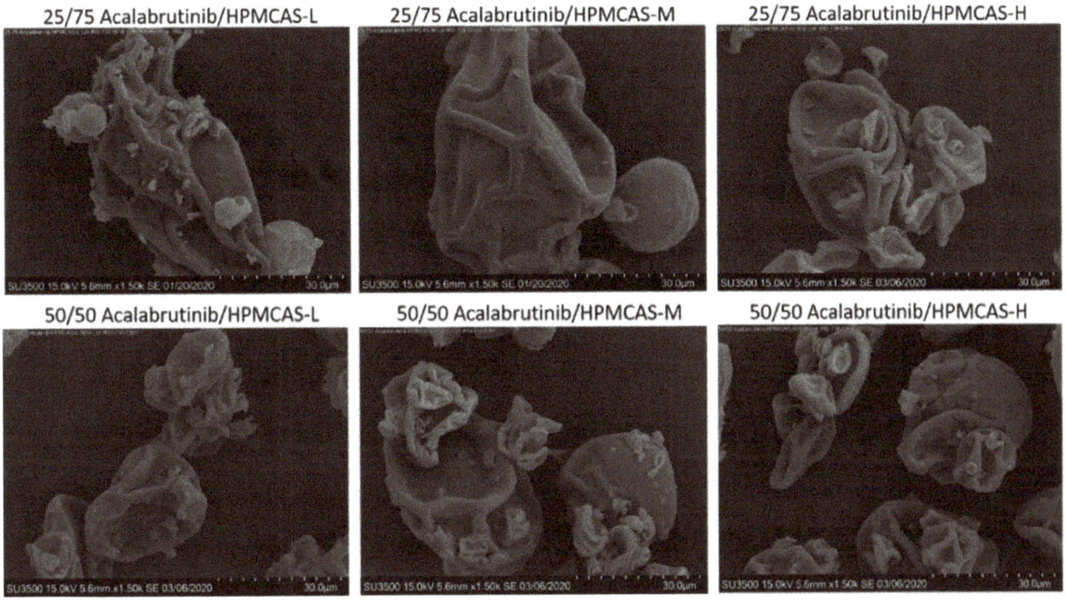

Figure A3. Scanning electron microscopy images of manufactured acalabrutinib ASDs.

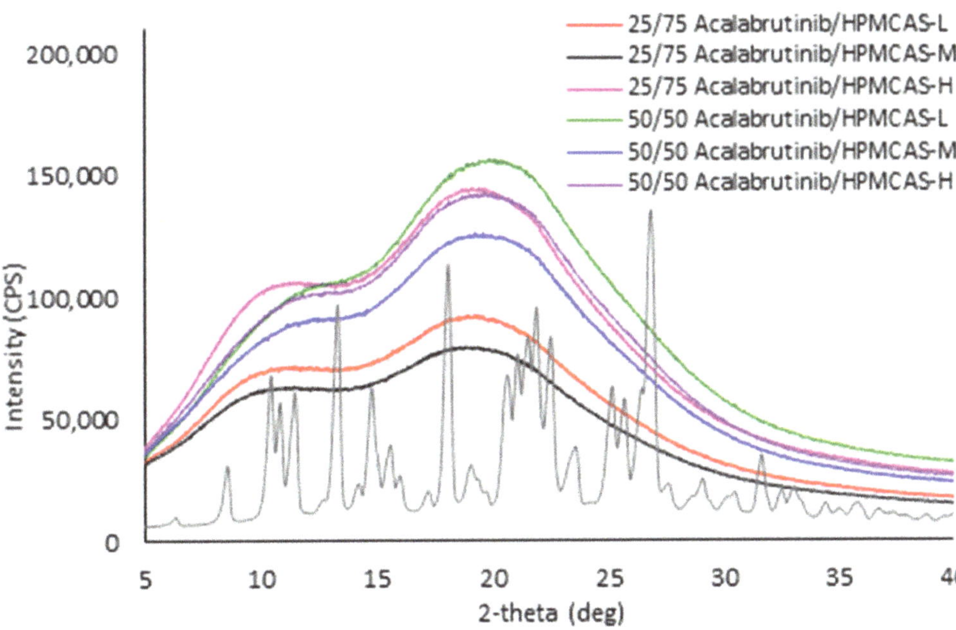

Figure A4. Powder X-ray diffractograms of manufactured acalabrutinib ASDs in comparison to acalabrutinib crystalline form I.

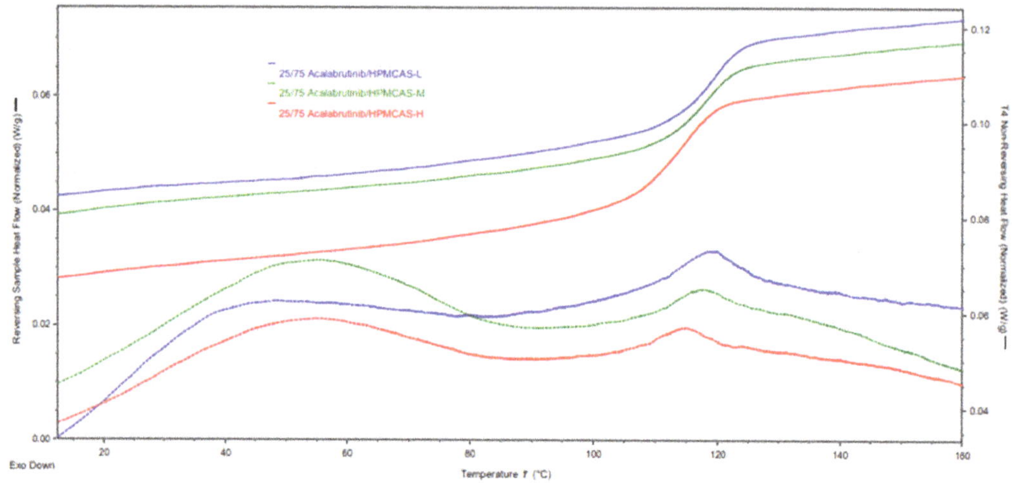

(**a**)

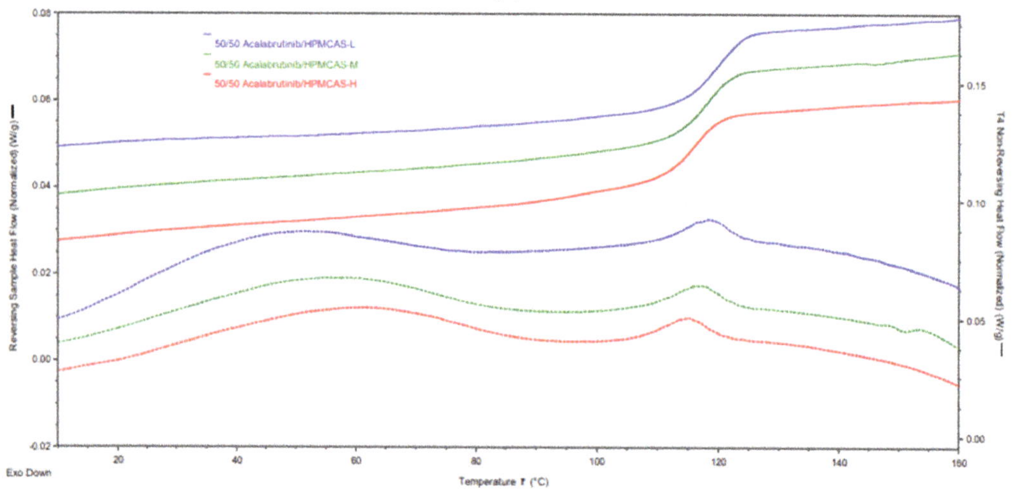

(**b**)

Figure A5. mDSC thermograms of manufactured acalabrutinib ASDs. Panel (**a**) 25 wt% drug loading ASDs. Panel (**b**) 50 wt% drug loading ASDs.

Table A9. Glass transition temperature (T_g) values measured via mDSC for manufactured acalabrutinib ASDs.

Formulation	T_g (°C)
25/75 Acalabrutinib/HPMCAS-L	118.7 ± 0.1
25/75 Acalabrutinib/HPMCAS-M	117.6 ± 0.2
25/75 Acalabrutinib/HPMCAS-H	114.1 ± 0.9
50/50 Acalabrutinib/HPMCAS-L	119.0 ± 0.0
50/50 Acalabrutinib/HPMCAS-M	118.0 ± 0.0
50/50 Acalabrutinib/HPMCAS-H	116.0 ± 1.0

Appendix A.8. In Vitro Dissolution Testing of ASD Intermediates Tabulated Results

Tabulated values for AUC enhancement of ASD intermediates during in vitro testing are shown in Table A10.

Table A10. AUC enhancements for ASD intermediates tested in gastric to intestinal transfer dissolution tests using a Pion µDiss Profiler.

ASD	AUC Enhancement [a] Gastric pH = 5	AUC Enhancement [a] Gastric pH = 6	AUC Enhancement [a] Gastric pH = 2
25/75 acalabrutinib/HPMCAS-H	7.2 (7.1–7.4)	6.8 (6.6–7.1)	1.1 (1.0–1.3)
25/75 acalabrutinib /HPMCAS-M	4.8 (3.9–5.7)	2.9 (2.4–3.4)	
25/75 acalabrutinib /HPMCAS-L	3.4 (3.2–3.6)	3.2 (2.9–3.5)	
50/50 acalabrutinib /HPMCAS-H	6.0 (5.1–6.9)	6.7 (6.1–7.3)	1.1 (0.9–1.2)
50/50 acalabrutinib /HPMCAS-M	2.4 (2.2–2.7)	2.9 (2.3–3.5)	0.8 (0.6–0.9)
50/50 acalabrutinib /HPMCAS-L	2.0 (1.3–2.6)	1.5 (1.5–1.5)	

[a] AUC enhancement = AUC(ASD)/AUC(crystalline acalabrutinib), reported as average (range), $n = 2$.

References

1. Brown, D.G.; Wobst, H.J. A decade of FDA-approved drugs (2010–2019): Trends and future directions. *J. Med. Chem.* **2021**, *64*, 2312–2338. [CrossRef]
2. Sawicki, E.; Schellens, J.H.; Beijnen, J.H.; Nuijen, B. Inventory of oral anticancer agents: Pharmaceutical formulation aspects with focus on the solid dispersion technique. *Cancer Treat. Rev.* **2016**, *50*, 247–263. [CrossRef]
3. Patel, D.; Bertz, R.; Ren, S.; Boulton, D.W.; Någård, M. A systematic review of gastric acid-reducing agent-mediated drug–drug interactions with orally administered medications. *Clin. Pharmacokinet.* **2020**, *59*, 447–462. [CrossRef] [PubMed]
4. Deisseroth, A.B. *Calquence FDA Label*; U.S. Food and Drug Administration: Silver Spring, MD, USA, 2019; Volume 2020. Available online: https://www.accessdata.fda.gov/drugsatfda_docs/label/2019/210259s006s007lbl.pdf (accessed on 15 March 2021).
5. Smelick, G.S.; Heffron, T.P.; Chu, L.; Dean, B.; West, D.A.; DuVall, S.L.; Lum, B.L.; Budha, N.; Holden, S.N.; Benet, L.Z.; et al. Prevalence of acid-reducing agents (ara) in cancer populations and ara drug–drug interaction potential for molecular targeted agents in clinical development. *Mol. Pharm.* **2013**, *10*, 4055–4062. [CrossRef] [PubMed]
6. Pepin, X.J.H.; Moir, A.J.; Mann, J.C.; Sanderson, N.J.; Barker, R.; Meehan, E.; Plumb, A.P.; Bailey, G.R.; Murphy, D.S.; Krejsa, C.M.; et al. Bridging in vitro dissolution and in vivo exposure for acalabrutinib. Part ii. A mechanistic pbpk model for ir formulation comparison, proton pump inhibitor drug interactions, and administration with acidic juices. *Eur. J. Pharm. Biopharm.* **2019**, *142*, 435–448. [CrossRef]
7. Zhang, L.; Wu, F.; Lee, S.C.; Zhao, H.; Zhang, L. Ph-dependent drug–drug interactions for weak base drugs: Potential implications for new drug development. *Clin. Pharmacol. Ther.* **2014**, *96*, 266–277. [CrossRef]
8. Mudie, D.M.; Stewart, A.M.; Rosales, J.A.; Biswas, N.; Adam, M.S.; Smith, A.; Craig, C.D.; Morgen, M.M.; Vodak, D.T. Amorphous solid dispersion tablets overcome acalabrutinib ph effect in dogs. *Pharmaceutics* **2021**, *13*, 557. [CrossRef] [PubMed]
9. Uchiyama, A.A.T.; Silva, P.A.I.A.; Lopes, M.S.M.; Yen, C.T.; Ricardo, E.D.; Mutão, T.; Pimenta, J.R.; Machado, L.M.; Shimba, D.S.; Peixoto, R.D. Proton pump inhibitors and oncologic treatment efficacy: A practical review of the literature for oncologists. *Curr. Oncol.* **2021**, *28*, 783–799. [CrossRef] [PubMed]
10. van Leeuwen, R.W.; van Gelder, T.; Mathijssen, R.H.; Jansman, F.G. Drug-drug interactions with tyrosine-kinase inhibitors: A clinical perspective. *Lancet Oncol.* **2014**, *15*, e315–e326. [CrossRef]
11. Blatter, F.; Ingallinera, T.; Barf, T.; Aret, E.; Krejsa, C.; Evarts, J. Solid Forms and Formuations of (s)-4->-Amino-3-(l-(but-2-ynoyl)pyrrolidin-2-yl)imidazo[1,5-ajpyrazin-1-yl)-n-(pyridin-2-yl)benzamide. 2016. Available online: https://patentimages.storage.googleapis.com/5e/4d/3f/0e3e4fb9e1a906/WO2017002095A1.pdf (accessed on 1 March 2021).
12. Almeida e Sousa, L.; Reutzel-Edens, S.M.; Stephenson, G.A.; Taylor, L.S. Assessment of the amorphous "solubility" of a group of diverse drugs using new experimental and theoretical approaches. *Mol. Pharm.* **2015**, *12*, 484–495. [CrossRef] [PubMed]
13. Stewart, A.; Yates, I.; Mudie, D.; Pivette, P.; Goodwin, A.; Sarmiento, A.; Winter, M.; Morgen, M.; Vodak, D. Mechanistic study of belinostat oral absorption from spray-dried dispersions. *J. Pharm. Sci.* **2019**, *108*, 326–336. [CrossRef]
14. Friesen, D.T.; Shanker, R.; Crew, M.; Smithey, D.T.; Curatolo, W.J.; Nightingale, J.A.S. Hydroxypropyl methylcellulose acetate succinate-based spray-dried dispersions: An overview. *Mol. Pharm.* **2008**, *5*, 1003–1019. [CrossRef] [PubMed]
15. Mudie, D.M.; Stewart, A.M.; Biswas, N.; Brodeur, T.J.; Shepard, K.B.; Smith, A.; Morgen, M.M.; Baumann, J.M.; Vodak, D.T. Novel high-drug-loaded amorphous dispersion tablets of posaconazole; in vivo and in vitro assessment. *Mol. Pharm.* **2020**, *17*, 4463–4472. [CrossRef] [PubMed]
16. Stewart, A.M.; Grass, M.E.; Brodeur, T.J.; Goodwin, A.K.; Morgen, M.M.; Friesen, D.T.; Vodak, D.T. Impact of drug-rich colloids of itraconazole and hpmcas on membrane flux in vitro and oral bioavailability in rats. *Mol. Pharm.* **2017**, *14*, 2437–2449. [CrossRef] [PubMed]

17. Mosquera-Giraldo, L.I.; Borca, C.H.; Meng, X.; Edgar, K.J.; Slipchenko, L.V.; Taylor, L.S. Mechanistic design of chemically diverse polymers with applications in oral drug delivery. *Biomacromolecules* **2016**, *17*, 3659–3671. [CrossRef]
18. Litou, C.; Vertzoni, M.; Goumas, C.; Vasdekis, V.; Xu, W.; Kesisoglou, F.; Reppas, C. Characteristics of the human upper gastrointestinal contents in the fasted state under hypo- and a-chlorhydric gastric conditions under conditions of typical drug–drug interaction studies. *Pharm. Res.* **2016**, *33*, 1399–1412. [CrossRef]
19. Prichard, P.J.; Yeomans, N.D.; Mihaly, G.W.; Jones, D.B.; Buckle, P.J.; Smallwood, R.A.; Louis, W.J. Omeprazole: A study of its inhibition of gastric ph and oral pharmacokinetics after morning or evening dosage. *Gastroenterology* **1985**, *88*, 64–69. [CrossRef]
20. Tutuian, R.; Katz, P.O.; Bochenek, W.; Castell, D.O. Dose-dependent control of intragastric ph by pantoprazole, 10, 20 or 40 mg, in healthy volunteers. *Aliment. Pharmacol. Ther.* **2002**, *16*, 829–836. [CrossRef]
21. Fancher, R.M.; Zhang, H.; Sleczka, B.; Derbin, G.; Rockar, R.; Marathe, P. Development of a canine model to enable the preclinical assessment of ph-dependent absorption of test compounds. *J. Pharm. Sci.* **2011**, *100*, 2979–2988. [CrossRef]
22. Koziolek, M.; Grimm, M.; Bollmann, T.; Schäfer, K.J.; Blattner, S.M.; Lotz, R.; Boeck, G.; Weitschies, W. Characterization of the gi transit conditions in beagle dogs with a telemetric motility capsule. *Eur. J. Pharm. Biopharm.* **2019**, *136*, 221–230. [CrossRef]
23. Arndt, M.; Chokshi, H.; Tang, K.; Parrott, N.J.; Reppas, C.; Dressman, J.B. Dissolution media simulating the proximal canine gastrointestinal tract in the fasted state. *Eur. J. Pharm. Biopharm.* **2013**, *84*, 633–641. [CrossRef]
24. Dressman, J.B. Comparison of canine and human gastrointestinal physiology. *Pharm. Res.* **1986**, *3*, 123–131. [CrossRef] [PubMed]
25. Smeets-Peeters, M.; Watson, T.; Minekus, M.; Havenaar, R. A review of the physiology of the canine digestive tract related to the development of in vitro systems. *Nutr. Res. Rev.* **1998**, *11*, 45–69. [CrossRef]
26. Wang, C.; Zhai, B.; Guo, H.; Wang, P.; Liu, Z.; Gu, H.; Ho, H.; Langguth, P.; Li, K.; Wang, C.; et al. In vivo measurement of gastric fluid volume in anesthetized dogs. *J. Drug Deliv. Sci. Technol.* **2020**, *55*, 101488. [CrossRef]
27. Takano, R.; Sugano, K.; Higashida, A.; Hayashi, Y.; Machida, M.; Aso, Y.; Yamashita, S. Oral absorption of poorly water-soluble drugs: Computer simulation of fraction absorbed in humans from a miniscale dissolution test. *Pharm. Res.* **2006**, *23*, 1144–1156. [CrossRef]
28. Serajuddin, A.T.; Jarowski, C.I. Effect of diffusion layer ph and solubility on the dissolution rate of pharmaceutical bases and their hydrochloride salts. I: Phenazopyridine. *J. Pharm. Sci.* **1985**, *74*, 142–147. [CrossRef]
29. Pepin, X.J.H.; Sanderson, N.J.; Blanazs, A.; Grover, S.; Ingallinera, T.G.; Mann, J.C. Bridging in vitro dissolution and in vivo exposure for acalabrutinib. Part i. Mechanistic modelling of drug product dissolution to derive a p-psd for pbpk model input. *Eur. J. Pharm. Biopharm.* **2019**, *142*, 421–434. [CrossRef]
30. Mudie, D.M.; Samiei, N.; Marshall, D.J.; Amidon, G.E.; Bergström, C.A.S. Selection of in vivo predictive dissolution media using drug substance and physiological properties. *AAPS J.* **2020**, *22*, 34. [CrossRef]
31. Ueda, K.; Moseson, D.E.; Pathak, V.; Taylor, L.S. Effect of polymer species on maximum aqueous phase supersaturation revealed by quantitative nuclear magnetic resonance spectroscopy. *Mol. Pharm.* **2021**, *18*, 1344–1355. [CrossRef]
32. Pepin, X.J.H.; Huckle, J.E.; Alluri, R.V.; Basu, S.; Dodd, S.; Parrott, N.; Emami Riedmaier, A. Understanding mechanisms of food effect and developing reliable pbpk models using a middle-out approach. *AAPS J.* **2021**, *23*, 12. [CrossRef]
33. Matsumura, N.; Hayashi, S.; Akiyama, Y.; Ono, A.; Funaki, S.; Tamura, N.; Kimoto, T.; Jiko, M.; Haruna, Y.; Sarashina, A.; et al. Prediction characteristics of oral absorption simulation software evaluated using structurally diverse low-solubility drugs. *J. Pharm. Sci.* **2020**, *109*, 1403–1416. [CrossRef]
34. Podoll, T.; Pearson, P.G.; Evarts, J.; Ingallinera, T.; Bibikova, E.; Sun, H.; Gohdes, M.; Cardinal, K.; Sanghvi, M.; Slatter, J.G. Bioavailability, biotransformation, and excretion of the covalent bruton tyrosine kinase inhibitor acalabrutinib in rats, dogs, and humans. *Drug Metab. Dispos.* **2019**, *47*, 145–154. [CrossRef]
35. Gala, U.H.; Miller, D.A.; Williams, R.O. Harnessing the therapeutic potential of anticancer drugs through amorphous solid dispersions. *Biochim. Biophys. Acta BBA Rev. Cancer* **2020**, *1873*, 188319. [CrossRef]
36. Chen, Y.; Pui, Y.; Chen, H.; Wang, S.; Serno, P.; Tonnis, W.; Chen, L.; Qian, F. Polymer-mediated drug supersaturation controlled by drug–polymer interactions persisting in an aqueous environment. *Mol. Pharm.* **2019**, *16*, 205–213. [CrossRef]
37. Ueda, K.; Higashi, K.; Yamamoto, K.; Moribe, K. The effect of hpmcas functional groups on drug crystallization from the supersaturated state and dissolution improvement. *Int. J. Pharm.* **2014**, *464*, 205–213. [CrossRef]
38. Kesisoglou, F.; Wang, M.; Galipeau, K.; Harmon, P.; Okoh, G.; Xu, W. Effect of amorphous nanoparticle size on bioavailability of anacetrapib in dogs. *J. Pharm. Sci.* **2019**, *108*, 2917–2925. [CrossRef]
39. Mitra, A.; Zhu, W.; Kesisoglou, F. Physiologically based absorption modeling for amorphous solid dispersion formulations. *Mol. Pharm.* **2016**, *13*, 3206–3215. [CrossRef]
40. Stewart, A.M.; Grass, M.E. Practical approach to modeling the impact of amorphous drug nanoparticles on the oral absorption of poorly soluble drugs. *Mol. Pharm.* **2020**, *17*, 180–189. [CrossRef]

pharmaceutics

MDPI

Article

Mutual Effects of Hydrogen Bonding and Polymer Hydrophobicity on Ibuprofen Crystal Inhibition in Solid Dispersions with Poly(*N*-vinyl pyrrolidone) and Poly(2-oxazolines)

Xiaoning Shan, Maryam A. Moghul, Adrian C. Williams and Vitaliy V. Khutoryanskiy *

Reading School of Pharmacy, University of Reading, Whiteknights, P.O. Box 224, Reading RG6 6AD, UK; Xiaoning.Shan@pgr.reading.ac.uk (X.S.); mmoghul.01@gmail.com (M.A.M.); a.c.williams@reading.ac.uk (A.C.W.)
* Correspondence: v.khutoryanskiy@reading.ac.uk; Tel.: +44-118-378-6119

Abstract: Poly(*N*-vinyl pyrrolidone) (PVP), poly(2-methyl-2-oxazoline) (PMOZ), poly(2-ethyl-2-oxazoline) (PEOZ), poly(2-*n*-propyl-2-oxazoline) (PnPOZ), and poly(2-isopropyl-2-oxazoline) (PiPOZ) were used to prepare solid dispersions with ibuprofen (IB), a model poorly-water soluble drug. Dispersions, prepared by solvent evaporation, were investigated using powder X-ray diffractometry, differential scanning calorimetry, and FTIR spectroscopy; hydrogen bonds formed between IB and all polymers in solid dispersions. PMOZ, the most hydrophilic polymer, showed the poorest ability to reduce or inhibit the crystallinity of IB. In contrast, the more hydrophobic polymers PVP, PEOZ, PnPOZ, and PiPOZ provided greater but similar abilities to reduce IB crystallinity, despite the differing polymer hydrophobicity and that PiPOZ is semi-crystalline. These results indicate that crystallinity disruption is predominantly due to hydrogen bonding between the drug molecules and the polymer. However, carrier properties affected drug dissolution, where PnPOZ exhibited lower critical solution temperature that inhibited the release of IB, whereas drug release from other systems was consistent with the degree of ibuprofen crystallinity within the dispersions.

Keywords: solid dispersions; hydrogen bonding; hydrophobicity; poly(*N*-vinyl pyrrolidone); poly(2-oxazolines); crystallinity; hydrophobic drug; amorphous; ibuprofen

check for updates

Citation: Shan, X.; Moghul, M.A.; Williams, A.C.; Khutoryanskiy, V.V. Mutual Effects of Hydrogen Bonding and Polymer Hydrophobicity on Ibuprofen Crystal Inhibition in Solid Dispersions with Poly(*N*-vinyl pyrrolidone) and Poly(2-oxazolines). *Pharmaceutics* **2021**, *13*, 659. https://doi.org/10.3390/pharmaceutics13050659

Academic Editor: Wouter L. J. Hinrichs

Received: 14 March 2021
Accepted: 28 April 2021
Published: 4 May 2021

1. Introduction

Whilst oral delivery remains the most common route for drug administration, most new active pharmaceutical ingredients are poorly water soluble and thus not well-absorbed after oral administration. Solid dispersion, defined as the dispersion of one or more active ingredient in a carrier or matrix at solid state, is an established platform technology to enhance the dissolution rate and improve the apparent solubility of a drug and, hence, increase the bioavailability of a range of poorly water soluble drugs [1–3]. Several classes of hydrophilic polymers have been used as carriers to prepare solid dispersions, including PVP [4–6] and its derivatives [7–9], polyethylene glycols [10,11], cellulose ethers [12,13] and poloxamers [14,15].

In solid dispersions, polymer–drug interactions can provide stability to the system by restricting the mobility of the drug molecules in the polymer matrix. Common interactions between drugs and polymers include ionic, hydrophobic, dipole–dipole, Van der Waals, and hydrogen bonding [16–18]. Hydrogen bonding is typically detected between drugs and polymers in solid dispersions, as reported extensively, for example, between IB and PVP [19,20], esomeprazole and hydroxypropyl methylcellulose (HPMC) [21], flurbiprofen and poly(ethylene oxide) [22], and for nifedipine and Eudragit® [23], indicating that this is a key mechanism in the successful formation of amorphous or semi-crystalline solid dispersions. In contrast, there are relatively few studies exploring the effects of carrier hydrophobicity on crystallization inhibition [16,24,25]. However, research typically focuses on hydrogen bonding or hydrophobicity, with little consideration given to the mutual effects of hydrogen bonding and polymer hydrophobicity on drug crystal inhibition.

Poly(2-oxazolines) have been reported as an alternative to PVP in solid dispersions for solubility and dissolution rate enhancement of poorly-water soluble drugs. For example, Fael et al. [26] found that a lower molecular weight of PEOZ (5000 g/mol) was superior to a higher molecular weight of the polymer (50,000 g/mol) in improving the dissolution behavior of glipizide. Boel et al. [27] showed that PEOZ maintained supersaturation of itraconazole and fenofibrate to a similar extent as PVP, poly(vinylpyrrolidone-*co*-vinyl acetate) (PVP-VA), and HPMC. Everaerts et al. [28] selected PEOZ, PnPOZ, poly(2-sec-butyl-2-oxazoline) (PsecBuOZ), and a combination of PEOZ with either PnPOZ or PsecBuOZ as carriers for amorphous solid dispersions with six drugs, and highlighted the potential of poly(2-oxazolines) as a novel polymer carrier to form amorphous solid dispersions.

In our previous work [29], PVP and a series of water-soluble poly(2-oxazolines) including PMOZ, PEOZ, PnPOZ, and PiPOZ were used to prepare solid dispersions with haloperidol. The effects of polymer hydrophobicity and their semi-crystalline nature on drug crystallinity were demonstrated. However, hydrogen bonding between haloperidol and poly(2-oxazolines) was almost absent due to the poor hydrogen bond donating ability of the haloperidol hydroxyl group.

In order to explore the impacts of both polymer hydrophobicity and drug–polymer hydrogen bonding, we selected IB, a hydrophobic crystalline drug and strong hydrogen bond donor (because of its carboxylic group), to prepare solid dispersions with poly(2-oxazolines) and PVP. Dispersions were prepared by solvent evaporation and characterized using FTIR spectroscopy, differential scanning calorimetry, and powder X-ray diffractometry. Solubility parameters and Flory–Huggins interaction parameters were calculated to predict drug–polymer miscibility, and drug dissolution studies were conducted to further explore the relationship between IB crystallization inhibition and release from the dispersions.

2. Materials and Methods

2.1. Materials

Poly(2-ethyl-2-oxazoline) (PEOZ), 50 kDa (polydispersity index, PDI 3-4); poly(*N*-vinyl pyrrolidone) (PVP), 55 kDa (K-value 30); and buffer tablets, pH 6.8 were from Sigma-Aldrich (Gillingham, UK). Poly(2-methyl-2-oxazoline) (PMOZ), poly(2-n-propyl-2-oxazoline) (PnPOZ), and poly(2-isopropyl-2-oxazoline) (PiPOZ) were synthesized according to our previously reported procedure [29]. Ibuprofen (IB) was from Tokyo Chemistry Industry (Japan). *N*, *N*-dimethylacetamide (DMA) was from Fisher Scientific (Loughborough, UK).

2.2. Preparation of Polymer–IB Solid Dispersions

Solid dispersions of polymer–IB were prepared in different repeating unit/drug molar ratios by solvent evaporation. DMA (1 mL) was used to dissolve 25 mg of IB with varying amounts of each polymer depending on the repeating unit/drug molar ratios. After dissolution, the solution was transferred to a Petri dish and the solvent was removed by evaporation at 50 °C on a heating base. The resultant solid was kept under vacuum for 72–96 h to remove residual DMA.

2.3. Characterization of Solid Dispersions

2.3.1. Powder X-ray Diffractometry (PXRD)

A small amount of each dry sample (~20 mg) was placed on a silica plate and analyzed in a Bruker D8 ADVANCE PXRD using Cu Kα radiation (λ = 1.5406 Å) over 5–64° for 1 h, with a step of 0.05°(2θ) and count time of 1.2 s at 40 mV, 40 mA, with the sample rotated at 30 rpm. The results were analyzed using EVA software and the background for each sample was removed.

2.3.2. Differential Scanning Calorimetry (DSC)

Thermal analysis of pure drug, polymers, and solid dispersions was performed using DSC (TA Instruments). Samples (3–5 mg) were loaded into pierced T_{zero} aluminum pans. The thermal behavior of each sample was investigated in a nitrogen atmosphere from 10 to a maximum of 220 °C at 10 °C/min. The degree of sample crystallinity was determined by the specific enthalpy (ΔH) of the drug melting peak using TA universal analysis software and was calculated as the ratio of ΔH of the drug in the solid dispersions to ΔH of pure IB (taken as 100% crystalline). Since the drug content in the dispersion was only a fraction of the sample weight, the degree of crystallinity was normalized according to the following equation:

$$\text{Crystallinity } (\%) = \left(\Delta H_s \times \frac{W_s}{W_i}\right) / \Delta H_i \times 100 \quad (1)$$

where ΔH_s is the ΔH of the drug in the solid dispersion, melting around 76 °C (melting point of IB), ΔH_i is the ΔH of pure IB, W_s is the weight of solid dispersions, and W_i is the weight of IB in solid dispersions.

2.3.3. Fourier Transform Infrared (FTIR) Spectroscopy

FTIR spectra were recorded on a Nicolet iS5 spectrometer using a diamond ATR (attenuated total reflection) accessory. After a background scan was collected, samples were placed on the crystal and scanned from 4000–400 cm^{-1} at a resolution of 4 cm^{-1} and with an average of 64 scans. OMINIC software was used for spectral analysis.

2.4. In Vitro Dissolution Studies

Dissolution of IB from solid dispersions ([polymer repeat unit]/[drug] = 1:1 mol/mol) used USP Apparatus II (paddle method) at 37 ± 0.5 °C with paddles at 50 rpm and simulated intestinal fluid (SIF) (PBS, pH = 6.8). A pharmaceutical grade empty vegan clear capsule size "0" filled with solid dispersion (equivalent to 100 mg drug) was placed in 900 mL SIF with a sinker. Samples (5 mL) were withdrawn at 2, 5, 10, 20, 40, 60, 80, 100, and 120 min, and filtered using a 0.45-µm syringe filter; an equal volume of SIF was added to the dissolution medium to maintain the volume. The drug was assayed by UV-visible spectroscopy at 265 nm. All dissolution studies were performed in triplicate.

2.5. Statistical Analysis

All solid dispersions for each polymer at all drug loadings were prepared three times independently. All analyses, PXRD, DSC, FTIR, and dissolution studies were in triplicate. Data are expressed as mean ± standard deviation.

3. Results and Discussion

3.1. Preparation and Characterization of Solid Dispersions

To evaluate the effects of different polymers on the crystallinity of IB, solid dispersions were prepared by solvent evaporation and were characterized by DSC, PXRD, and FTIR, with DSC used to calculate the crystallinity of IB in the dispersions. The X-ray diffractogram of IB (Figure 1) shows multiple distinctive peaks, notably at 6.2°, 12.3°, 14.1°, 14.9°, 16.8°, 17.9°, 19.6°, 20.3°, 22.5°, 25.3°, 28.4°, and 28.8°, in agreement with the literature [30] and demonstrating the crystalline nature of pure IB.

Clearly, the drug is diluted when included in the polymer dispersion, and so peak intensities fall in all solid dispersions with PVP, PEOZ, PnPOZ, and PiPOZ at molar ratio [polymer]/[IB] = 0.3:1. In these systems, the drug is in excess of the polymer hydrogen bond acceptor repeat units and so is expected to remain largely crystalline. However, when the molar ratio of polymer repeat unit to IB is 1:1, the X-ray data in Figure 1 indicate that IB crystallinity is completely lost (although PMOZ systems showed some anomalous results, discussed below). It is also interesting to note that PiPOZ alone is semi-crystalline and presents a feature at 8.14° 2θ, but this is lost at both 0.3:1 (where IB is in excess) and at a 1:1 mole ratio with the drug (highlighted in Figure 1d). This demonstrates that whilst solid dispersion studies typically focus on disruption of drug crystallinity, clearly the interactions between drug and carrier can also affect the nature and properties of the polymeric dispersant.

(a)

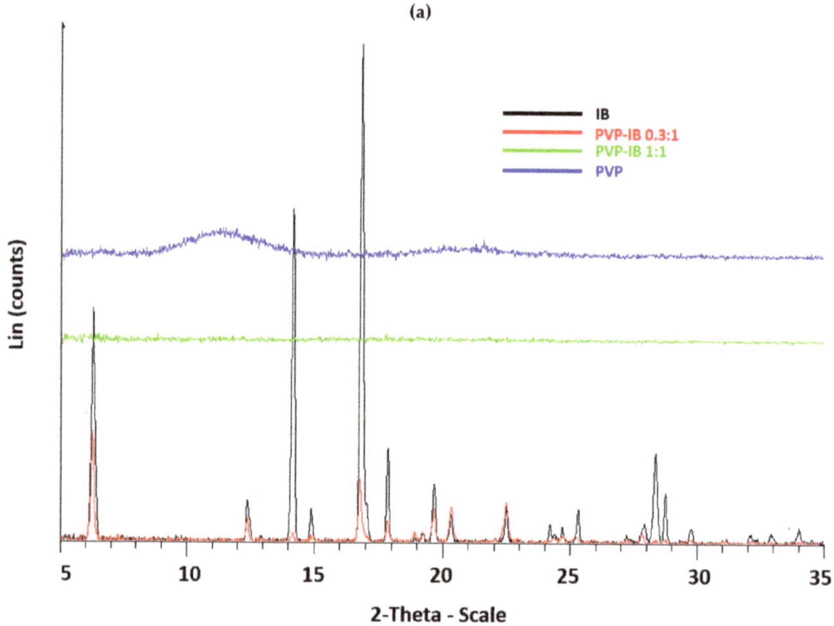

(b)

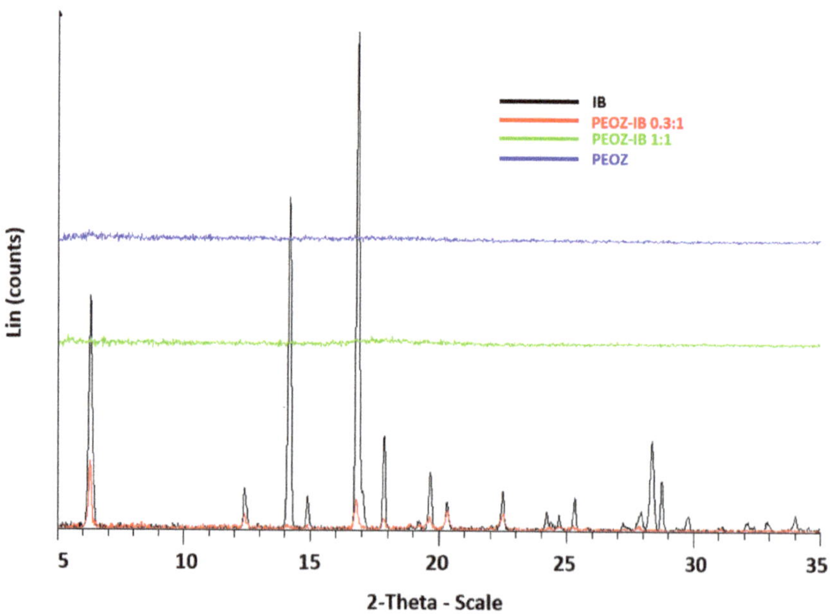

Figure 1. *Cont.*

(c)

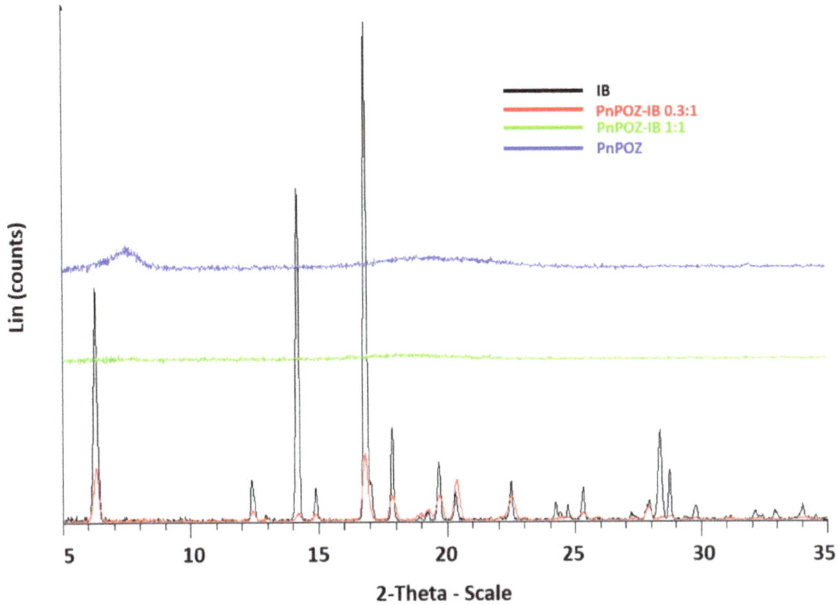

(d)

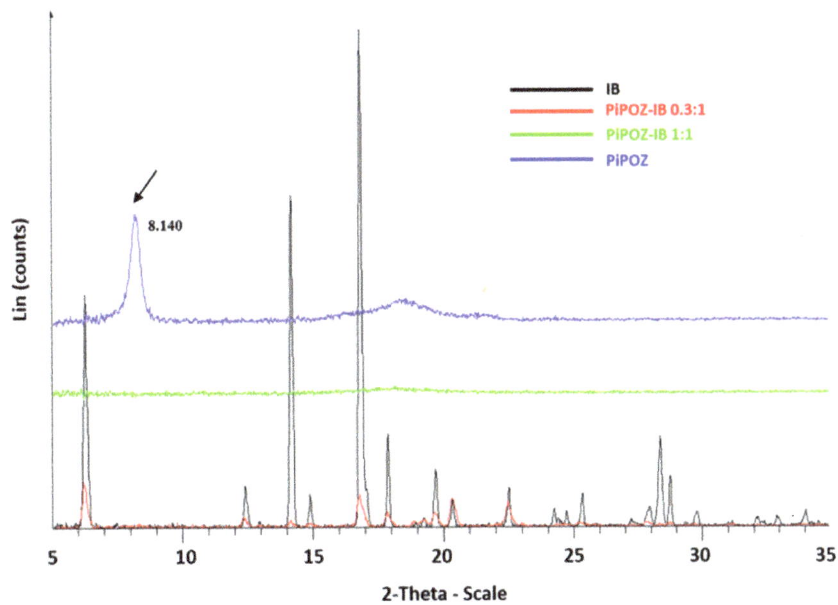

Figure 1. *Cont.*

(e)

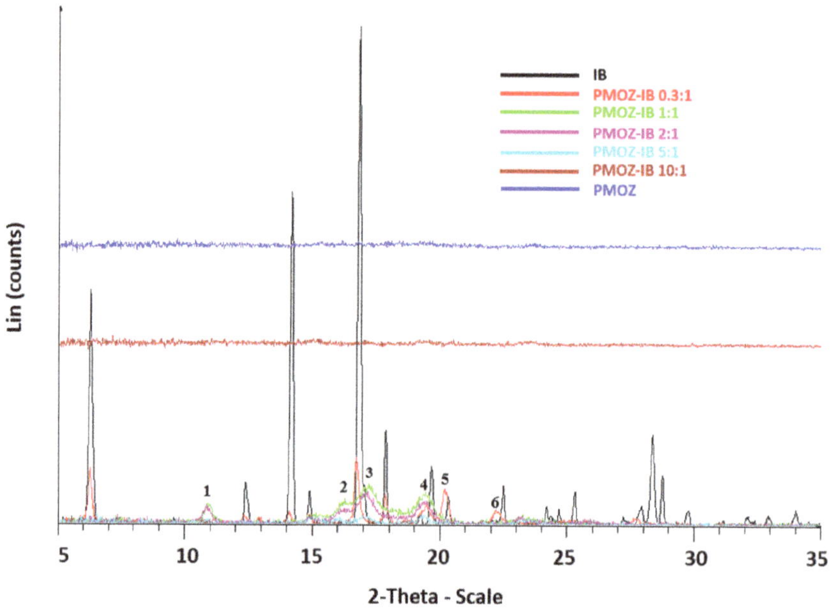

Figure 1. X-ray diffraction diagrams of PVP-IB SDs (**a**), PEOZ-IB SDs (**b**), PnPOZ-IB SDs (**c**), PiPOZ-IB SDs (**d**), and PMOZ-IB SDs (**e**).

When dispersed in PMOZ, the X-ray data indicate different behavior for ibuprofen. As above, at a 0.3:1 molar ratio, where the drug is in excess, there is a reduction in the intensity of the peaks from ibuprofen due to the dilution effect. Some peaks observed in samples at this ratio were slightly shifted from their positions seen with pure ibuprofen at 19.6°, 20.3°, and 22.5° (denoted as 4, 5, and 6, respectively, in Figure 1e) while the characteristic peak of IB at 16.8° 2θ did not significantly move, indicating that IB was predominantly in its original form. However, in contrast with the other polymers, it is clear that ibuprofen has some structure in PMOZ at 1:1, 1:2, and 1:5 drug:polymer compositions, with a series of broader diffraction features seen between 16 and 20°; again, these features show reduced intensities as the drug is further diluted in the polymer. For these systems, the original strong diffraction peak from the initial crystalline ibuprofen at 16.8° 2θ is lost, and broader features at 16.1°, 17.2°, and 19.5° 2θ are seen (peaks 2, 3, and 4, respectively, in Figure 1e). Further, a new feature at 10.5° is seen in these systems, which is absent from the pure ibuprofen diffractogram (peak 1). The changes to the diffraction peak positions and breadth of the new features are consistent with a semi-crystalline structure, but one that differs from the initial ibuprofen crystal lattice. Racemic ibuprofen is known to be polymorphic, but the diffraction peaks seen for the drug when dispersed in PMOZ are not consistent with the reported form II polymorph [31]. Thus, in our system, it is feasible that a semi-crystalline PMOZ-IB complex is formed. Finally, at a 1:10 drug:polymer composition, all diffraction peaks were lost, suggesting the formation of an amorphous dispersion.

FTIR spectra were recorded from the components and dispersions to probe molecular interactions. Functional groups of particular interest are the carboxyl group of IB [−C=O(OH)], where the −OH acts as a proton donor, and the carbonyl group (–C=O) and the nitrogen atom of polymers, which act as proton acceptors. In addition, correlation between frequency shifts and intermolecular interaction between drugs and polymers in solid dispersions is well known [32–34], and so was used to investigate hydrogen bonding in our polymer–drug solid dispersion systems.

The FTIR spectra of IB, polymers, and polymer–IB solid dispersions are shown in Figures S1 and S2. For clarity, the spectra are expanded between 1800–1550 cm^{-1} in Figure 2. IB vibrational frequencies and their assignments are given in Table S1 and agree with the literature [35]. Briefly, absorption bands between 3100–2800 cm^{-1} are attributed to C−H stretching modes, with peak intensities gradually reducing as the drug quantity falls in the solid dispersions (Figure S2). Two medium intensity features, appearing at 2725 cm^{-1} and 2633 cm^{-1} in the spectrum of IB, can be assigned to the stretching vibration of the cyclic dimerized hydroxyl groups, which is subjected to intermolecular hydrogen bonding [19,36] (Figure S1). However, these bands are lost in the spectra of amorphous solid dispersions, indicating that the drug dimeric structure is lost as a result of interaction with the polymers.

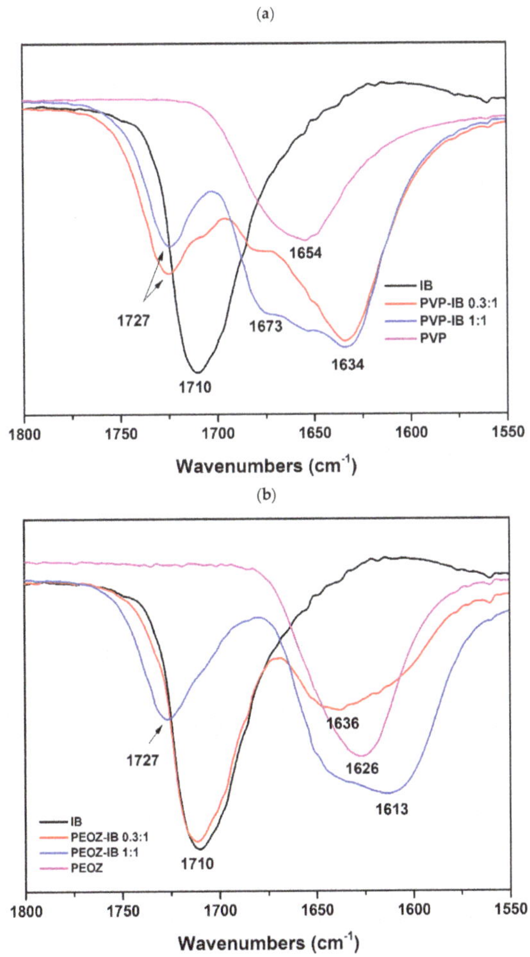

Figure 2. *Cont.*

(c)

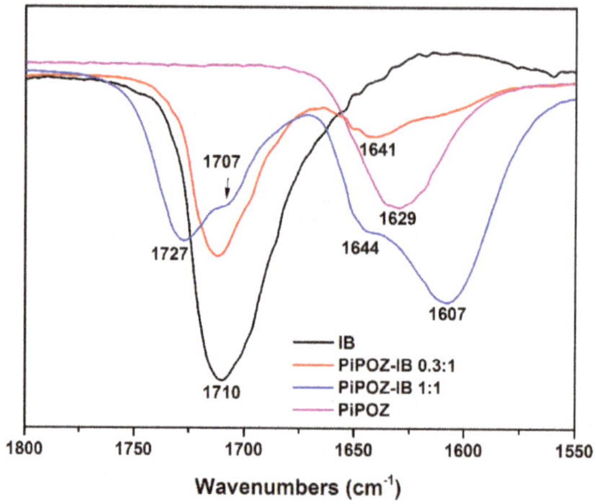

(d)

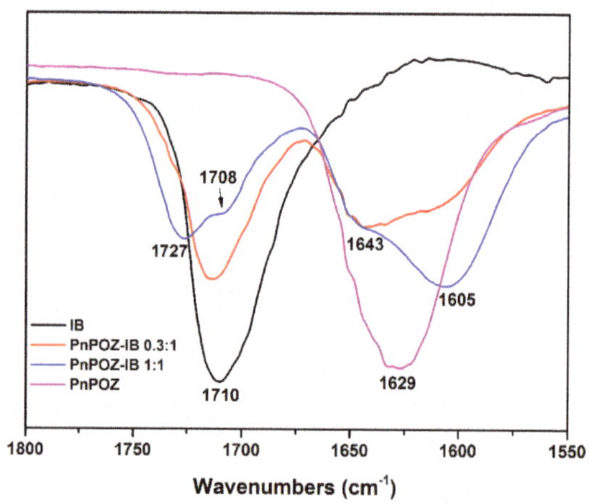

(e)

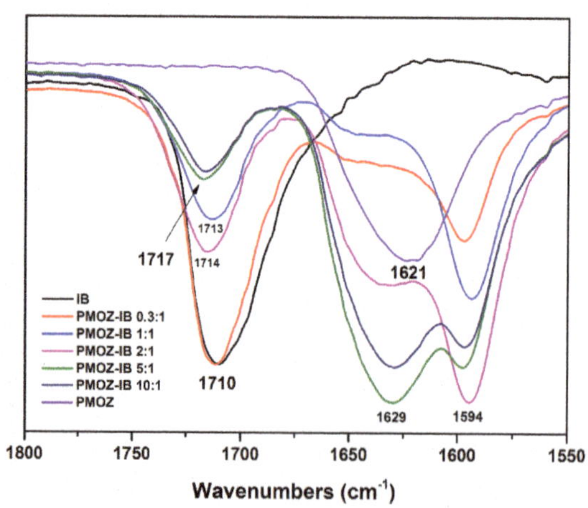

Figure 2. FTIR spectra of PVP-IB SDs (**a**), PEOZ-IB SDs (**b**), PnPOZ-IB SDs (**c**), PiPOZ-IB SDs (**d**), and PMOZ-IB SDs (**e**) in the range of 1800–1550 cm^{-1}.

The FTIR spectrum of IB shows a strong carbonyl stretching mode at 1710 cm^{-1} (Figure 2), which shifted to higher wavenumbers when dispersed in the polymers, and especially at ratios where X-ray diffraction showed no drug crystallinity (i.e., 1:1 ratios). These red shifts are summarized in Table 1. In contrast to the polyoxazoline carriers, the spectra of 0.3:1 mol PVP:IB showed that the IB carbonyl stretching mode shifted from 1710 cm^{-1} to 1727 cm^{-1}, despite the excess of IB to polymer monomer units, indicating the strong hydrogen bonding between IB with PVP may have consequential disruption to the IB crystal lattice. Furthermore, the carbonyl stretching mode for PVP at 1654 cm^{-1} was replaced by two peaks at 1634 cm^{-1} and 1673 cm^{-1}, with this latter peak strengthening at [PVP]/[IB] = 1:1 (Figure 2a).

Table 1. The red shift of the carbonyl stretching mode from the carboxylic acid of IB at 1710 cm^{-1} in (polymer)/(IB) = 0.3 mol and 1:1 mol solid dispersions.

Polymer–Drug	Wavenumbers (cm^{-1})			
	0.3:1 mol	**Red Shift**	**1:1 mol**	**Red Shift**
PVP-IB	1727	17	1727	17
PMOZ-IB	1710	0	1713	3
PEOZ-IB	1711	1	1727	17
PnPOZ-IB	1713	3	1727	17
PiPOZ-IB	1712	2	1727	17

The 1:1 ratio of IB in dispersions with PEOZ, PnPOZ, and PiPOZ showed similar infrared spectra, in agreement with that for PVP. The carbonyl stretching mode of IB showed a consistent red shift of 17 cm^{-1} from 1710 cm^{-1} to 1727 cm^{-1}, and the carbonyl stretch in the polymers split from the single peak at 1626 or 1629 cm^{-1} to give features at both higher and lower wavenumbers. At higher drug loadings (polymer: drug 0.3:1), the data suggest some drug–polymer interactions occurred, but these are somewhat obscured by the "free" excess IB within the systems.

As with the X-ray investigation, dispersions with PMOZ showed different molecular interactions than for the other polyoxazolines. A modest red shift in the IB carbonyl feature of 3 cm^{-1} was seen at a 1:1 stoichiometry, and when excess polymer was employed (10:1 polymer repeat unit: IB), the shift was still modest at 7 cm^{-1}. The PMOZ carbonyl mode was seen at 1621 cm^{-1} in the polymer alone (Figure 2e). There is again evidence for this mode splitting in the dispersion with peaks consistently at ~1594 cm^{-1} and ~1650 cm^{-1} in the samples at 0.3:1 and 1:1 mole ratios. This peak apparently moves towards ~1629 cm^{-1} as the polymer content increases, but in fact is due to the increased contribution of the "excess" (or "free") PMOZ carbonyl peak intensity, which overlaps and obscured the carbonyl group of PMOZ that interacts with IB. The weaker interaction of PMOZ with IB compared with that in other polyoxazoline dispersions can be attributed to PMOZ's relatively high hydrophilicity, which inhibits its ability to disorder the hydrophobic drug molecules [29].

From the IR data, there is no evidence for hydrogen bonding between the carboxylic groups of IB and nitrogen atoms in the polymers, given the invariant C-N stretching mode (Figure S3). Although PVP can form hydrogen bonds either through the nitrogen or carbonyl group [37], steric hindrance constrains the involvement of nitrogen atom in intermolecular interactions, so the carbonyl group is more favorable for hydrogen bonding [38,39].

Overall, the changes in the carbonyl band of IB and polymers indicate a modified carbonyl environment caused by the hydrogen bonding between the carboxylic groups of IB and carbonyl groups of the polymers. The relatively high red shift of the carbonyl mode of IB at [PVP]/[IB] = 0.3:1 mol confirmed strong hydrogen bonding between the drug and PVP, and at 1:1 mole ratio, the dispersions with PVP, PEOZ, PnPOZ, and PiPOZ all showed similar red shifts of this feature, suggesting near equivalent hydrogen bond interaction strengths. In contrast, the interaction between PMOZ and IB was relatively weak as a result of PMOZ's high hydrophilicity, but no new spectral features were found to demonstrate the presence of a novel drug: PMOZ complex.

DSC experiments were used to investigate the thermal behavior of the solid dispersions and to estimate drug crystallinity within the dispersions. The DSC thermogram of pure IB showed a single characteristic melting peak at 76 °C, confirming its crystalline nature (Figure 3) and in agreement with previous reports [40,41]. In all dispersions at 0.3:1 mole ratio, the excess IB was seen to melt at a lower temperature, and the broadening of the melting event is consistent with disorder being introduced into the crystal lattice and interactions with the polymer occurring. At 1:1 mole ratio, the drug melting peak was lost in all dispersions except in dispersion with PMOZ, in agreement with the X-ray data.

With PMOZ, a second endothermic peak appeared at 121.2 °C in (PMOZ)/(IB) = 0.3:1, at 137.7 °C in (PMOZ)/(IB) = 1:1, at 137.9 °C in (PMOZ)/(IB) = 2:1, and at 128.3 °C in (PMOZ)/(IB) = 5:1, potentially due to semi-crystalline IB or a PMOZ-IB complex. In addition, the melting peak seen at 203.5 °C for semi-crystalline PiPOZ was lost in (PiPOZ)/(IB) solid dispersions; whilst reports tend to focus on the disruption to drug crystallinity in solid dispersions, clearly the drug also has the potential to disrupt the structure of the polymeric carrier, as indicated here.

Drug crystallinity in polymer–IB solid dispersions was calculated from the specific enthalpy of the melting peak. As can be seen from Figure 4, IB crystallinity was reduced in all dispersions with PVP, PEOZ, PnPOZ, and PiPOZ, and the drug was essentially amorphous at a molar ratio of 1:1. The crystallinity of IB in dispersions with PMOZ could not be quantified by this approach due to the formation of new thermal features and the potential formation of a complex with this polymer.

(a)

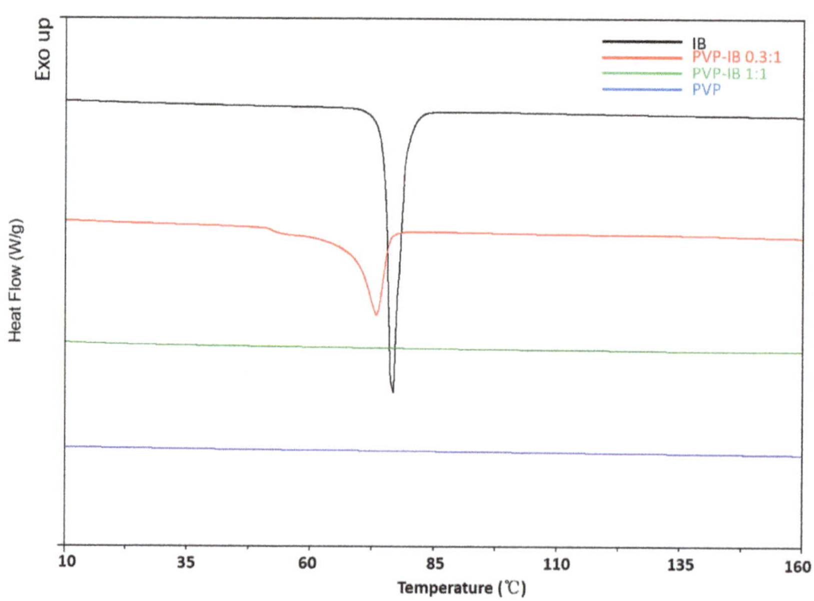

Figure 3. *Cont.*

(b)

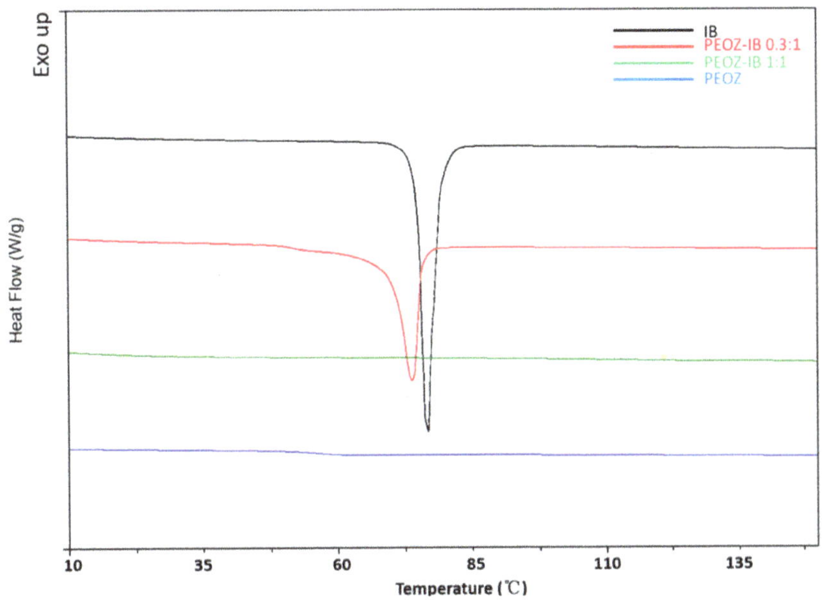

(c)

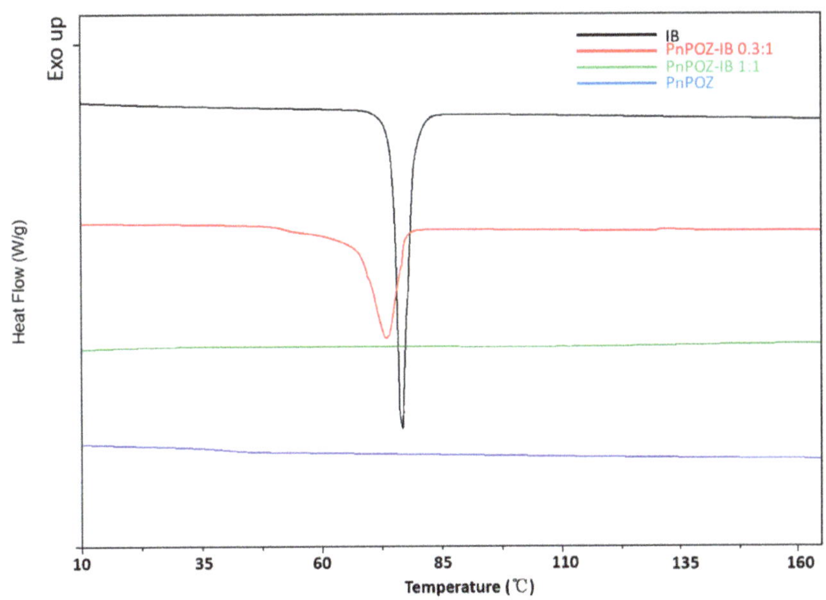

Figure 3. *Cont.*

(**d**)

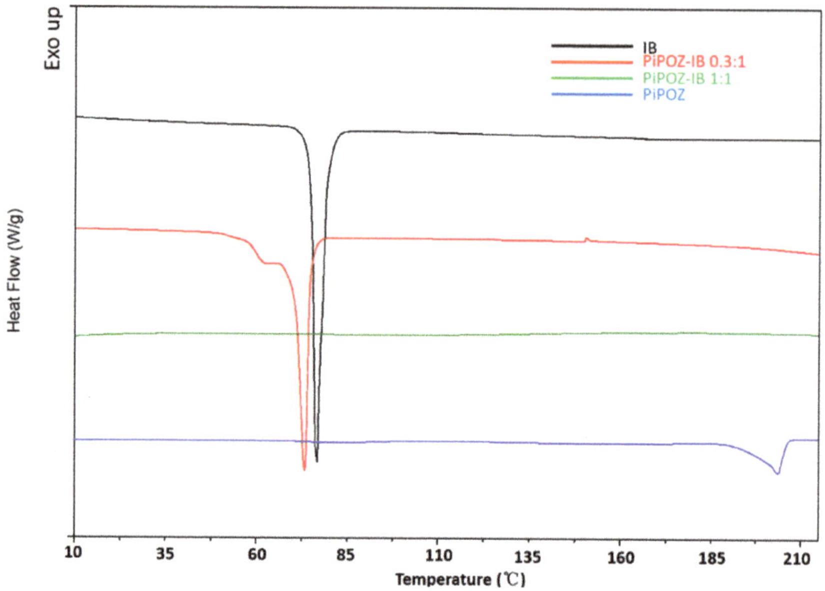

(**e**)

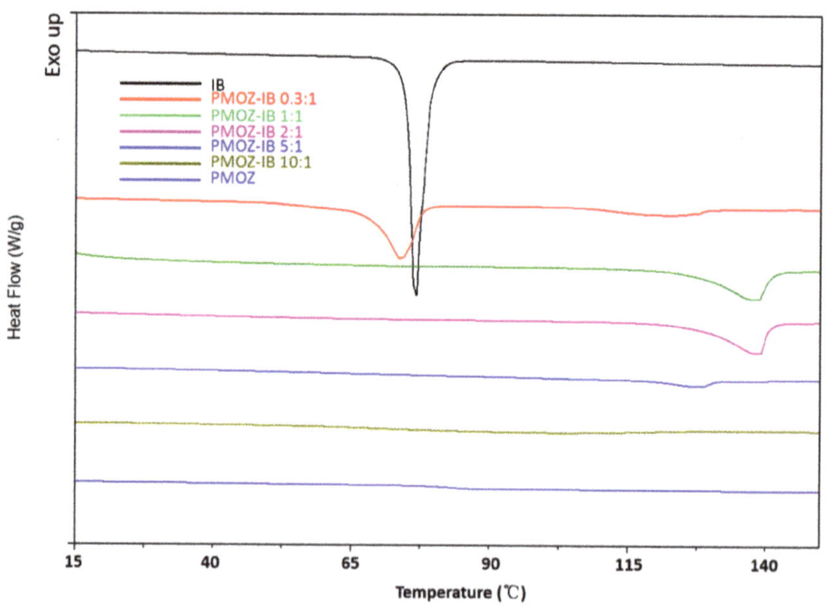

Figure 3. DSC thermograms of PVP-IB SDs (**a**), PEOZ-IB SDs (**b**), PnPOZ-IB SDs (**c**), PiPOZ-IB SDs (**d**), and PMOZ-IB SDs (**e**).

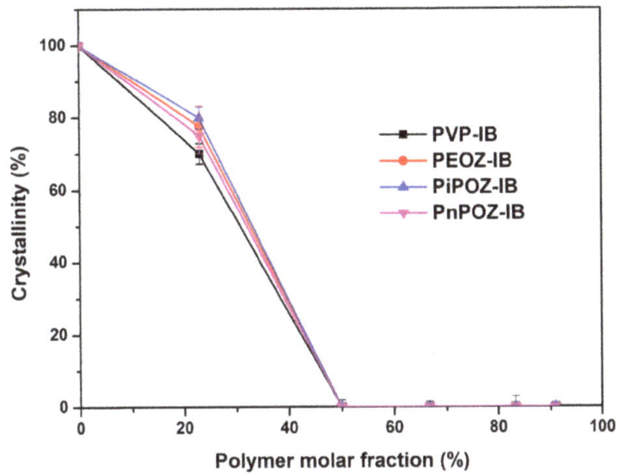

Figure 4. Crystallinity of polymer–IB solid dispersions as a function of polymer molar fraction.

3.2. Theoretical Evaluation of Drug-Polymer Miscibility

3.2.1. Solubility Parameters

The solubility parameter is a measure of cohesive energy density (CED: the cohesive energy per unit volume) of a material. The cohesive energy represents the total attractive forces within a condensed state material and can be defined as the quantity of energy needed to separate the atoms/molecules of a solid or liquid to a distance where the atoms or molecules possess no potential energy, that is, no interactions occur between atoms and molecules [30]. Consequently, solubility parameters have been used to predict the solubility/miscibility of one component into/with another component [42]. For this study, the solubility parameters were calculated using the Van Krevelen method [43], rather than the Fedors method [44,45], since the former considers hydrogen bonding. The Van Krevelen method provides:

$$\delta = \sqrt{{\delta_d}^2 + {\delta_p}^2 + {\delta_h}^2} \tag{2}$$

$$\text{where } \delta_d = \frac{\sum F_{di}}{V} \delta_p = \frac{\sqrt{\sum {F_{pi}}^2}}{V} \delta_h = \sqrt{\frac{\sum E_{hi}}{V}}$$

where δ is the total solubility parameter; δ_d, the contribution from dispersion forces; δ_p, the contribution from polar forces; δ_h, the contribution of hydrogen bonding; F_{di}, the molar attraction constant due to dispersion component; F_{pi}, the molar attraction constant due to polar component; E_{hi}, the hydrogen bonding energy; and V, the molar volume. For various groups, the values of F_{di}, F_{pi}, E_{hi}, and V (molar volume) are given in the literature [43,45]. The solubility parameters of these five polymers were taken from our previous study [29] and the solubility parameters calculated for IB and PVP are in good agreement with the literature [46].

Compounds with similar values of δ are likely to be miscible because the energy required to break interactions within each component is balanced by the energy released by interaction between the components. Greenhalgh et al. [30] classified dispersions based on the difference between the solubility parameters of excipients and drugs ($\Delta\delta$). The authors demonstrated that compounds with $\Delta\delta < 7.0$ MPa$^{1/2}$ are likely to be miscible. However, compounds with $\Delta\delta > 10.0$ MPa$^{1/2}$ are likely to be immiscible. The calculated solubility parameters for IB, PVP, PMOZ, PEOZ, PnPOZ, and PiPOZ are summarized in Table 2.

It can be seen from Table 2 that all the polymers are expected to be miscible with IB with $\Delta\delta$ values ranging from 3.1 to 6.9, except for PMOZ ($\Delta\delta$ = 7.6). The rank order values for $\Delta\delta$ miscibility (PiPOZ; PnPOZ; PEOZ; PVP) are inconsistent with their ability to disrupt ibuprofen crystallinity, which may be explained by confounding factors such as the stronger hydrogen bonding seen between PVP and IB, as suggested from the FTIR data (Table 1). Although widely used, this approach has limitations and tends to be most widely applicable for drug–polymer systems where Van der Waals interactions play a major role, whereas for drug–polymer mixtures forming highly directional interactions such as hydrogen bonds or long range forces such as ionic interactions, this method can yield erroneous results [1,47].

Table 2. Solubility parameters of drug and polymers.

Drug and Polymers	Solubility Parameters (δ) ($MPa^{1/2}$)		Group Classification
	Van Krevelen Method	Δδ	
IB	19.4		
PVP	26.3	6.9	Miscible
PMOZ	27.0	7.6	Not miscible
PEOZ	24.5	5.1	Miscible
PnPOZ	22.9	3.5	Miscible
PiPOZ	22.5	3.1	Miscible

3.2.2. Flory–Huggins Interaction Parameter

Flory–Huggins theory considers melting point depression as an indicator of miscibility. According to this [48], the relationship between the melting temperature of the pure drug ($T_m{}^0$) and the depressed melting point of the drug in the drug–polymer system (T_m) can be described by the following equation [49–51]:

$$\frac{1}{T_m} - \frac{1}{T_m^0} = -\frac{R}{\Delta H}\left(\ln\Phi + \left(1 - \frac{1}{m}\right)(1-\Phi) + \chi(1-\Phi)^2\right) \quad (3)$$

where R is the gas constant (8.31 J/mol·K), ΔH is the heat of fusion of the pure drug, Φ is the volume fraction of the drug in the solid dispersion (i.e., drug loading), m is the volume ratio between polymer and drug, and χ is the drug–polymer interaction parameter representing the difference between the drug–polymer contact interaction and the average self-contact interactions of drug–drug and polymer–polymer [49]. A negative χ value indicates that the interaction between a polymer and a drug is stronger than the attraction within polymer–polymer and drug–drug pairs. More negative values of χ indicate better affinity between the polymer and the drug and, for example, could be caused by hydrogen bonding between the drug and the polymer. Positive χ values indicate that drug molecules and polymer segments have stronger affinity to interact with those of their own kind rather than interacting with each other [50].

Given that all the polymer–IB solid dispersion systems showed depressed drug melting points at (polymer)/(drug) = 0.3:1 mol, the χ values of these dispersions were calculated and are listed in Table 3. Again, the PMOZ-IB system could not be investigated by this method.

Table 3. Flory–Huggins interaction parameters of polymer–IB solid dispersion systems at the molar ratio of 0.3:1.

Polymer–Drug	$V_{Polymer\ Repeat\ Unit}$ [a] (cm^3/mol)	$V_{Polymer}$ [b] (cm^3/mol)	V_{Drug} [c] (cm^3/mol)	M [d]	T_m (°C)	χ
PVP-IB	80.0	40,000	195.5	204.60	73.27	−3.71
PEOZ-IB	74.1	37,050		189.51	73.80	−3.85
PnPOZ-IB	90.2	45,100		230.69	73.50	−3.32
PiPOZ-IB	90.5	45,250		231.46	73.34	−3.52

a is the molecular volume of polymer repeating unit, calculated from the literature [43,45]. *b* is the molecular volume of polymer, calculated by multiplying $V_{polymer\ repeat\ unit}$ by the repeat unit number. *c* is the molecular volume of IB, calculated from the literature [43,45], and is in agreement with the value taken from [52]. *d* is the volume ratio between the polymer and the drug.

As can be seen, the drug–polymer interaction parameters are all negative and broadly similar, ranging from −3.85 for PEOZ-IB to −3.32 for PnPOZ-IB. Interestingly, the Flory–Huggins approach suggests a rank order of (greatest interactions to least) of PEOZ > PVP > PiPOZ > PnPOZ, whereas the rank order of solubility parameter miscibility was PiPOZ > PnPOZ > PEOZ > PVP. Clearly, both approaches provide approximations (almost a "yes/no" guide), rather than a predictive ability to develop the optimal solid dispersion, since other factors influence the polymer's ability to disrupt the drug's crystallinity.

The X-ray, thermal, and infrared studies showed that PMOZ has a lower propensity to disorder ibuprofen than the other polymers. Its solubility parameter difference to ibuprofen ($\Delta\delta$) was 7.6 MPa$^{1/2}$, and thus beyond the notional value of 7 for miscibility but close to the borderline value 6.9 MPa$^{1/2}$ calculated for PVP, which has the greatest tendency to disorder the drug. An alternative explanation is that the hydrophobic–hydrophilic balance (HHB) value for PMOZ (3.95) demonstrates that it is highly hydrophilic, and so the hydrophobic IB molecules will be less likely to molecularly disperse into the hydrophilic domains of PMOZ, consistent with our earlier findings on the non-hydrogen bonding dispersions with haloperidol where, again, PMOZ showed reduced interactions compared with the more hydrophobic carriers [29]. The importance of polymer hydrophobicity for crystal growth inhibitors has previously been reported [16].

3.3. *In Vitro* Dissolution Studies

The dissolution profiles of IB and polymer–IB (all 1:1 mol/mol) solid disperisons are shown in Figure 5. The dissolution of pure IB within 60 min was below 70%, with ~50% released in the first 20 min. As expected from the crystallinity data (Figure 4), dissolution was rapid from solid dispersions with PVP, PEOZ, and PiPOZ, where over 80% of the drug was released in the first 20 min. Drug release from solid dispersion with PMOZ was slower compared to PVP, PEOZ, and PiPOZ, but faster than IB alone, with ~70% of the drug released in 20 min, consistent with the analytical and theoretical discussions above. Despite the reduction of drug crystallinity and the system being essentially amorphous as determined by XRD and DSC, dispersions formed with PnPOZ showed slower dissolution than pure crystalline IB, with less than 30% released in 20 min. This result is consistent with our previous study [29] and can be explained by this polymer's lower critical solution temperature (LCST) of ~25 °C [53,54], which is much lower than the temperature used in the dissolution studies (37 °C). Under these conditions, PnPOZ remains insoluble in the dissolution medium, which limits drug release from these solid dispersions. Further detailed dissolution studies of these formulations will be of interest in the future, for example, in assessing release below 25 °C and evaluating the extent of IB supersaturation on the evolution of kinetic solubility profiles [55].

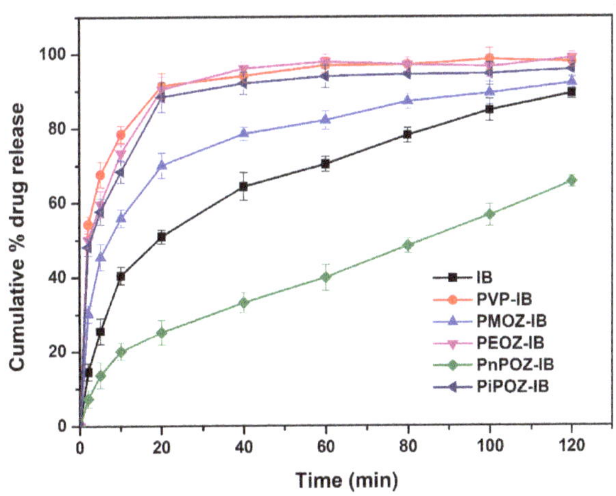

Figure 5. Dissolution profiles of pure IB and from different polymer–IB solid dispersions ((polymer repeat unit)/(drug) = 1:1 mol/mol). Cumulative % drug release with standard error of mean has been plotted against time.

4. Conclusions

Solid dispersions of IB were prepared using the poly(2-oxazolines) and PVP. Physical characterization of the dispersions showed that the polymers were able to disrupt the ibuprofen crystallinity, forming apparently amorphous dispersion at 1:1 mole ratios, and that hydrogen bonding was the prime mechanism for the interaction; however, the interactions between PMOZ and ibuprofen were more complex and hydrogen bonding was less prominent. The theoretical approach using the differences in solubility parameters between the drug and carrier or calculating the Flory–Huggins interaction parameters suggested compatibility between the drug and carriers, but the rank order of the predicted interactions varied between the two approaches. The purpose of generating solid dispersions is to enhance the dissolution rate of a poorly water soluble drug, and our studies demonstrated that the dispersions were able to significantly increase ibuprofen dissolution. However, our studies also show that other factors can significantly impact the performance of a solid dispersion. Physical characterization, for example, XRD showing that the drug is amorphous, can be assumed to result in enhanced dissolution. However, we show that not only is the crystallinity of a drug affected by dispersion, but so too is the structure of a semi-crystalline polymer (PiPOZ). The hydrophilicity of a carrier may reduce interactions with a hydrophobic drug, and so HHB may be an additional factor. Furthermore, the solution behavior of the carrier can also influence performance; physical characterization and theoretical models implied that dispersions with PnPOZ would be as effective as the other carriers, but the lower critical solution temperature (~25 °C) meant that this amorphous dispersion performed worse than the ibuprofen alone in the dissolution studies. Thus, both physical interactions, such as hydrogen bonding, and polymer properties, such as hydrophobicity, need to be considered when selecting carriers for solid dispersions.

Supplementary Materials: The following are available online at https://www.mdpi.com/article/10.3390/pharmaceutics13050659/s1, Figure S1. FTIR of ibuprofen. Figure S2: FTIR full spectra of PVP-IB SDs (**a**), PEOZ-IB SDs (**b**), PnPOZ-IB SDs (**c**), PiPOZ-IB SDs (**d**) and PMOZ-IB SDs (**e**) Figure S3: FTIR spectra of PVP-IB SDs and POZ-IB SDs in the range of 1400~900 cm^{-1}. The peaks (marked with an arrow) are attributed to C-N mode. Table S1: FTIR spectral data of ibuprofen (s—strong; w—weak; sym—symmetrical; asym—asymmetrical; str—stretching; m—medium; vs—very strong; vw—very weak.).

Author Contributions: Conceptualization, V.V.K. and A.C.W.; methodology, X.S. and M.A.M.; software, X.S. and M.A.M.; validation, X.S. and M.A.M.; formal analysis, X.S.; investigation, X.S.; resources, X.S.; data curation, X.S.; writing—original draft preparation, X.S.; writing—review and editing, V.V.K. and A.C.W.; visualization, X.S.; supervision, V.V.K. and A.C.W.; project administration, funding acquisition, V.V.K. All authors have read and agreed to the published version of the manuscript.

Funding: This research received no external funding.

Institutional Review Board Statement: Not applicable

Informed Consent Statement: Not applicable

Data Availability Statement: Data is contained within the article and in supplementary material.

Acknowledgments: The authors are grateful to the University of Reading and the China Scholarship Council (201707040071) for funding the PhD studentship of X.S. The assistance of staff at the Chemical Analysis Facility (CAF, University of Reading) with DSC and PXRD experiments is also acknowledged. The authors are also grateful to Kenneth Shankland for his valuable advice in interpreting PXRD and DSC data.

Conflicts of Interest: The authors declare no conflict of interest.

References

1. Baghel, S.; Cathcart, H.; O'Reilly, N.J. Polymeric amorphous solid dispersions: A review of amorphization, crystallization, stabilization, solid-state characterization, and aqueous solubilization of biopharmaceutical classification system class ii drugs. *J. Pharm. Sci.* **2016**, *105*, 2527–2544. [CrossRef] [PubMed]
2. Kim, K.T.; Lee, J.Y.; Lee, M.Y.; Song, C.K.; Choi, J.H.; Kim, D.D. Solid dispersions as a drug delivery system. *J. Pharm. Investig.* **2011**, *41*, 125–142. [CrossRef]
3. Tekade, A.R.; Yadav, J.N. A review on solid dispersion and carriers used therein for solubility enhancement of poorly water soluble drugs. *Adv. Pharm. Bull.* **2020**, *10*, 359–369. [CrossRef] [PubMed]
4. Biswal, S.; Sahoo, J.; Murthy, P.N. Physicochemical properties of solid dispersions of gliclazide in polyvinylpyrrolidone k90. *AAPS Pharm. Sci. Tech* **2009**, *10*, 329–334. [CrossRef]
5. Franco, P.; De Marco, I. The use of poly(*N*-vinyl pyrrolidone) in the delivery of drugs: A review. *Polymers* **2020**, *12*, 1114. [CrossRef]
6. Rumondor, A.C.; Ivanisevic, I.; Bates, S.; Alonzo, D.E.; Taylor, L.S. Evaluation of drug-polymer miscibility in amorphous solid dispersion systems. *Pharm. Res.* **2009**, *26*, 2523–2534. [CrossRef]
7. Indulkar, A.S.; Lou, X.; Zhang, G.G.Z.; Taylor, L.S. Insights into the dissolution mechanism of ritonavir-copovidone amorphous solid dispersions: Importance of congruent release for enhanced performance. *Mol. Pharm.* **2019**, *16*, 1327–1339. [CrossRef]
8. Panini, P.; Rampazzo, M.; Singh, A.; Vanhoutte, F.; Van den Mooter, G. Myth or truth: The glass forming ability class iii drugs will always form single-phase homogenous amorphous solid dispersion formulations. *Pharmaceutics* **2019**, *11*, 529. [CrossRef]
9. Que, C.; Lou, X.; Zemlyanov, D.Y.; Mo, H.; Indulkar, A.S.; Gao, Y.; Zhang, G.G.Z.; Taylor, L.S. Insights into the dissolution behavior of ledipasvir-copovidone amorphous solid dispersions: Role of drug loading and intermolecular interactions. *Mol. Pharm.* **2019**, *16*, 5054–5067. [CrossRef]
10. Biswal, S.; Sahoo, J.; Murthy, P.N.; Giradkar, R.P.; Avari, J.G. Enhancement of dissolution rate of gliclazide using solid dispersions with polyethylene glycol 6000. *AAPS Pharm. Sci. Tech* **2008**, *9*, 563–570. [CrossRef]
11. Urbanetz, N.A. Stabilization of solid dispersions of nimodipine and polyethylene glycol 2000. *Eur. J. Pharm. Sci.* **2006**, *28*, 67–76. [CrossRef]
12. Ghosh, I.; Snyder, J.; Vippagunta, R.; Alvine, M.; Vakil, R.; Tong, W.Q.; Vippagunta, S. Comparison of hpmc based polymers performance as carriers for manufacture of solid dispersions using the melt extruder. *Int. J. Pharm.* **2011**, *419*, 12–19. [CrossRef]
13. Li, B.; Harich, K.; Wegiel, L.; Taylor, L.S.; Edgar, K.J. Stability and solubility enhancement of ellagic acid in cellulose ester solid dispersions. *Carbohydr. Polym.* **2013**, *92*, 1443–1450. [CrossRef]
14. Ali, W.; Williams, A.C.; Rawlinson, C.F. Stochiometrically governed molecular interactions in drug: Poloxamer solid dispersions. *Int. J. Pharm.* **2010**, *391*, 162–168. [CrossRef]
15. Newa, M.; Bhandari, K.H.; Oh, D.H.; Kim, Y.R.; Sung, J.H.; Kim, J.O.; Woo, J.S.; Choi, H.G.; Yong, C.S. Enhanced dissolution of ibuprofen using solid dispersion with poloxamer 407. *Arch. Pharm. Res.* **2008**, *31*, 1497–1507. [CrossRef]
16. Ilevbare, G.A.; Liu, H.; Edgar, K.J.; Taylor, L.S. Understanding polymer properties important for crystal growth inhibition—impact of chemically diverse polymers on solution crystal growth of ritonavir. *Cryst. Growth Des.* **2012**, *12*, 3133–3143. [CrossRef]
17. Li, Y.; Pang, H.; Guo, Z.; Lin, L.; Dong, Y.; Li, G.; Lu, M.; Wu, C. Interactions between drugs and polymers influencing hot melt extrusion. *J. Pharm. Pharmacol.* **2014**, *66*, 148–166. [CrossRef]
18. Tran, T.T.D.; Tran, P.H.L. Molecular interactions in solid dispersions of poorly water-soluble drugs. *Pharmaceutics* **2020**, *12*, 745. [CrossRef]
19. Niemczyk, A.I.; Williams, A.C.; Rawlinson-Malone, C.F.; Hayes, W.; Greenland, B.W.; Chappell, D.; Khutoryanskaya, O.; Timmins, P. Novel polyvinylpyrrolidones to improve delivery of poorly water-soluble drugs: From design to synthesis and evaluation. *Mol. Pharm.* **2012**, *9*, 2237–2247. [CrossRef]
20. Rawlinson, C.F.; Williams, A.C.; Timmins, P.; Grimsey, I. Polymer-mediated disruption of drug crystallinity. *Int. J. Pharm.* **2007**, *336*, 42–48. [CrossRef]
21. Van Nguyen, H.; Baek, N.; Lee, B.J. Enhanced gastric stability of esomeprazole by molecular interaction and modulation of microenvironmental ph with alkalizers in solid dispersion. *Int. J. Pharm.* **2017**, *523*, 189–202. [CrossRef]
22. Ozeki, T.; Yuasa, H.; Kanaya, Y. Application of the solid dispersion method to the controlled release of medicine. Ix. Difference in the release of flurbiprofen from solid dispersions with poly(ethylene oxide) and hydroxypropylcellulose and the interact. *Int. J. Pharm.* **1997**, *155*, 209–217. [CrossRef]
23. Huang, J.; Wigent, R.J.; Schwartz, J.B. Drug-polymer interaction and its significance on the physical stability of nifedipine amorphous dispersion in microparticles of an ammonio methacrylate copolymer and ethylcellulose binary blend. *J. Pharm. Sci.* **2008**, *97*, 251–262. [CrossRef] [PubMed]
24. Rasenack, N.; Hartenhauer, H.; Müller, B.W. Microcrystals for dissolution rate enhancement of poorly water-soluble drugs. *Int. J. Pharm.* **2003**, *254*, 137–145. [CrossRef]
25. Zimmermann, A.; Millqvist-Fureby, A.; Elema, M.R.; Hansen, T.; Mullertz, A.; Hovgaard, L. Adsorption of pharmaceutical excipients onto microcrystals of siramesine hydrochloride: Effects on physicochemical properties. *Eur. J. Pharm. Biopharm.* **2009**, *71*, 109–116. [CrossRef]
26. Fael, H.; Rafols, C.; Demirel, A.L. Poly(2-ethyl-2-oxazoline) as an alternative to poly(vinylpyrrolidone) in solid dispersions for solubility and dissolution rate enhancement of drugs. *J. Pharm. Sci.* **2018**, *107*, 2428–2438. [CrossRef] [PubMed]

27. Boel, E.; Smeets, A.; Vergaelen, M.; De la Rosa, V.R.; Hoogenboom, R.; Van den Mooter, G. Comparative study of the potential of poly(2-ethyl-2-oxazoline) as carrier in the formulation of amorphous solid dispersions of poorly soluble drugs. *Eur. J. Pharm. Biopharm.* **2019**, *144*, 79–90. [CrossRef]
28. Everaerts, M.; Tigrine, A.; de la Rosa, V.R.; Hoogenboom, R.; Adriaensens, P.; Clasen, C.; Van den Mooter, G. Unravelling the miscibility of poly(2-oxazoline)s: A novel polymer class for the formulation of amorphous solid dispersions. *Molecules* **2020**, *25*, 3587. [CrossRef]
29. Shan, X.; Williams, A.C.; Khutoryanskiy, V.V. Polymer structure and property effects on solid dispersions with haloperidol: Poly(n-vinyl pyrrolidone) and poly(2-oxazolines) studies. *Int. J. Pharm.* **2020**, *590*, 119884. [CrossRef]
30. Greenhalgh, D.J.; Williams, A.C.; Timmins, P.; York, P. Solubility parameters as predictors of miscibility in solid dispersions. *J. Pharm. Sci.* **1999**, *88*, 1182–1190. [CrossRef] [PubMed]
31. Dudognon, E.; Danede, F.; Descamps, M.; Correia, N.T. Evidence for a new crystalline phase of racemic ibuprofen. *Pharm. Res.* **2008**, *25*, 2853–2858. [CrossRef]
32. Rostkowska, H.; Nowak, M.J.; Lapinski, L.; Adamowicz, L. Ir spectral and theoretical characterization of intramolecular hydrogen bonds closing five-membered rings. *Phys. Chem. Chem. Phys.* **2001**, *3*, 3012–3017. [CrossRef]
33. Rumondor, A.C.; Wikstrom, H.; Van Eerdenbrugh, B.; Taylor, L.S. Understanding the tendency of amorphous solid dispersions to undergo amorphous-amorphous phase separation in the presence of absorbed moisture. *AAPS Pharm. Sci. Tech* **2011**, *12*, 1209–1219. [CrossRef]
34. Teberekidis, V.I.; Sigalas, M.P. Theoretical study of hydrogen bond interactions of felodipine with polyvinylpyrrolidone and polyethyleneglycol. *J. Mol. Struct.* **2007**, *803*, 29–38. [CrossRef]
35. Ramukutty, S.; Ramachandran, E. Growth, spectral and thermal studies of ibuprofen crystals. *Cryst. Res. Technol.* **2012**, *47*, 31–38. [CrossRef]
36. Bogdanova, S.; Pajeva, I.; Nikolova, P.; Tsakovska, I.; Müller, B. Interactions of poly(vinylpyrrolidone) with ibuprofen and naproxen: Experimental and modeling studies. *Pharm. Res.* **2005**, *22*, 806–815. [CrossRef]
37. Iannuccelli, V.; Coppi, G.; Leo, E.; Fontana, F.; Bernabei, M.T. Pvp solid dispersions for the controlled release of furosemide from a floating multiple-unit system. *Drug Dev. Ind. Pharm.* **2000**, *26*, 595–603. [CrossRef]
38. Chadha, R.; Kapoor, V.K.; Kumar, A. Analytical techniques used to characterize drug-polyvinylpyrrolidone systems in solid and liquid states—An overview. *J. Sci. Ind. Res.* **2006**, *65*, 459–469.
39. Hosono, T.; Tsuchiya, S.; Matsumaru, H. Model of interaction of ajmaline with polyvinylpyrrolidone. *J. Pharm. Sci.* **1980**, *69*, 824–826. [CrossRef]
40. Sekizaki, H.; Danjo, K.; Eguchi, H.; Yonezawa, Y.; Sunada, H.; Otsuka, A. Solid-state interaction of ibuprofen with polyvinylpyrrolidone. *Chem. Pharm. Bull.* **1995**, *43*, 988–993. [CrossRef]
41. Williams, A.C.; Timmins, P.; Lu, M.; Forbes, R.T. Disorder and dissolution enhancement: Deposition of ibuprofen on to insoluble polymers. *Eur. J. Pharm. Sci.* **2005**, *26*, 288–294. [CrossRef] [PubMed]
42. Jankovic, S.; Tsakiridou, G.; Ditzinger, F.; Koehl, N.J.; Price, D.J.; Ilie, A.R.; Kalantzi, L.; Kimpe, K.; Holm, R.; Nair, A.; et al. Application of the solubility parameter concept to assist with oral delivery of poorly water-soluble drugs—A PEARRL review. *J. Pharm. Pharmacol.* **2019**, *71*, 441–463. [CrossRef] [PubMed]
43. Krevelen, D.W.V. *Properties of Polymers*; Elsevier: Amsterdam, The Netherlands, 1990; pp. 189–221.
44. Chokshi, R.J.; Sandhu, H.K.; Iyer, R.M.; Shah, N.H.; Malick, A.W.; Zia, H. Characterization of physico-mechanical properties of indomethacin and polymers to assess their suitability for hot-melt extrusion processes as a means to manufacture solid dispersion/solution. *J. Pharm. Sci.* **2005**, *94*, 2463–2474. [CrossRef] [PubMed]
45. Fedors, R.F. A method for estimating both the solubility parameters and molar volumes of liquids. *Polym. Eng. Sci.* **1974**, *14*, 147–154. [CrossRef]
46. Kitak, T.; Dumicic, A.; Planinsek, O.; Sibanc, R.; Srcic, S. Determination of solubility parameters of ibuprofen and ibuprofen lysinate. *Molecules* **2015**, *20*, 21549–21568. [CrossRef]
47. Barton, A.F.M. Solubility parameters. *Chem. Rev.* **1975**, *75*, 731–753. [CrossRef]
48. Flory, P.J. *Principles of Polymer Chemistry*; Cornell University Press: Ithaca, NY, USA, 1953.
49. Huang, Y.; Dai, W.G. Fundamental aspects of solid dispersion technology for poorly soluble drugs. *Acta Pharm. Sin. B* **2014**, *4*, 18–25. [CrossRef]
50. Lin, D.; Huang, Y. A thermal analysis method to predict the complete phase diagram of drug-polymer solid dispersions. *Int. J. Pharm.* **2010**, *399*, 109–115. [CrossRef]
51. Moseson, D.E.; Taylor, L.S. The application of temperature-composition phase diagrams for hot melt extrusion processing of amorphous solid dispersions to prevent residual crystallinity. *Int. J. Pharm.* **2018**, *553*, 454–466. [CrossRef]
52. Kong, C.Y.; Sugiura, K.; Funazukuri, T.; Miyake, K.; Okajima, I.; Badhulika, S.; Sako, T. The retention factors and partial molar volumes of ibuprofen at infinite dilution in supercritical carbon dioxide at t = (308.15, 313.15, 323.15, 333.15, 343.15 and 353.15) k. *J. Mol. Liq.* **2019**, *296*, 111849. [CrossRef]
53. Bouten, P.; Lava, K.; van Hest, J.; Hoogenboom, R. Thermal properties of methyl ester-containing poly(2-oxazoline)s. *Polymers* **2015**, *7*, 1998–2008. [CrossRef]
54. Park, J.S.; Kataoka, K. Comprehensive and accurate control of thermosensitivity of poly(2-alkyl-2-oxazoline)s via well-defined gradient or random copolymerization. *Macromolecules* **2007**, *40*, 3599–3609. [CrossRef]
55. Han, Y.R.; Lee, P.I. Effect of extent of supersaturation on the evolution of kinetic solubility profiles. *Mol. Pharm.* **2017**, *14*, 206–220. [CrossRef]

pharmaceutics

MDPI

Article

Amorphous Solid Dispersion Tablets Overcome Acalabrutinib pH Effect in Dogs

Deanna M. Mudie [1,*], Aaron M. Stewart [1], Jesus A. Rosales [1,2], Nishant Biswas [1], Molly S. Adam [1], Adam Smith [1], Christopher D. Craig [1], Michael M. Morgen [1] and David T. Vodak [1]

[1] Global Research & Development, Lonza, Bend, OR 97703, USA; aaron.stewart@lonza.com (A.M.S.); rosaleja@uw.edu (J.A.R.); nishantbb@gmail.com (N.B.); molly.adam@lonza.com (M.S.A.); adam.smith@lonza.com (A.S.); chris.craig@lonza.com (C.D.C.); michael.morgen@lonza.com (M.M.M.); david.vodak@lonza.com (D.T.V.)

[2] Pharmaceutics Department, University of Washington, Seattle, WA 98195, USA

* Correspondence: deanna.mudie@lonza.com

Abstract: Calquence® (crystalline acalabrutinib), a commercially marketed tyrosine kinase inhibitor (TKI), exhibits significantly reduced oral exposure when taken with acid-reducing agents (ARAs) due to the low solubility of the weakly basic drug at elevated gastric pH. These drug–drug interactions (DDIs) negatively impact patient treatment and quality of life due to the strict dosing regimens required. In this study, reduced plasma drug exposure at high gastric pH was overcome using a spray-dried amorphous solid dispersion (ASD) comprising 50% acalabrutinib and 50% hydroxypropyl methylcellulose acetate succinate (HPMCAS, H grade) formulated as an immediate-release (IR) tablet. ASD tablets achieved similar area under the plasma drug concentration–time curve (AUC) at low and high gastric pH and outperformed Calquence capsules 2.4-fold at high gastric pH in beagle dogs. In vitro multicompartment dissolution testing conducted a priori to the in vivo study successfully predicted the improved formulation performance. In addition, ASD tablets were 60% smaller than Calquence capsules and demonstrated good laboratory-scale manufacturability, physical stability, and chemical stability. ASD dosage forms are attractive for improving patient compliance and the efficacy of acalabrutinib and other weakly basic drugs that have pH-dependent absorption.

Keywords: acalabrutinib; amorphous solid dispersion; acid-reducing agent; bioavailability enhancement; kinase inhibitor; pH effect; proton pump inhibitor; spray drying

Citation: Mudie, D.M.; Stewart, A.M.; Rosales, J.A.; Biswas, N.; Adam, M.S.; Smith, A.; Craig, C.D.; Morgen, M.M.; Vodak, D.T. Amorphous Solid Dispersion Tablets Overcome Acalabrutinib pH Effect in Dogs. *Pharmaceutics* **2021**, *13*, 557. https://doi.org/10.3390/pharmaceutics13040557

Academic Editors: Vitaliy Khutoryanskiy and Hisham Al-Obaidi

Received: 29 March 2021
Accepted: 13 April 2021
Published: 15 April 2021

1. Introduction

Calquence® is a commercially marketed kinase inhibitor used as second-line therapy for adult patients seeking treatment for mantle cell lymphoma, chronic lymphocytic leukemia, or small lymphocytic lymphoma [1]. The active pharmaceutical ingredient (API) in Calquence is the crystalline Form I of acalabrutinib, which is a diprotic weak base with pK_a values of 3.5 and 5.8 [2]. It is a Biopharmaceutics Classification System (BCS) Class 2 drug with low intrinsic solubility and a moderate log P [2,3]. The chemical structure and compound properties of acalabrutinib are shown in Figure 1 and Table 1, respectively.

Calquence has clinically meaningful drug–drug interactions (DDIs) with acid-reducing agents (ARAs) such as proton pump inhibitors (PPIs), where concomitant use results in reduced area under the plasma drug concentration–time curve (AUC) values [4]. For example, coadministration of Calquence with 40 mg of the PPI omeprazole for 5 days decreased AUC by 43% in healthy subjects [1]. According to the Food and Drug Administration (FDA) label for Calquence, patients must avoid coadministration with PPIs and stagger dosing with histamine H_2 receptor antagonists (H_2RAs) and antacids [1]. Because ARAs are commonly prescribed to cancer patients, these DDIs can undermine efficacy and patient compliance due to the complex dosing schedules required [5].

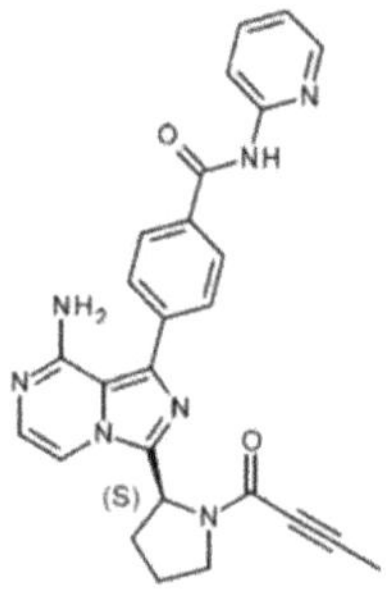

Figure 1. Acalabrutinib chemical structure.

The mechanism of reduced performance of acalabrutinib crystalline Form I when taken with ARAs is decreased solubility at the elevated gastric pH levels that occur when ARAs are administered, referred to here as the "ARA effect" [4,6]. The reduced solubility of the weakly basic drug decreases dissolution, which decreases absorption across the intestinal membrane and increases the extent of dose metabolized in humans [2,7]. This gastric pH-dependent mechanism that occurs with weakly basic drugs is the most common DDI for ARAs. However, natural variations in gastric pH also cause undesirable variability in Calquence performance, irrespective of the use of ARAs [4,8–10].

Table 1. Acalabrutinib Form I physicochemical properties.

Compound Property	Value
Molecular weight (g/mol)	465.5
pK_as in physiological range	3.5 (basic) and 5.8 (basic) [a]
log P	2.0 [a]
Melting temperature (T_m) (°C)	214 [b]
Glass-transition temperature (T_g) (°C)	133 [c]
Crystalline intrinsic solubility (μg/mL)	48 [a]

[a] Reference [2]. [b] Reference [11]. [c] Measured in house using modulated differential scanning calorimetry (mDSC). Refer to Appendix A.1 for method information.

To overcome the ARA effect, a spray-dried amorphous solid dispersion (ASD) was developed and formulated as an immediate-release (IR) tablet. ASD technologies are used to enhance solubilization and increase the bioavailability of poorly soluble drugs [12]. The amorphous drug form has higher solubility than the crystalline drug form due to its higher activity [13]. Since the amorphous form is thermodynamically unstable, it is typically formulated as an ASD with a polymer to stabilize the high free energy state [14].

Since the ARA effect is caused by the decreased solubility of acalabrutinib crystalline Form I at elevated gastric pH, it was hypothesized that an ASD of acalabrutinib could increase API solubility enough to provide rapid and high extent of gastrointestinal (GI) dissolution of the API at high and low gastric pH values (i.e., in the presence and absence of ARAs). Furthermore, achieving complete absorption in the presence of ARAs would provide limited potential for increased absorption at low, fasted-state gastric pH where the API is more soluble. Achieving similar AUC values under both conditions could thereby mitigate the ARA effect and potentially negate the DDI seen with crystalline Form I acalabrutinib.

The goal of this study was to demonstrate mitigation of the ARA effect in a beagle dog model using acalabrutinib ASD IR tablets at a 100 mg dose. A 50/50 (% *w*/*w*) ASD containing acalabrutinib and hydroxypropyl methylcellulose acetate succinate (HPMCAS, HF grade) was prepared using a laboratory (kilogram)-scale spray dryer and incorporated into ASD IR tablets. In vivo tests were conducted in fasted beagle dogs (1) pretreated with pentagastrin to increase gastric acid secretion and lower stomach pH or (2) pretreated with famotidine (an ARA) to decrease gastric acid secretion and elevate stomach pH. ASD IR tablets and commercially available Calquence capsules were administered to the dogs and the results were compared.

In addition to the in vivo study, ASD tablets and Calquence capsules were tested and compared in an in vitro multicompartment dissolution apparatus designed to reflect average gastrointestinal physiology in beagle dogs pretreated with pentagastrin or famotidine. Further, physical and chemical stability of the ASD, and chemical stability of the ASD tablet were assessed to demonstrate the integrity of amorphous acalabrutinib in the ASD tablet.

2. Materials and Methods

2.1. Material Sourcing

Acalabrutinib (CAS 1420477-60-6, >98% purity) was purchased from LC Laboratories (Woburn, MA, USA). HPMCAS-HF (Aqoat, HF grade) was purchased from Shin-Etsu Chemical Co., Ltd. (Tokyo, Japan). Hydrochloric acid (HCl), sodium acetate, sodium phosphate, potassium phosphate, and sodium chloride (NaCl) were purchased from Sigma Aldrich Chemical Company (St. Louis, MO, USA). Ammonium acetate was purchased from Thermo Fisher Scientific (Waltham, MA, USA). Fasted-state simulated intestinal fluid (FaSSIF) powder was purchased from Biorelevant.com Ltd. (London, UK). Methanol (HPLC grade) was purchased from Honeywell (Morris Plains, NJ, USA). Calquence capsules were purchased from Drug World (Cold Spring, NY, USA). Avicel PH-101 (microcrystalline cellulose) was purchased from FMC Corporation (Philadelphia, PA, USA). Pearlitol 25 (mannitol) was purchased from Roquette America (Geneva, IL, USA). Ac-Di-Sol (croscarmellose sodium) was purchased from Dupont (Wilmington, DE, USA). Cab-O-Sil M5P (fumed silica) was purchased from Cabot Corporation (Alpharetta, GA, USA). Magnesium stearate was purchased from Macron Fine Chemicals/Avantor (Radnor, PA, USA). Covance Laboratories (Madison, WI, USA) purchased pentagastrin from Sigma Aldrich Chemical Company and famotidine from Kirkland Signature (Kirkland, WA, USA) for use in in vivo testing.

2.2. ASD Manufacturing and Characterization

A 32 g batch of 50/50 (% *w*/*w*) acalabrutinib/HPMCAS-HF grade (Aqoat®, Shin-Etsu Chemical Co., Ltd., Tokyo, Japan) ASD was manufactured using a customized laboratory-scale spray dryer (0.3 m chamber diameter) capable of drying-gas flow rates of up to 35 kg/h. This ASD composition was selected as the lead after screening several different dispersion polymers and one additional drug loading (data to be provided in a future publication). Methanol was used as the spray solvent. The total solids loading in the spray solution was 5.4 wt%. Solutions were sprayed using a Schlick 2.0 pressure-swirl nozzle (Model 121, 200-μm orifice, Schlick Americas, Bluffton, SC, USA) at an outlet temperature of 45 to 50 °C and an inlet temperature of 142 to 150 °C.

After material was collected in a cyclone, residual solvent was removed in a secondary drying step using a vacuum dryer (Model TVO-2, Cascade TEK, Cornelius, OR, USA) for 20 h at 40 °C with a nitrogen sweep gas (−60 cmHg, 3 standard liters per minute). Solvent removal was confirmed to be below International Council for Harmonization (ICH) thresholds for methanol (<3000 ppm) using gas chromatography (GC). The ASD was analyzed using modulated differential scanning calorimetry (mDSC) to ensure a single T_g as an indication of drug-polymer homogeneity. Powder X-ray diffraction (PXRD) was used to verify the ASD was amorphous. Scanning electron microscopy (SEM) was used to visually identify potential changes in morphology or the presence of surface crystals. Method details can be found in Appendix A.1.

2.3. ASD Tablet Manufacturing and Characterization

The ASD IR tablets, which had a 100 mg unit dosage strength, a 400 mg total mass and a drug loading of 25 wt%, were manufactured using the 50/50 (% *w*/*w*) acalabrutinib/HPMC AS-H ASD. ASD tablet information is provided in Table 2.

Table 2. Detailed ASD tablet formulation and composition.

Function	Ingredient	Tablet Fraction (wt%) [a]
	Intragranular	
ASD	50/50 acalabrutinib/Aqoat (HPMCAS-HF)	50.0
Ductile filler	Avicel® PH-101	14.3
Brittle filler	Pearlitol® 25	28.7
Disintegrant	Ac-Di-Sol®	6.0
Glidant	Cab-O-Sil® (M5P)	0.25
Lubricant	Magnesium stearate	0.25
	Extragranular	
Glidant	Cab-O-Sil	0.25
Lubricant	Magnesium stearate	0.25
	Total tablet mass:	400 mg

[a] Tablets were made using 11 mm standard round concave (SRC) tooling.

ASD tablets were manufactured using a small-scale, semi-manual dry-granulation process. The ASD and intragranular excipients (see Table 2) were blended in a Turbula blender (Glen Mills Inc., Clifton, NJ, USA). The intragranular blend was compressed into slugs using a Manesty F3 single-station tablet press (Manesty Ltd., Knowsley, UK) with half-inch flat-faced tooling. Slugs were milled using a 1Zpresso Pro coffee grinder (1Zpresso, New Taipei City, Taiwan) at a setting of 4.5. The milled slugs were blended with extragranular excipients (see Table 2) to create a final blend. The final blend was compressed to a target tensile strength of 2 MPa using a Manesty F3 single-station tablet press with 11 mm SRC tooling. Complete manufacturing details can be found in Appendix A.2.

Tablet disintegration rates were determined in HCl (pH 2) and phosphate-buffered saline (pH 6, 67 mM phosphate) in a U.S. Pharmacopeia (USP) disintegration apparatus (ZT-71 disintegration tester, Erweka, Heusenstamm, Germany) using three replicates.

2.4. Stability

The ASD was stored at 40 °C/75% relative humidity (RH) to assess physical and chemical stability under accelerated storage conditions. Initial ASD samples and aged ASDs were analyzed for crystallinity using PXRD, thermal properties using mDSC, changes in morphology or presence of surface crystals using SEM, and chemical degradation by reverse-phase high-performance liquid chromatography (RP-HPLC). The ASD and ASD tablets were stored at ambient and refrigerated conditions to assess the chemical stability at nonaccelerated conditions. Initial and aged ASD samples were analyzed for total related substances using RP-HPLC. See Appendix A.3 for method details.

2.5. In Vitro Dissolution Testing

The in vitro dissolution performance of the ASD tablet and the commercially available Calquence capsule was evaluated using a controlled-transfer dissolution (CTD) apparatus containing stomach, duodenum, and jejunum/waste compartments [15,16]. The in vitro testing parameters were selected based on acalabrutinib physicochemical properties; the dose; and the average physiology of fasted beagle dogs (1) pretreated with pentagastrin to increase gastric acid secretion and lower stomach pH and (2) pretreated with famotidine to decrease gastric acid secretion and elevate stomach pH [17–23]. Due to its basic pK_as (3.5 and 5.8), acalabrutinib is expected to dissolve to a high extent in low-pH gastric environments and supersaturate upon emptying into the upper intestine [2]. A CTD test was used to evaluate the dissolution behavior of acalabrutinib in various gastric environments and maintain supersaturation upon emptying into an intestinal compartment. CTD test parameters are summarized in Table 3.

Table 3. Summary of CTD in vitro testing parameters.

Parameter	Value
Dose (mg)	100
Dosing volume (mL)	50
Dosing medium	Milli-Q water
Gastric resting medium	HCl (pH 2) and 34 mM NaCl
Gastric secretion medium	HCl (pH 6) and 34 mM NaCl
Gastric resting volume (mL)	50
Gastric secretion rate (mL/min)	2.4
Gastric emptying half-life (monoexponential) (min)	15
Duodenal resting and secretion medium	Phosphate (pH 6.5) and FaSSIF powder [a]
Duodenal fluid volume (mL)	50
Duodenal secretion rate (mL/min)	2.4
Duodenal emptying rate	Gastric emptying and duodenal secretion
Jejunal medium	Gastric and duodenal composition
Jejunal volume	Starts at 0 and increases to 419 mL at 90 min

[a] Consisted of 67 mM phosphate-buffered saline containing 1% (*w/w*) (13.4 mM) FaSSIF powder and 82 mM NaCl.

Fiber-optic UV probe detection was used to measure acalabrutinib concentrations in the stomach and duodenum compartments of the CTD apparatus. Before the experiment, unique calibration curves were generated for each UV probe (2 mm path length) by delivering aliquots of a known amount of stock acalabrutinib solution (10 mg/mL acalabrutinib in methanol) to 50 mL of gastric or intestinal medium held at 37 °C. To begin dosing, a single 100 mg ASD tablet or 100 mg Calquence capsule was placed in a hanging basket and submerged in the stomach compartment along with 50 mL water (for a 100 mL total volume in the stomach compartment). This achieves a dose concentration of 1 mg/mL acalabrutinib in the stomach compartment at the onset of the experiment. The ASD tablet was agitated for 1 min and the Calquence capsule was agitated for 10 min before initiating gastric emptying to allow dosage-form disintegration and dispersal. Each compartment was stirred at 150 rpm and held at 37 °C by circulating water through jacketed glass vessels. The dissolution performance in each compartment was monitored using Pion Rainbow™ UV probes (Pion Inc., Billerica, MA, USA) at 360 to 364 nm in the stomach and the duodenum compartments. The apparent concentrations measured consisted of (1) drug dissolved in aqueous medium, (2) drug partitioned into bile-salt micelles (when present) as micelle-bound drug. Drug concentrations were monitored for 90 min after gastric emptying started. All samples were analyzed in duplicate.

2.6. Pharmacokinetic (PK) Study

ASD tablets and Calquence capsules were evaluated for in vivo performance in beagle dogs ($n = 6$) at a 100 mg dose strength. The study was conducted by Covance Laboratories Inc. (Madison, WI, USA) in accordance with the protocol, protocol amendment, and Covance standard operating procedures (SOPs). All procedures were in compliance with the Animal Welfare Act Regulations (9 CFR 3).

Calquence capsules were stored at room temperature according to package directions and ASD tablets were stored at 2 to 8 °C prior to dosing. The study design is shown in Table 4. Dogs were pretreated with intramuscular pentagastrin in Phases 1 and 3 to increase gastric acid secretion and lower stomach pH (representing fasted human gastric pH) [17]. Dogs were pretreated with oral famotidine in Phases 2 and 4 to decrease gastric acid creation and raise stomach pH (representing the gastric pH of fasted humans taking ARAs) [17,18].

The purebred, non-naive beagle dogs used in the study weighed 8.9 to 10.7 kg and were 1 to 2 years in age. For each phase, animals were fasted overnight before dose administration. Food was returned to the animals at approximately 4 h after dosing.

For each phase, a flush with approximately 10 mL water was administered to facilitate swallowing of the tablet or capsule dosage form.

Table 4. In vivo study design.

Phase [a]	Test Article	Pretreatment	Dosage Form per Animal
1	Acalabrutinib ASD	Pentagastrin [b]	1 tablet
2	Acalabrutinib ASD	Famotidine [c]	1 tablet
3	Calquence	Pentagastrin [b]	1 capsule
4	Calquence	Famotidine [c]	1 capsule

[a] Each phase consisted of 6 dogs at a 100 mg target acalabrutinib dose level with a 7 day washout between phases. [b] For Phases 1 and 3, animals received a subcutaneous injection of pentagastrin (6 mg/kg/60 mg/mL/0.1 mL/kg) approximately 30 min before test article administration. [c] For Phases 2 and 4, animals received a 40 mg oral dose of famotidine (two 20 mg tablets) approximately 60 min before test article administration.

For each phase, blood (approximately 1 mL) was collected from each animal from a jugular vein into tubes containing K_2EDTA before dosing at 0.25, 0.5, 1, 2, 4, 8, 12, and 24 h after dosing. Blood was maintained in chilled cryogenic racks before centrifugation to obtain plasma. Centrifugation began within 1 h of collection. Plasma was placed into 96-well tubes with barcode labels. Plasma was maintained on dry ice before storage at approximately −70 °C.

Acalabrutinib concentrations in dog plasma were measured by liquid chromatography–tandem mass spectrometry (LC–MS/MS) using a Sciex API-5000 system (Framingham, MA, USA) equipped with positive-ionization turbo ion spray. The concentrations of the samples were quantitated using a linear calibration curve that was prepared by diluting a stock solution of acalabrutinib in dimethylformamide at a concentration of 20 µg/mL. The stock solution was diluted to a range of 1–1000 ng/mL in 1:1 acetonitrile:water and spiked with blank protein. An internal standard, consisting of labetalol dissolved at 200 ng/mL in acetonitrile, was used to ensure system suitability. The dog plasma samples were prepared for analysis by precipitating the protein in acetonitrile in a 96-well plate and diluting the supernatant in 1 equivalent of acetonitrile. HPLC analysis was conducted using a 50 mm × 2 mm, 2.5 µm pore size Phenomenex Luna C18(2) high-speed-technology (HST) column (Phenomenex, Torrance, CA, USA) at a flow rate of 0.6 mL/min. The mobile phases consisted of 15 mM ammonium formate in water containing 0.1 % formic acid and 0.1 % formic acid in acetonitrile. During analysis, the column temperature was maintained at 50 °C, whereas the sample compartment was maintained at 5 °C.

Pharmacokinetic parameters including maximum drug plasma concentration (C_{max}), time to maximum plasma concentration (T_{max}), AUC through 24 h ($AUC_{0–24}$), and the AUC extrapolated to infinity ($AUC_{0\text{-}inf}$) were determined for each individual subject using Microsoft Excel (Microsoft Corporation, Seattle, WA, USA). $AUC_{0\text{-}24}$ and $AUC_{0\text{-}inf}$ were calculated using a linear trapezoidal method. Results of each individual subject were averaged to obtain the reported averages and standard deviations. $AUC_{0\text{-}inf}$ ratios (famotidine/pentagastrin) were calculated for ASD tablets and Calquence capsules. In addition, the $AUC_{0\text{-}inf}$ values for ASD tablets and Calquence capsules were compared with those of Calquence capsules after pentagastrin pretreatment, as were statistical *p*-values using a one-way analysis of variance (ANOVA) test for AUC relative to Calquence after famotidine pretreatment and Calquence after pentagastrin pretreatment.

3. Results

3.1. ASD and Tablet Manufacturing and Characterization

A 50/50 (% *w*/*w*) acalabrutinib/HPMCAS-H ASD was spray dried with a high yield (97% after secondary drying). The ASD contained 100 ppm of residual methanol, well below the ICH threshold of 3000 ppm. Furthermore, the ASD was confirmed to be amorphous and homogenous by mDSC as evident by a single T_g centered at 115.6 ± 0.3 °C. PXRD showed no evidence of crystallinity, based on the absence of sharp diffraction peaks characteristic of crystalline acalabrutinib. SEM demonstrated that ASD particles were primarily collapsed spheres with no evidence of surface crystals. Complete characterization results can be found in Appendix A.4.

ASD tablets were successfully manufactured to a target tensile strength of 2 MPa. The compression pressure needed to achieve a 2 MPa tensile strength was 78 MPa, with an average solid fraction of 0.80. On average, the tablets were 4.9 mm thick and had a tablet volume of 370 mm^3. This tablet volume is 60% smaller than the dosage form size of Calquence capsules, which have a calculated Size 1 capsule volume of ~870 mm^3.

Tablets disintegrated rapidly in the USP disintegration tester, with a 32 ± 2 s disintegration time (average ± standard deviation) at pH 2 and a 37 ± 4 s disintegration time at pH 6.

3.2. Stability

The ASD was stored at 40 °C/75% RH to assess physical and chemical stability when stored under accelerated conditions. The ASD showed good physical stability when stored at 40 °C/75% RH/open for 6 months, as indicated by a lack of sharp diffraction peaks by PXRD, a lack of particle fusion or other particle surface morphology changes by SEM micrographs, and a single T_g by mDSC after aging at accelerated conditions. Although the ASD showed good physical stability at 40 °C/75% RH, significant chemical degradation of the ASD was observed after storage for 12 weeks at 40 °C/75% RH, with HPLC results showing a 780% increase in impurities.

The ASD and ASD tablet were stored at ambient and refrigerated conditions to assess chemical stability. Storage at refrigerated conditions with and without desiccant and at 25 °C/60% RH/sealed with desiccant mitigated chemical degradation of the ASD and ASD tablet. For example, refrigerated storage for 12 weeks resulted in no increase in impurities for the ASD or ASD tablet. Storage at 25 °C/60% RH/sealed with desiccant for 12 weeks resulted in no increase in impurities for the ASD and only a 6% increase in impurities (total impurities = 0.69 ± 0.00 area%) for the ASD tablet. Results and discussion can be found in Appendix A.5.

3.3. Tablet In Vitro Dissolution Performance

ASD tablets and Calquence capsules were tested in the CTD apparatus using 0.01 N HCl (pH 2) as the gastric medium. As the dissolution results in Figure 2 show, dissolution of the ASD tablets and Calquence capsules were practically identical under conditions simulating fasted dogs treated with pentagastrin. Acalabrutinib is a weakly basic compound with pK_as of 3.5 and 5.8. As a result of ionization, acalabrutinib has a high crystalline solubility at pH 2. This high crystalline drug solubility drives rapid dissolution of both dosage forms in the gastric compartment and, upon transfer to the duodenal compartment, results in supersaturated drug concentrations and identical duodenal AUC values.

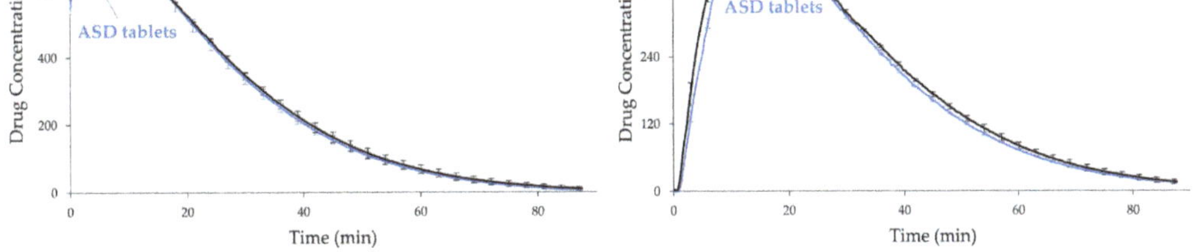

Figure 2. Dissolution profiles of acalabrutinib ASD tablets and Calquence capsules in CTD apparatus, using pH 2 gastric conditions to simulate fasted dogs treated with pentagastrin (n = 2).

The dissolution profiles from the CTD apparatus using 1×10^{-6} N HCl (pH 6) as the gastric medium are shown in Figure 3. In contrast to the pH 2 results, the ASD tablet outperformed the Calquence capsule in conditions that simulated fasted dogs treated with ARAs such as famotidine. The crystalline solubility of acalabrutinib is orders of magnitude lower at pH 6 than at pH 2. As such, the crystalline drug in the Calquence capsules is solubility-limited in the gastric compartment and does not supersaturate to the same extent when transferred to the duodenal compartment. In contrast, the ASD supersaturates in both the gastric and duodenal compartments, resulting in a 3.4-fold calculated enhancement in duodenal AUC for the ASD tablet relative to the Calquence capsules, as shown in Table 5.

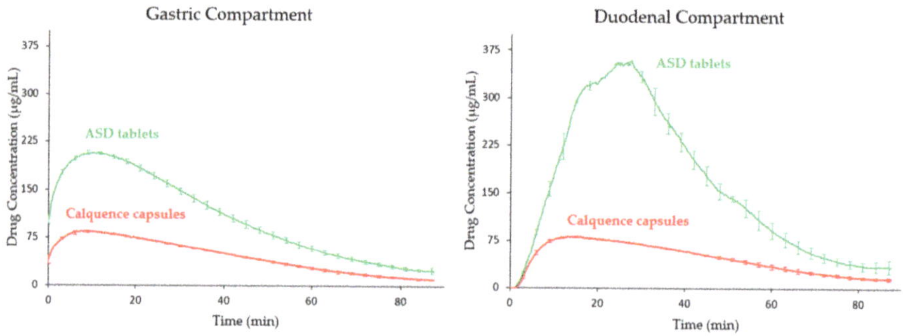

Figure 3. Dissolution profiles of acalabrutinib ASD tablets and Calquence capsules in the CTD apparatus under pH 6 gastric conditions to simulate fasted dogs treated with famotidine (ARAs) (n = 2).

Table 5. Calculated in vitro AUC in the duodenal compartment during CTD testing of ASD tablets and Calquence capsules.

Test Article with Simulated Pretreatment	Average In Vitro Duodenal AUC (µg-min/mL)	In Vitro AUC Ratio (Famotidine/ Pentagastrin)
ASD tablet, pentagastrin (pH 2 stomach)	16,946	0.83
ASD tablet, famotidine (pH 6 stomach)	14,102	-
Calquence, pentagastrin (pH 2 stomach)	17,617	0.24
Calquence, famotidine (pH 6 stomach)	4173	-

3.4. PK Study

ASD tablets and Calquence capsules were dosed to dogs pretreated with pentagastrin or famotidine. Profiles for blood plasma concentration versus time for each formulation treatment from 0 to 12 h are plotted in Figure 4, with tabulated noncompartmental PK results in Table 6. Profiles of blood plasma drug concentration versus time from 0 to 24 h and additional noncompartmental PK results can be found in Appendix A.6. The performance of the Calquence capsules and ASD tablets was similar after pentagastrin pretreatment. This result is in line with expectations prior to study initiation and supports historical performance data in dogs and humans, where acalabrutinib is well absorbed when administered to subjects with low stomach pH [2,6].

For the famotidine pretreatment condition, the ASD tablet achieved a 2.4-fold higher AUC than the Calquence capsules, as well as approximately 93% of the AUC observed for the pentagastrin-treated dogs, overcoming the impact of stomach pH. As expected, the performance of the Calquence capsules suffered at high stomach pH, resulting in roughly a 3-fold decrease in $AUC_{0\text{-inf}}$ compared to the pentagastrin-treated dogs. Again, this is in line with expectations based on historical performance data for Calquence dosed with ARAs. The $AUC_{0\text{-}inf}$ values for the ASD tablets in pentagastrin- and famotidine-treated dogs were statistically equivalent to each other, and to that of the Calquence capsules for pentagastrin-treated dogs. However, the $AUC_{0\text{-}inf}$ values for the ASD tablet in pentagastrin- and famotidine-treated dogs were statistically higher than the $AUC_{0\text{-}inf}$ values for Calquence capsules in famotidine-treated dogs.

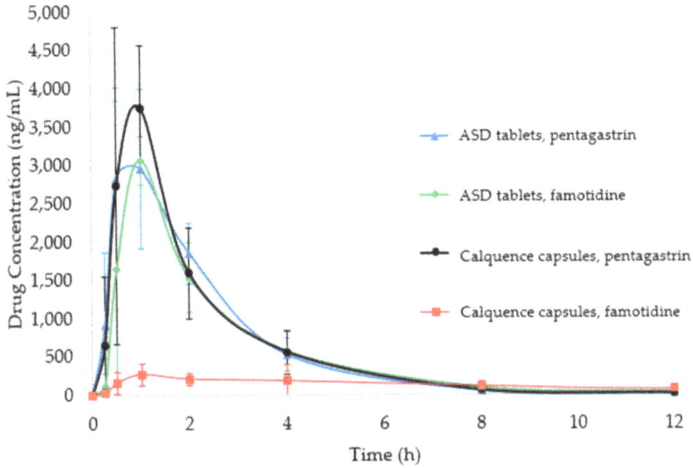

Figure 4. Profiles for blood plasma concentration versus time from 0 to 12 h measured in beagle dogs for ASD tablets and Calquence capsules (n = 6).

Table 6. Noncompartmental PK results from the acalabrutinib beagle dog study. Data are reported as the mean with the standard deviation in parentheses.

Test Article, Pretreatment	ASD Tablet, Pentagastrin	ASD Tablet, Famotidine	Calquence Capsule, Pentagastrin	Calquence Capsule, Famotidine
$AUC_{0\text{-}inf}$ (ng-h/mL)	8161 (1364) [a]	7579 (1423) [a]	8365 (1201)	3112 (1415) [b]
C_{max} (ng/mL)	3332 (769)	3443 (996)	4480 (516)	355 (121)
T_{max} (h)	0.9 (0.5)	0.9 (0.2)	0.8 (0.2)	1.6 (1.2)
AUC ratio compared to Calquence capsule, pentagastrin	0.98	0.91	1.00	0.37

[a] Statistically equivalent to Calquence capsule, pentagastrin ($p > 0.05$). [b] Statistically different from Calquence capsule, pentagastrin ($p < 0.05$).

4. Discussion

4.1. ASDs for Improving Low-Solubility, Weakly Basic Drugs

ASD dosage forms are attractive for enhancing the oral exposure of poorly soluble (e.g., BCS Class 2 and 4) drugs. The ASD tablets in this study boost AUC values 2.4-fold compared to that of Calquence capsules in beagle dogs at high gastric pH levels. Based on this result, ASD tablets are a promising alternative to Calquence capsules for improving patient compliance and efficacy [5]. Currently, patients are instructed to avoid co-administration with PPIs and to stagger dosing with H_2RAs and antacids, so these ASD tablets present the potential for co-administration with all types of ARAs [1]. Furthermore, these ASD tablets are 60% smaller than Calquence capsules at the same unit dosage strength, which should make them easier for patients to swallow [24,25]. While the ASD tablets were manufactured with a round, convex shape for this dog study, the size reduction is expected to translate to other tablet shapes (e.g., oblong) that may further facilitate swallowability in humans [24].

The results of this study are generally applicable to other small-molecule protein kinase inhibitors besides acalabrutinib, as well as weakly basic drugs in general. Several weakly basic, oral oncologic drugs on the market have shown evidence of decreased absorption as a result of high gastric pH when taken with ARAs [26,27]. Weakly basic drugs made up 78% of new molecular entities approved between 2003 and 2013 that showed a clinical DDI with ARAs [4]. Since small-molecule protein kinase inhibitors are prevalent in pharmaceutical pipelines and many are poorly soluble, ASD dosage forms have the potential to enable effective delivery and improve the experience of many cancer patients [28–31]. In addition to removing the ARA effect as highlighted in this study, ASD dosage forms can also mitigate food effects, decrease plasma variability and reduce dose [30,32].

4.2. In Vitro Versus In Vivo Performance

The in vitro CTD methodology forecast relative in vivo performance between the ASD tablet and Calquence capsules a priori to the dog study. As shown in Figure 5, the in vivo AUC values of ASD tablets relative to that of Calquence capsules in the pentagastrin pretreatment phase showed the same rank-ordering as in vitro duodenal AUC values relative to those of Calquence capsules at pH 2 (the initial gastric pH condition).

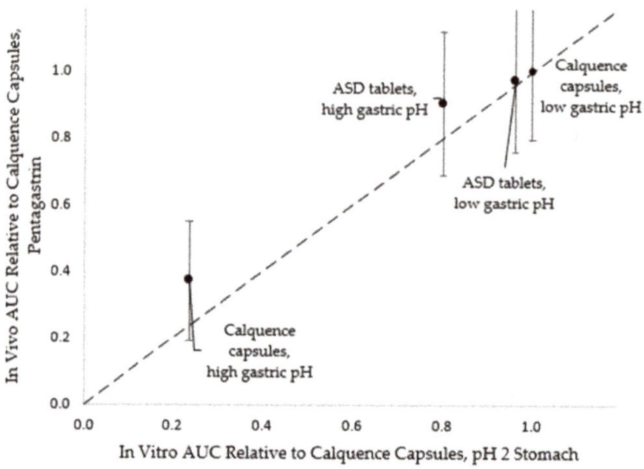

Figure 5. AUC values relative to those of Calquence capsules (pentagastrin pretreatment) in vivo versus duodenal AUC values relative to those of Calquence capsules (pH 2 initial gastric pH) in vitro. Relative AUC is calculated using average AUC. Error bars represent fractional uncertainties, where uncertainty in AUC is the standard deviation. Dashed line = 1:1 correlation line.

On average, the ASD tablets performed somewhat better in vivo than they did in vitro relative to the Calquence capsules at low initial gastric pH. This result is not surprising, because high-permeability drugs such as acalabrutinib often perform better in vivo than in vitro using dissolution test apparatuses that lack an absorption compartment [33,34]. Coupling in vitro testing with in silico modeling is a useful strategy for capturing the sensitivity of formulations to the multitude of important physiological variables [35,36]. Nonetheless, in this study, the CTD methodology proved useful in forecasting the in vivo performance of the ASD tablets and Calquence capsules.

4.3. Dog Model for Studying ARA Effect

In this study, a dog model was chosen to demonstrate the ability of an ASD dosage form to mitigate the acalabrutinib ARA effect. Beagle dogs are a common preclinical species and surrogate for humans due to their ability to swallow intact dosage forms and the relative ease in maintaining the animals. Although beagle dogs can be useful surrogates for humans, several differences exist between human and beagle dog physiology, including gastric pH [23,37,38].

It is common practice to modulate gastric pH in dogs when evaluating oral drug product performance. Pretreatment with intramuscular pentagastrin or oral famotidine has provided gastric pH values in beagle dogs that were more consistent and stable than those of dogs that had not been pretreated—in the range of pH 1–2 (for pentagastrin) or pH 7 (for famotidine) [17,18]. These findings make intramuscular pentagastrin and oral famotidine pretreatments useful for studying the impact of a wide range in gastric pH on weakly basic drugs, as was done in this study [17,18,39–41].

Famotidine represents a worst-case scenario for weakly basic drugs since a pH of 7 falls at the high end of the pH range, where drug solubility is lowest. Measured gastric pH in humans after taking ARAs can range from approximately pH 3 to 7 depending upon type of ARA, dose, duration of treatment, and individual response [42–44]. Therefore, the fact that the difference in AUC of the ASD tablet was only 7% over such a wide pH range (i.e., ~1 to 7), suggests low pH sensitivity and high-performance robustness for different ARA treatments. Further, it suggests high robustness of the ASD tablet to the natural pH variations that can be found even in patients who are not taking ARAs [8].

In attempting to use the current dog study results to forecast the utility of ASDs for overcoming pH effects in humans, it should be noted that acalabrutinib metabolism differs substantially between humans and dogs. In humans, approximately 50% of acalabrutinib is metabolized by CYP3A4 in the gut and 50% is metabolized by first-pass liver extraction [2]. In contrast, in dogs, acalabrutinib undergoes less-extensive metabolism upon first pass through the gut and liver, with reports of bioavailability in the 70% to 80% range at the human prescribed dose (100 mg) [2,45]. In humans, lower solubility at high gastric pH purportedly reduces absorption across the GI membrane and increases the extent of metabolized drug, whereas in dogs, the lower solubility at higher pH is expected to primarily reduce absorption [2]. Despite these metabolism differences, in both dogs and humans, the ASD tablet increases dissolved drug concentrations at elevated gastric pH levels. Therefore, success in dogs suggests that the ASD tablets will also overcome the ARA effect in humans.

5. Conclusions

This study demonstrates that ASD tablets are an effective enabling technology for overcoming reduction in AUC of the weakly basic drug acalabrutinib when co-administered with a gastric ARA. In beagle dogs, ASD tablets achieved similar AUC values at low and high gastric pH conditions and outperformed Calquence capsules 2.4-fold at high gastric pH. Relative formulation performance was successfully forecast using a multicompartment in vitro CTD apparatus.

The ASD was easy to manufacture at high yield on laboratory-scale spray-drying equipment and the resulting ASD tablets were 60% smaller than Calquence capsules. The ASD had good physical stability and the ASD and ASD tablet showed good chemical stability when stored refrigerated or at room temperature with a desiccant. An ASD dosage form represents a useful strategy for improving patient compliance and efficacy of acalabrutinib. This strategy could be extended to other small-molecule drug products and problem statements requiring bioavailability enhancement to drive improved in vivo performance.

Author Contributions: Conceptualization, D.M.M., A.M.S., N.B., M.M.M., and D.T.V.; methodology, D.M.M., A.M.S., J.A.R., M.S.A., C.D.C., and M.M.M.; formal analysis, D.M.M., A.M.S., and J.A.R.; investigation, J.A.R., N.B., M.S.A., and A.S.; data curation, D.M.M., A.M.S., and J.A.R.; writing—original draft preparation, D.M.M., A.M.S., and J.A.R.; writing—review and editing, D.M.M., A.M.S., J.A.R., M.S.A., and M.M.M.; visualization, D.M.M., A.M.S., and J.A.R.; supervision, D.M.M. and D.T.V.; project administration, D.M.M. All authors have read and agreed to the published version of the manuscript.

Funding: This research received no external funding.

Institutional Review Board Statement: The dog study was performed at Covance Laboratories. All procedures were in compliance with the Animal Welfare Act Regulations (9 CFR 3) as mandated by Animal and Plant Health Inspection Service, United States Department of Agriculture.

Informed Consent Statement: Not applicable.

Data Availability Statement: The data in this study in the form of Microsoft Excel worksheets are available from the corresponding author upon request.

Acknowledgments: We would like to thank Covance Laboratories (Madison, WI, USA) for conducting the in vivo study. We would like to thank Henny Zijlstra for identifying the Calquence ARA-effect problem statement. We would also like to thank Dwayne T. Friesen for his scientific advice in developing the in vivo study and Terry Podoll for his correspondence about acalabrutinib metabolism in beagle dogs. We would like to thank Jonathan Cape for his scientific advice on the chemical stability analysis, and we would like to thank Lauren Switala and Kimberly Shepard for assistance with analytical characterization. We would also like to thank Ann Malkin for her review and editing of the manuscript. No financial support or funding was provided for this work.

Conflicts of Interest: The authors declare no conflict of interest.

Appendix A

The following is available in the Appendix A.1 API, ASD and ASD tablet characterization methods, Appendix A.2 ASD tablet manufacturing methods, Appendix A.3 physical and chemical stability methods, Appendix A.4 ASD and API characterization results, Appendix A.5 physical and chemical stability results and discussion, and Appendix A.6 PK study results.

Appendix A.1. Active Pharmaceutical Ingredient (API), Amorphous Solid Dispersion (ASD) and ASD Tablet Characterization Methods

Appendix A.1.1. Modulated Differential Scanning Calorimetry (mDSC)

API was analyzed to determine the glass-transition temperature (T_g) using a Mettler Toledo DSC 3+ instrument (Mettler Toledo, LLC, Columbus, OH, USA). Samples were prepared as loose powder, loaded into a hermetically sealed pan perforated just before scanning. The instrument was run in alternating DSC mode at a scan rate of 2.5 °C/min, modulation of ± 1.5 °C/min, and scan range of 25–160 °C (n = 2).

A 50/50 ASD was spray dried as described in the text using acalabrutinib and the HF grade of hydroxypropyl methylcellulose acetate succinate (HPMCAS-HF) (Aqoat). The ASDs were analyzed to confirm that they were homogeneous (as evidenced by a single T_g) using a TA Instruments Q2000 mDSC instrument (TA Instruments-Waters LLC, New Castle, DE, USA). Samples were prepared as loose powder, loaded into a T_{zero} nonhermetically sealed pan (TA Instruments). The instrument was run in modulated mode at a scan rate of 2.5 °C/min, modulation of ± 1.5 °C/min, and scan range of 0–160 °C (n = 3). As-received HPMCAS-HF was also analyzed using mDSC as a reference using the same method used for the ASD (n = 2).

Appendix A.1.2. Powder X-ray Diffraction (PXRD)

ASD samples were analyzed using PXRD to confirm they were amorphous, as evidenced by the lack of sharp Bragg diffraction peaks in the X-ray pattern, using a Rigaku MiniFlex600 X-ray diffractometer (Rigaku, The Woodlands, TX, USA) equipped with a Cu-Kα source. The scan rate was set to 2.5°/min with a 0.02° step size from 3° to 40° on the 2Θ scale.

Appendix A.1.3. Scanning Electron Micrography (SEM)

As-received API and ASD samples were imaged using a Hitachi SU3500 scanning electron microscope (Hitachi High Technologies America Inc., Schaumburg, IL, USA) to assess morphology and confirm the absence of crystals. Samples were spread onto a post using double-sided tape and sputter-coated for 8 minutes at 8 V and 20 mA using an Anatech Hummer 6.2 sputter coater (Anatech USA, Hayward, CA, USA).

Appendix A.1.4. Reverse-Phase High-Performance Liquid Chromatography (RP-HPLC)

ASD and ASD tablet samples were analyzed for total related substances using RP-HPLC using a Waters XBridge C8 (150 × 3 mm, 3.5 μm pores) column (Waters Corporation, Milford, MO, USA) at a temperature of 40 °C. Mobile phase was pumped into the column at a flow rate of 0.7 mL/min using a gradient method. 10 mM ammonium acetate was used as mobile phase A and methanol was used as mobile phase B supplied according to the following sequence: 0 min: 20% B, 30 min: 70% B, 40 min: 70% B, 41 min: 20% B, 45. Samples were diluted with 8/2 (*v*/*v*) methanol/water and injection volume was 5 μL. The sample tray was held at uncontrolled room temperature. Analysis wavelength was 230 nm and 4 nm bandwidth.

Appendix A.1.5. RP-HPLC–Mass Spectrometry (MS)

RP-HPLC–MS was performed as a first step in identifying acalabrutinib degradation products using a Sciex Qtrap-4000 triple quadrupole mass spectrometer (Sciex, Framingham, MA, USA) interfaced with an Agilent 1200 HPLC (Agilent Technologies, Santa Clara, CA, USA). The RP-HPLC method was modified from the RP-HPLC method described above to adapt the method to the lower pressure limit of the 1100 Agilent HPLC instrument. A Waters XBridge C8 (150 3 mm, 3.5 μm pores) column (Waters Corporation, Milford, MO, USA) was used for the separation at a column oven temperature of 45 °C. Mobile phase was pumped into the column at a flow rate of 0.5 mL/min using a gradient method. 10 mM ammonium acetate was used as mobile phase A and methanol was used as mobile phase B supplied according to the following sequence: 0 min: 20% B, 30 min: 70% B, 40 min: 70% B, 41 min: 20% B, 45. Samples were diluted with 8/2 (*v*/*v*) methanol/water and injection volume was 5 μL. The sample tray was held at uncontrolled room temperature. The eluent was first passed through the UV detector prior to entering the MS. The analysis wavelength was 230 nm and 4 nm bandwidth. MS parameters were as follows: Positive mode ionization, 3500 V ESI voltage, 250 °C heater temperature, 12 psi gas source 1, 15 psi gas source 2, 60 V declustering potential, 10 V exit potential. Q1 MS scans were conducted from 100 to 1000 m/z with a 500 ms cycle time.

Appendix A.2. ASD Tablet Manufacturing Methods

ASD tablets were made using a small-scale, semi-manual manufacturing process. ASD and intragranular excipients (see Table 2 in main text) were passed through a #20 mesh screen. The ASD and all intragranular excipients excluding magnesium stearate were blended in a Turbula blender (Glen Mills Inc., Clifton, NJ, USA) for 15 min at 32 rpm. Magnesium stearate was added to the blender and the mixture was blended for 4 min at 32 rpm. The intragranular blend was compressed into slugs using a Manesty F3 single-station tablet press (Manesty Ltd., Knowsley, Great Britain) equipped with half-inch flat-faced tooling. The compression pressure was adjusted to achieve tensile strengths in the range of 0.5–0.7 MPa. Tensile strength was calculated using Equation (A1), where TS is tensile strength, F is breaking force, D is tablet diameter, and H is tablet thickness.

$$\mathrm{TS} = \frac{2\,\mathrm{F}}{\pi\,\mathrm{D}\,\mathrm{H}} \tag{A1}$$

Breaking force was determined using a Schleuniger hardness tester (Sotax, Westborough, MA, USA).

Slugs were milled using a 1Zpresso Pro coffee grinder (1Zpresso, New Taipei City, Taiwan) at a setting of 4.5. The extragranular Cab-O-Sil was passed through a #20 mesh screen and added to a Turbula blender together with Ac-Di-Sol and the intragranular blend and blended for 15 minutes at 32 rpm. The magnesium stearate was screened through a #20 mesh screen and added to the Turbula blender for an additional 4 min of blending. The final blend was then compressed into tablets of a target tensile strength of 2 MPa using the Manesty F3 press equipped with 11 mm standard round concave (SRC) tooling.

Appendix A.3. Physical and Chemical Stability Methods

The ASD was stored at 40 °C/75% relative humidity (RH) to assess physical and chemical stability when stored under accelerated conditions. Approximately 100 mg of ASD was placed in a single 4 mL glass vial, with enough glass vials prepared for each storage condition, timepoint and replicate. Each vial was covered with perforated aluminum foil and transferred to a temperature/humidity-controlled oven (Model ES2000, Environmental Specialties International Inc., Baton Rouge, LA, USA) at 40 °C/75% RH and allowed to stand undisturbed for 4, 12, and 24 weeks. At each timepoint, samples were removed from the oven and transferred to a vacuum desiccator for up to 18 h to remove water. After drying, the samples were removed from the vacuum desiccator and stored at 5 ± 3 °C. Initial ASD samples and aged ASDs were analyzed for crystallinity using SEM and PXRD, thermal properties using mDSC (in triplicate) and total related substances using RP-HPLC (duplicate analysis).

Additional ASD samples were stored at ambient and refrigerated conditions to assess the chemical stability upon storage at nonaccelerated conditions. The storage conditions are detailed in Table A1. Approximately 100 mg of ASD was placed in a single 4 mL glass vial, with enough glass vials prepared for each storage condition, timepoint and replicate. A subset of vials was placed in sealed 40 cc high-density polyethylene (HDPE) bottles with and without 1 g of desiccant. These vials were subsequently stored at 5 °C for 2, 5, and 12 weeks. A second subset of vials was placed in sealed 40 cc HDPE bottles with desiccant as well as 40 cc HDPE bottles with perforated aluminum seals (without desiccant). These vials were subsequently transferred to a temperature- and humidity-controlled oven held at 25 °C/60% RH and allowed to stand undisturbed for 2, 5, and 12 weeks. At each timepoint, samples were removed from storage and transferred to a vacuum desiccator for up to 18 hours to remove water. After drying, the samples were removed from the vacuum desiccator and stored at 5 °C. Initial and aged ASD samples were analyzed for total related substances using RP-HPLC (duplicate analysis).

ASD tablets were stored at ambient and refrigerated conditions to assess chemical stability (See Table A1). A subset of tablets were placed in sealed 40 cc HDPE bottles with and without 1 g of desiccant. These bottles were subsequently stored at 5 °C ±3 °C for 2, 5, and 12 weeks. A second subset of tablets were placed in sealed 40 cc HDPE bottles with desiccant as well as 40 cc HDPE bottles with perforated aluminum seals (without desiccant). These bottles were then transferred to a temperature- and humidity-controlled oven held at 25 °C/60% RH and allowed to stand undisturbed for 2, 5, and 12 weeks. At each timepoint, samples were removed from storage and stored at 5 °C. Initial and aged tablets were analyzed for total related substances using RP-HPLC.

Table A1. Storage conditions for the evaluation of ASD and ASD tablet stability. Checkmarks indicate the conditions that were tested for each sample.

No.	Storage Conditions	ASD Physical Stability	ASD Chemical Stability	ASD Tablet Chemical Stability
1	5 °C/sealed with desiccant		✔	✔
2	5 °C/sealed		✔	✔
3	25 °C/60% RH/sealed with desiccant		✔	✔
4	25 °C/60% RH/open, perforated seal		✔	✔
5	40 °C/75% RH/open, perforated seal	✔		

Appendix A.4. ASD and API Characterization Results

The ASD was confirmed to be a homogenous single-phase amorphous solid by PXRD, mDSC, and SEM. The PXRD diffraction patterns of the ASD showed no evidence of crystallinity, demonstrated by the absence of sharp diffraction peaks characteristic of pure crystalline acalabrutinib, as shown in Figure A1. SEM images showed typical SDD surface morphology with no evidence of surface crystals, as shown in Figure A2, panel (a). The mDSC thermograms showed a single, average dry T_g of 115.6 ± 0.3 °C, as shown in Figure A3. As-received API showed blade-like morphology, as shown in Figure A2, panel (b).

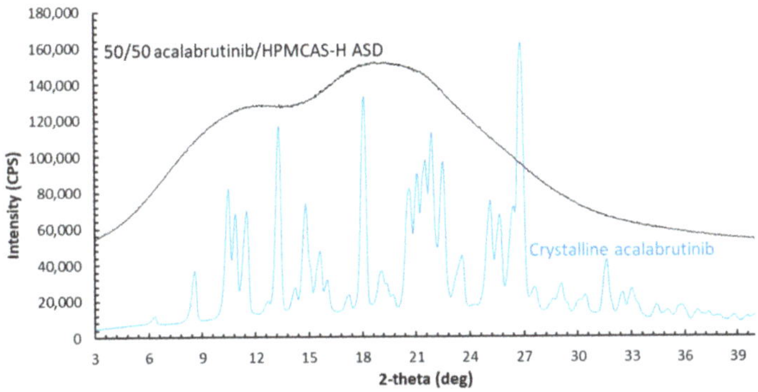

Figure A1. PXRD diffractograms for the ASD and crystalline drug.

(a)

(b)

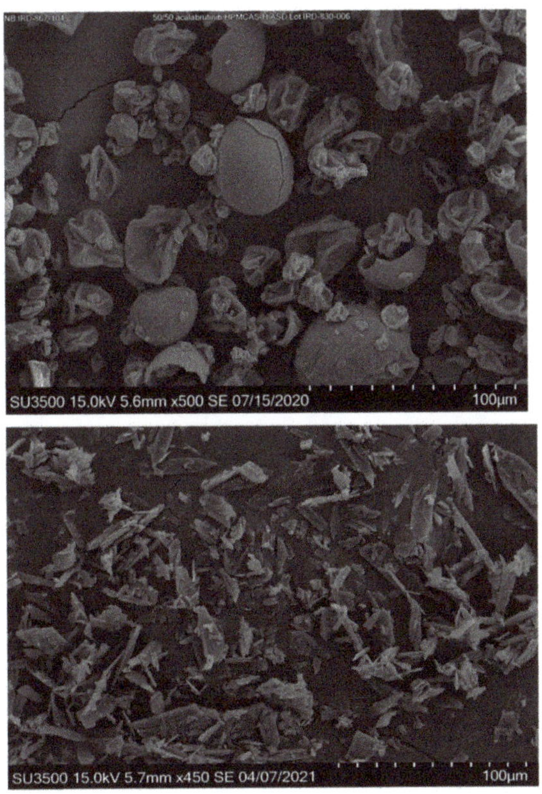

Figure A2. Panel (**a**): SEM micrograph of the ASD, panel (**b**); SEM micrograph of as-received acalabrutinib.

(**a**)

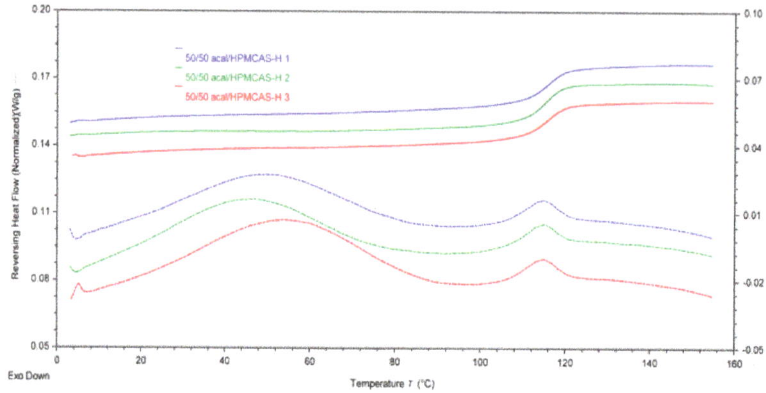

(**b**)

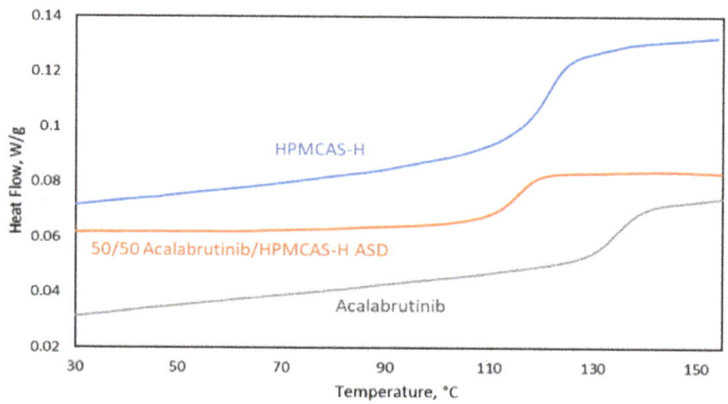

Figure A3. Panel (**a**): mDSC thermograms of the ASD samples, showing a single T_g at 115.6 °C. Panel (**b**): single replicate thermograms of ASD sample compared to acalabrutinib API (T_g = 133 °C) and HPMCAS-H (T_g = 121 °C) samples for reference.

Appendix A.5. Physical and Chemical Stability Results and Discussion

The ASD remained physically stable when stored at 40 °C/75% RH open for 6 months. Good physical stability was indicated by a lack of sharp diffraction peaks by PXRD (see Figure A4), a lack of particle fusion or other particle surface morphology changes by SEM micrography (see Figure A5, and a single T_g by mDSC (see Figure A6) after aging at accelerated conditions.

Testing demonstrated the ASD and ASD tablets had good chemical stability after 12 weeks of storage at refrigerated conditions with and without desiccant, and at 25 °C/60% RH sealed with desiccant (see Tables A2 and A3). When refrigerated, both the ASD and ASD tablets showed no increase in impurities after 12 weeks of storage. When stored at 25 °C/60% RH sealed with desiccant for 12 weeks, the ASD showed no increase in impurities and the ASD tablets showed a small (6%) increase in impurities. However, this increase was 4-fold lower than that observed for ASD tablets stored at 25 °C/60% RH/open (with a perforated seal) without desiccant. Overall, the ASD and the ASD tablets showed a similar growth in impurities across all conditions.

Quantifying total impurities rather than identifying each impurity was the main focus of the chemical stability analysis. However, RP-HPLC–MS analysis provided a first step toward identifying impurities, as shown in Table A4. The degradant that increased the most upon storage was Ad4, with the proposed degradation pathway being alkyne hydration. Confidently identifying each degradant would require additional experimentation, which was outside the scope of this study.

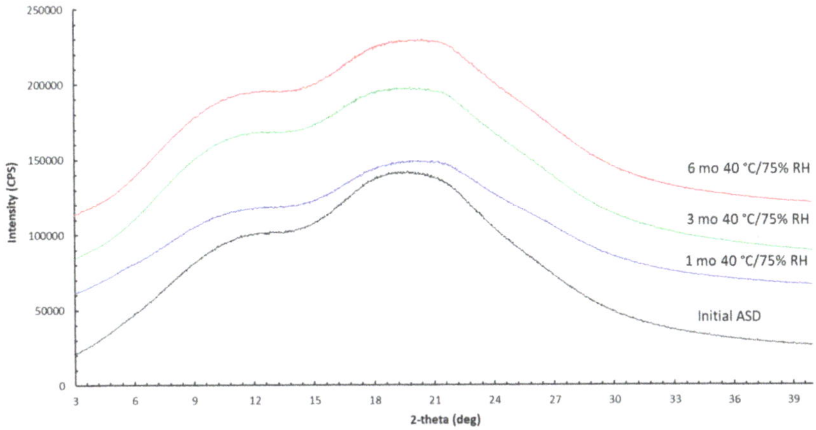

Figure A4. Diffractograms of ASD samples before and after storage at 40 °C/75% RH/open.

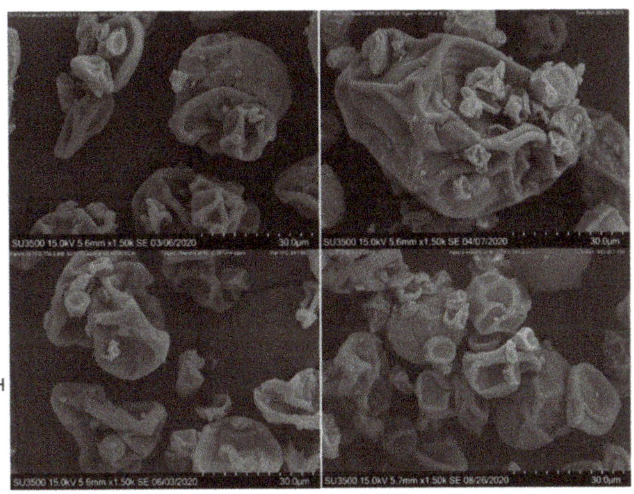

Figure A5. SEM micrographs of ASD samples before and after storage at 40 °C/75% RH/open.

Table A2. Total impurities of ASD and ASD tablets before and after storage for 12 weeks, as measured by HPLC.

	ASD			ASD Tablet		
Storage Condition	**Initial Impurities (area%)**	**Impurities after 12 wk (area%)**	**Change after 12 wk (%) [a]**	**Initial Impurities (area%)**	**Impurities after 12 wk (area%)**	**Change after 12 wk (%) [a]**
5 °C/sealed with desiccant	0.89 ± 0.01	0.85 ± 0.00	−4.5	0.65 ± 0.01	0.63 ± 0.01	−3.1
5 °C/sealed	0.89 ± 0.01	0.89 ± 0.01	0.0	0.65 ± 0.01	0.63 ± 0.01	−3.1
25 °C/60% RH/ sealed with desiccant	0.89 ± 0.01	0.89 ± 0.01	0.0	0.65 ± 0.01	0.69 ± 0.00	6.2
25 °C/60% RH/open, perforated seal	0.89 ± 0.01	1.15 ± 0.07	29.2	0.65 ± 0.01	0.82 ± 0.05	26.2
40 °C/75% RH/ open, perforated seal	0.60 [b] ± 0.01	5.3 ± 0.08	783.3	-	-	-

[a] Average % change = (final − initial)/initial × 100%; [b] A different batch of ASD was used for the study conducted at 40 °C/75% RH/open.

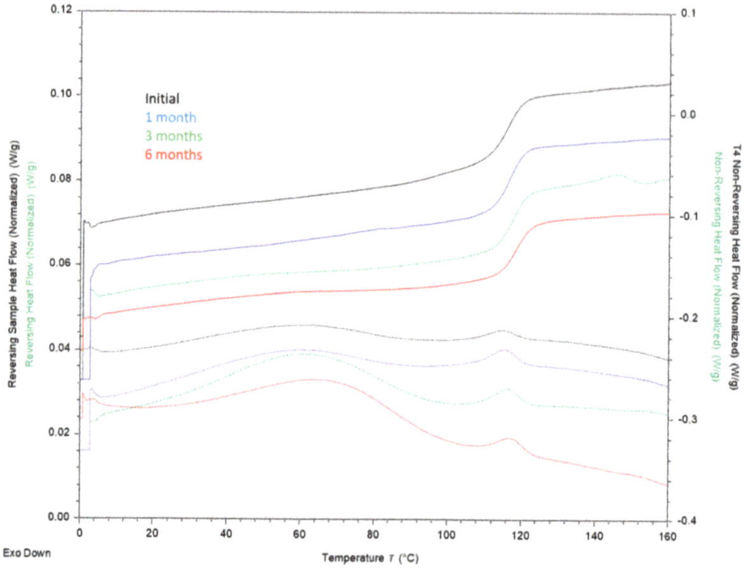

Figure A6. mDSC thermograms of ASD samples before and after storage at 40 °C/75% RH/open, demonstrating no change in thermal properties after storage at accelerated conditions.

Table A3. Total impurities of ASD, ASD tablets, and crystalline drug before and after storage for up to 12 weeks as measured by HPLC.

	Total Impurities (area %)				
Storage Time	**Initial**	**2 wk**	**4 wk**	**5 wk**	**12 wk**
			ASD		
5 °C/sealed with desiccant	0.89 ± 0.01	0.87 ± 0.10	–	0.87 ± 0.07	0.85 ± 0.00
5 °C/sealed	0.89 ± 0.01	0.82 ± 0.08	–	0.89 ± 0.00	0.89 ± 0.01
25 °C/60% RH/sealed with desiccant	0.89 ± 0.01	0.84 ± 0.06	–	0.81 ± 0.05	0.89 ± 0.01
25 °C/60% RH/open, perforated seal	0.89 ± 0.01	0.90 ± 0.07	–	0.95 ± 0.08	1.15 ± 0.07
40 °C/75% RH/open, perforated seal	0.60 ± 0.01	–	1.25 ± 0.03	–	5.27 ± 0.08
			ASD Tablets		
5 °C/sealed with desiccant	0.65 ± 0.01	0.65 ± 0.02	–	0.65 ± 0.06	0.63 ± 0.01
5 °C/sealed	0.65 ± 0.01	0.66 ± 0.06	–	0.70 ± 0.15	0.63 ± 0.01
25 °C/60% RH/sealed with desiccant	0.65 ± 0.01	0.70 ± 0.02	–	0.62 ± 0.00	0.69 ± 0.00
25 °C/60% RH/open, perforated seal	0.65 ± 0.01	0.70 ± 0.05	–	0.74 ± 0.08	0.82 ± 0.05
			Crystalline Drug		
40 °C/75% RH/open, perforated seal	0.59 ± 0.00	–	0.70 ± 0.04	–	0.94 ± 0.05

Table A4. Proposed degradants as measured by RP-HPLC–MS analysis.

Peak Retention Time (min)	Relative Retention Time (min)	Degradant	Characteristic Ion Mass-to-Charge Ratio (m/z)	Proposed Degradation Pathway	Formula Weight of Proposed Product (Da) [a]
11.05	0.65	Ad1	442.6	Transamidation with acetate at pyrrolidine ring	441.4
14.79	0.87	Ad2	400.6	Amide cleavage at pyrrolidine ring	399.4
15.09	0.89	Ad3	550.8	Putative addition product	
16.87	0.99	Ad4	484.6	Alkyne hydration	483.5
17.45	1.03	Ad5	532.7	Amidation of free amine with 2-butyne-4-one fragment	532
20.23	1.19	Ad6	546/548	Succinamide formation at free amine	548
23.07	1.36	Ad7	457.7	Unknown	
25.8	1.52	Ad8	932.2/466.7	Addition and dimerization	
27.9	1.64	Ad9	888/444	Dimerization	

[a] Formula weight is typically ~ 1 unit lower than the characteristic ion mass-to-charge ratio (m/z) because each ion picks up an extra H+ during ionization.

Podoll and coworkers have previously reported that the primary mechanisms of degradation for acalabrutinib in the solid state is hydrolysis of the amide and alkyne groups [2]. In the case of the ASD, the rate of hydrolysis may be aided by the increase in water activity in the amorphous form of acalabrutinib relative to the crystal form, coupled with an acidic environment within the ASD particle. As such, protecting the ASD tablets from water by storing them under refrigerated conditions or at room temperature conditions (25 °C/40% RH) with desiccant mitigates the chemical instability observed when storing the ASD at 40 °C/75% RH. Although these packaging configurations are less ideal than storage in uncontrolled environments, they are not uncommon for pharmaceutical drug products. Recommended storage conditions for Calquence are storage at 20 to 25 °C with excursions permitted to 15 to 30 °C in packaging with no desiccants [1].

Tablets were assayed using the RP-HPLC method upon manufacturing and shortly after the in vivo study (upon receipt of retain tablets from the Covance Laboratories study site), achieving 100–101 %label claim (n = 2) and 102–103% label claim (n = 2), respectively.

Appendix A.6. Pharmacokinetic (PK) Study Results

Profiles of blood plasma drug concentration versus time are shown in Figure A7 for each formulation treatment from 0 to 24 h. Noncompartmental PK results from the acalabrutinib beagle dog study are shown in Table A5.

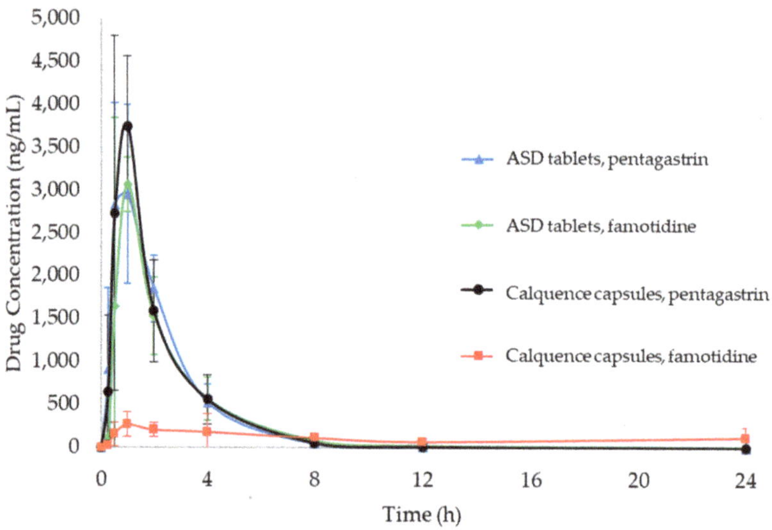

Figure A7. Profiles of blood plasma drug concentration versus time from 0 to 24 h measured in beagle dogs for ASD tablets and Calquence capsules.

Table A5. Noncompartmental PK results from the acalabrutinib beagle dog study. Data are reported as the mean with a single standard deviation in parentheses.

Test Article, Pretreatment	ASD Tablet, Pentagastrin	ASD Tablet, Famotidine	Calquence Capsules, Pentagastrin	Calquence Capsules, Famotidine
$AUC_{0\text{-}24}$ (ng h/mL)	8157 (1363)	7558 (1429)	8363 (1199)	2875 (1284)
$AUC_{0\text{-}inf}$ (ng h/mL)	8161 (1364)	7579 (1423)	8365 (1201)	3112 (1415)
$AUC_{0\text{-}inf}$ *p*-value (compared to Calquence, famotidine)	<0.05	<0.05	<0.05	n/a [a]
$AUC_{0\text{-}inf}$ *p*-value (compared to Calquence capsules, pentagastrin)	0.81	0.37	n/a	<0.05

[a] n/a = not applicable.

References

1. Deisseroth, A.B. *Calquence FDA Label*; U.S. Food and Drug Administration: Silver Spring, MD, USA, 2019; Volume 2020. Available online: https://www.accessdata.fda.gov/drugsatfda_docs/label/2019/210259s006s007lbl.pdf (accessed on 15 March 2021).
2. Pepin, X.J.H.; Moir, A.J.; Mann, J.C.; Sanderson, N.J.; Barker, R.; Meehan, E.; Plumb, A.P.; Bailey, G.R.; Murphy, D.S.; Krejsa, C.M.; et al. Bridging in vitro dissolution and in vivo exposure for acalabrutinib. Part ii. A mechanistic pbpk model for ir formulation comparison, proton pump inhibitor drug interactions, and administration with acidic juices. *Eur. J. Pharm. Biopharm.* **2019**, *142*, 435–448. [PubMed]
3. Center For Drug Evaluation and Research. *Product Quality Review*; Application Number: 210259orig1s000; Center For Drug Evaluation and Research: Silver Spring, MD, USA, 2017.
4. Patel, D.; Bertz, R.; Ren, S.; Boulton, D.W.; Någård, M. A systematic review of gastric acid-reducing agent-mediated drug–Drug interactions with orally administered medications. *Clin. Pharmacokinet.* **2020**, *59*, 447–462. [PubMed]
5. Smelick, G.S.; Heffron, T.P.; Chu, L.; Dean, B.; West, D.A.; DuVall, S.L.; Lum, B.L.; Budha, N.; Holden, S.N.; Benet, L.Z.; et al. Prevalence of acid-reducing agents (ara) in cancer populations and ara drug–Drug interaction potential for molecular targeted agents in clinical development. *Mol. Pharm.* **2013**, *10*, 4055–4062. [PubMed]
6. Pepin, X.J.H.; Sanderson, N.J.; Blanazs, A.; Grover, S.; Ingallinera, T.G.; Mann, J.C. Bridging in vitro dissolution and in vivo exposure for acalabrutinib. Part i. Mechanistic modelling of drug product dissolution to derive a p-psd for pbpk model input. *Eur. J. Pharm. Biopharm.* **2019**, *142*, 421–434.
7. Zhang, L.; Wu, F.; Lee, S.C.; Zhao, H.; Zhang, L. Ph-dependent drug–Drug interactions for weak base drugs: Potential implications for new drug development. *Clin. Pharmacol. Ther.* **2014**, *96*, 266–277.

8. Hens, B.; Tsume, Y.; Bermejo, M.; Paixao, P.; Koenigsknecht, M.J.; Baker, J.R.; Hasler, W.L.; Lionberger, R.; Fan, J.; Dickens, J.; et al. Low buffer capacity and alternating motility along the human gastrointestinal tract: Implications for in vivo dissolution and absorption of ionizable drugs. *Mol. Pharm.* **2017**, *14*, 4281–4294.
9. Riethorst, D.; Mols, R.; Duchateau, G.; Tack, J.; Brouwers, J.; Augustijns, P. Characterization of human duodenal fluids in fasted and fed state conditions. *J. Pharm. Sci.* **2016**, *105*, 673–681.
10. Shih, G.L.; Brensinger, C.; Katzka, D.A.; Metz, D.C. Influence of age and gender on gastric acid secretion as estimated by integrated acidity in patients referred for 24-h ambulatory ph monitoring. *Am. J. Gastroenterol.* **2003**, *98*, 1713–1718.
11. Blatter, F.; Ingallinera, T.; Barf, T.; Aret, T.; Krejsa, C.; Evarts, J. Solid Forms and Formuations of (s)-4->-Amino-3-(l-(but-2-ynoyl)pyrrolidin-2-yl)imidazo[1,5-ajpyrazin- 1-yl)-n-(pyridin-2-yl)benzamide. 2016. Available online: https://patentimages.storage.googleapis.com/5e/4d/3f/0e3e4fb9e1a906/WO2017002095A1.pdf (accessed on 1 March 2021).
12. Van den Mooter, G. The use of amorphous solid dispersions: A formulation strategy to overcome poor solubility and dissolution rate. *Drug Discov. Today Technol.* **2012**, *9*, e79–e85.
13. Almeida e Sousa, L.; Reutzel-Edens, S.M.; Stephenson, G.A.; Taylor, L.S. Assessment of the amorphous "solubility" of a group of diverse drugs using new experimental and theoretical approaches. *Mol. Pharm.* **2015**, *12*, 484–495.
14. Baghel, S.; Cathcart, H.; O'Reilly, N.J. Polymeric amorphous solid dispersions: A review of amorphization, crystallization, stabilization, solid-state characterization, and aqueous solubilization of biopharmaceutical classification system class ii drugs. *J. Pharm. Sci.* **2016**, *105*, 2527–2544.
15. Mudie, D.M.; Buchanan, S.; Stewart, A.M.; Smith, A.; Shepard, K.B.; Biswas, N.; Marshall, D.; Ekdahl, A.; Pluntze, A.; Craig, C.D.; et al. A novel architecture for achieving high drug loading in amorphous spray dried dispersion tablets. *Int. J. Pharm. X* **2020**, *2*, 100042.
16. Mudie, D.M.; Stewart, A.M.; Biswas, N.; Brodeur, T.J.; Shepard, K.B.; Smith, A.; Morgen, M.M.; Baumann, J.M.; Vodak, D.T. Novel high-drug-loaded amorphous dispersion tablets of posaconazole; in vivo and in vitro assessment. *Mol. Pharm.* **2020**, *17*, 4463–4472.
17. Fancher, R.M.; Zhang, H.; Sleczka, B.; Derbin, G.; Rockar, R.; Marathe, P. Development of a canine model to enable the preclinical assessment of ph-dependent absorption of test compounds. *J. Pharm. Sci.* **2011**, *100*, 2979–2988.
18. Koziolek, M.; Grimm, M.; Bollmann, T.; Schäfer, K.J.; Blattner, S.M.; Lotz, R.; Boeck, G.; Weitschies, W. Characterization of the gi transit conditions in beagle dogs with a telemetric motility capsule. *Eur. J. Pharm. Biopharm.* **2019**, *136*, 221–230.
19. Meyer, J.H.; Dressman, J.; Fink, A.; Amidon, G. Effect of size and density on canine gastric emptying of nondigestible solids. *Gastroenterology* **1985**, *89*, 805–813.
20. Dressman, J.B. Comparison of canine and human gastrointestinal physiology. *Pharm. Res.* **1986**, *3*, 123–131.
21. Arndt, M.; Chokshi, H.; Tang, K.; Parrott, N.J.; Reppas, C.; Dressman, J.B. Dissolution media simulating the proximal canine gastrointestinal tract in the fasted state. *Eur. J. Pharm. Biopharm.* **2013**, *84*, 633–641.
22. Wang, C.; Zhai, B.; Guo, H.; Wang, P.; Liu, Z.; Gu, H.; Ho, H.; Langguth, P.; Li, K.; Wang, C.; et al. In vivo measurement of gastric fluid volume in anesthetized dogs. *J. Drug Deliv. Sci. Technol.* **2020**, *55*, 101488.
23. Smeets-Peeters, M.; Watson, T.; Minekus, M.; Havenaar, R. A review of the physiology of the canine digestive tract related to the development of in vitro systems. *Nutr. Res. Rev.* **1998**, *11*, 45–69.
24. Drumond, N.; Stegemann, S. Better medicines for older patients: Considerations between patient characteristics and solid oral dosage form designs to improve swallowing experience. *Pharmaceutics* **2021**, *13*, 32.
25. Raber-Durlacher, J.E.; Brennan, M.T.; Verdonck-de Leeuw, I.M.; Gibson, R.J.; Eilers, J.G.; Waltimo, T.; Bots, C.P.; Michelet, M.; Sollecito, T.P.; Rouleau, T.S.; et al. Swallowing dysfunction in cancer patients. *Support. Care Cancer* **2012**, *20*, 433–443.
26. van Leeuwen, R.W.; van Gelder, T.; Mathijssen, R.H.; Jansman, F.G. Drug-drug interactions with tyrosine-kinase inhibitors: A clinical perspective. *Lancet Oncol.* **2014**, *15*, e315–e326.
27. Uchiyama, A.A.T.; Silva, P.A.I.A.; Lopes, M.S.M.; Yen, C.T.; Ricardo, E.D.; Mutão, T.; Pimenta, J.R.; Machado, L.M.; Shimba, D.S.; Peixoto, R.D. Proton pump inhibitors and oncologic treatment efficacy: A practical review of the literature for oncologists. *Curr. Oncol.* **2021**, *28*, 783–799.
28. Sawicki, E.; Schellens, J.H.; Beijnen, J.H.; Nuijen, B. Inventory of oral anticancer agents: Pharmaceutical formulation aspects with focus on the solid dispersion technique. *Cancer Treat. Rev.* **2016**, *50*, 247–263.
29. Herbrink, M.; Nuijen, B.; Schellens, J.H.; Beijnen, J.H. Variability in bioavailability of small molecular tyrosine kinase inhibitors. *Cancer Treat. Rev.* **2015**, *41*, 412–422.
30. Gala, U.H.; Miller, D.A.; Williams, R.O. Harnessing the therapeutic potential of anticancer drugs through amorphous solid dispersions. *Biochim. Biophys. Acta (BBA) Rev. Cancer* **2020**, *1873*, 188319.
31. Roskoski, R. Properties of fda-approved small molecule protein kinase inhibitors: A 2021 update. *Pharmacol. Res.* **2021**, *165*, 105463. [PubMed]
32. Veerman, G.D.M.; Hussaarts, K.; Jansman, F.G.A.; Koolen, S.W.L.; van Leeuwen, R.W.F.; Mathijssen, R.H.J. Clinical implications of food-drug interactions with small-molecule kinase inhibitors. *Lancet Oncol.* **2020**, *21*, e265–e279. [PubMed]
33. Pathak, S.M.; Schaefer, K.J.; Jamei, M.; Turner, D.B. Biopharmaceutic ivive-mechanistic modeling of single- and two-phase in vitro experiments to obtain drug-specific parameters for incorporation into pbpk models. *J. Pharm. Sci.* **2019**, *108*, 1604–1618.
34. Xu, H.; Vela, S.; Shi, Y.; Marroum, P.; Gao, P. In vitro characterization of ritonavir drug products and correlation to human in vivo performance. *Mol. Pharm.* **2017**, *14*, 3801–3814.

35. Hens, B.; Talattof, A.; Paixão, P.; Bermejo, M.; Tsume, Y.; Löbenberg, R.; Amidon, G.L. Measuring the impact of gastrointestinal variables on the systemic outcome of two suspensions of posaconazole by a pbpk model. *AAPS J.* **2018**, *20*, 57.
36. Jamei, M.; Abrahamsson, B.; Brown, J.; Bevernage, J.; Bolger, M.B.; Heimbach, T.; Karlsson, E.; Kotzagiorgis, E.; Lindahl, A.; McAllister, M.; et al. Current status and future opportunities for incorporation of dissolution data in pbpk modeling for pharmaceutical development and regulatory applications: Orbito consortium commentary. *Eur. J. Pharm. Biopharm.* **2020**, *155*, 55–68.
37. Grignard, E.; Taylor, R.; McAllister, M.; Box, K.; Fotaki, N. Considerations for the development of in vitro dissolution tests to reduce or replace preclinical oral absorption studies. *Eur. J. Pharm. Sci.* **2017**, *99*, 193–201.
38. Kararli, T.T. Comparison of the gastrointestinal anatomy, physiology, and biochemistry of humans and commonly used laboratory animals. *Biopharm. Drug Dispos.* **1995**, *16*, 351–380.
39. Zane, P.; Guo, Z.; MacGerorge, D.; Vicat, P.; Ollier, C. Use of the pentagastrin dog model to explore the food effects on formulations in early drug development. *Eur. J. Pharm. Sci.* **2014**, *57*, 207–213.
40. Mitra, A.; Kesisoglou, F. Impaired drug absorption due to high stomach ph: A review of strategies for mitigation of such effect to enable pharmaceutical product development. *Mol. Pharm.* **2013**, *10*, 3970–3979.
41. Mitra, A.; Kesisoglou, F.; Beauchamp, M.; Zhu, W.; Chiti, F.; Wu, Y. Using absorption simulation and gastric ph modulated dog model for formulation development to overcome achlorhydria effect. *Mol. Pharm.* **2011**, *8*, 2216–2223.
42. Litou, C.; Vertzoni, M.; Goumas, C.; Vasdekis, V.; Xu, W.; Kesisoglou, F.; Reppas, C. Characteristics of the human upper gastrointestinal contents in the fasted state under hypo- and a-chlorhydric gastric conditions under conditions of typical drug—Drug interaction studies. *Pharm. Res.* **2016**, *33*, 1399–1412.
43. Tutuian, R.; Katz, P.O.; Bochenek, W.; Castell, D.O. Dose-dependent control of intragastric ph by pantoprazole, 10, 20 or 40 mg, in healthy volunteers. *Aliment. Pharmacol. Ther.* **2002**, *16*, 829–836.
44. Prichard, P.J.; Yeomans, N.D.; Mihaly, G.W.; Jones, D.B.; Buckle, P.J.; Smallwood, R.A.; Louis, W.J. Omeprazole: A study of its inhibition of gastric ph and oral pharmacokinetics after morning or evening dosage. *Gastroenterology* **1985**, *88*, 64–69. [PubMed]
45. Podoll, T.; Pearson, P.G.; Evarts, J.; Ingallinera, T.; Bibikova, E.; Sun, H.; Gohdes, M.; Cardinal, K.; Sanghvi, M.; Slatter, J.G. Bioavailability, biotransformation, and excretion of the covalent bruton tyrosine kinase inhibitor acalabrutinib in rats, dogs, and humans. *Drug Metab. Dispos.* **2019**, *47*, 145–154. [PubMed]

pharmaceutics

Article

Green Nanotechnology in the Formulation of a Novel Solid Dispersed Multilayered Core-Sheath Raloxifene-Loaded Nanofibrous Buccal Film; In Vitro and In Vivo Characterization

Sara Nageeb El-Helaly [1,2], Eman Abd-Elrasheed [3], Samar A. Salim [4], Rania H. Fahmy [1,3], Salwa Salah [1] and Manal M. EL-Ashmoony [1,*]

1 Department of Pharmaceutics and Industrial Pharmacy, Faculty of Pharmacy, Cairo University, Cairo 11562, Egypt; sara.elhelaly@pharma.cu.edu.eg (S.N.E.-H.); rania.fahmy@pharma.cu.edu.eg (R.H.F.); salwa.salah@pharma.cu.edu.eg (S.S.)
2 Department of Pharmaceutics and Industrial Pharmacy, Faculty of Pharmacy, New Giza University, NewGiza, Km 22 Cairo-Alex Road, Giza 12588, Egypt
3 Department of Pharmaceutics and Industrial Pharmacy, Faculty of Pharmacy, Al-Ahram Canadian University, 6th of October City 12556, Egypt; emansaadahmed136@gmail.com
4 Nanotechnology Research Center (NTRC), The British University in Egypt (BUE), El-Sherouk City, Cairo 11837, Egypt; Samar.Salim@bue.edu.eg
* Correspondence: manal.elashmoony@pharma.cu.edu.eg; Tel.: +20-12-2362-0458

Citation: Nageeb El-Helaly, S.; Abd-Elrasheed, E.; Salim, S.A.; Fahmy, R.H.; Salah, S.; EL-Ashmoony, M.M. Green Nanotechnology in the Formulation of a Novel Solid Dispersed Multilayered Core-Sheath Raloxifene-Loaded Nanofibrous Buccal Film; In Vitro and In Vivo Characterization. *Pharmaceutics* **2021**, *13*, 474. https://doi.org/10.3390/pharmaceutics13040474

Academic Editors: Vitaliy Khutoryanskiy and Hisham Al-Obaidi

Received: 26 February 2021
Accepted: 24 March 2021
Published: 1 April 2021

Abstract: Green nanotechnology utilizes the principles of green chemistry to formulate eco-friendly nanocarrier systems to mitigate patients and environment hazards. Raloxifene (RLX) demonstrates poor aqueous solubility (BCS class II) and low bioavailability, only 2% (extensive first-pass metabolism). The aim of this study is to enhance RLX solubility and bioavailability via development of novel solid dispersed multilayered core-sheath RLX-loaded nanofibers (RLX-NFs) without the involvement of organic solvents. A modified emulsion electrospinning technique was developed. Electrospinning of an RLX-nanoemulsion (RLX-NE) with polymer solution (poly vinyl alcohol (PVA), hydroxypropyl methylcellulose (HPMC), and chitosan (CS) in different volume ratios (1:9, 2:8, and 4:6) using D-optimal response surface methodology was adopted. In vitro characterization of RLX-loaded NFs was performed; scanning electron microscope (SEM), thermal analysis, drug content, release studies, and bioadhesion potential. The optimum NFs formula was evaluated for morphology using high-resolution transmission electron microscopy (HRTEM), and ex vivo drug permeation. The superiority of E2 (comprising RLX-NE and PVA (2:8)) over other NF formulae was statistically observed with respect to Q60 (56.048%), Q240 (94.612%), fiber size (594.678 nm), mucoadhesion time 24 h, flux (5.51 μg/cm^2/h), and enhancement ratio (2.12). RLX pharmacokinetics parameters were evaluated in rabbits following buccal application of NF formula E2, relative to RLX oral dispersion. E2 showed significantly higher Cmax (53.18 ± 4.56 ng/mL), and relative bioavailability (≈2.29-fold).

Keywords: green nanotechnology; raloxifene; nanoemulsion; core-sheath nanofibers; response surface methodology; ex vivo drug permeation; in vivo pharmacokinetic studies

1. Introduction

The postmenopausal state is characterized by a sharp fall in the estrogen levels circulating in plasma. Such a fall is responsible for various symptoms: Hot flushes, mood swings, sleep disorders, as well as metabolic changes, including cardiovascular diseases and osteoporosis. These symptoms collectively affect postmenopausal women's quality of life. Consequently, they add up to be the foremost cause of morbidity and mortality in postmenopausal women [1–4]. Raloxifene hydrochloride (RLX) is a selective estrogen receptor modulator (SERM) that has shown to preserve the beneficial effects of estrogen mainly increased bone mineral density (BMD) [5] and decreased LDL-cholesterol [6], and, at the same time, has an antiestrogen effect on mammary tissues and the uterus [7,8]. RLX prevents the transcriptional activation of genes containing the estrogen response element in reproductive tissue. In vitro studies showed RLX inhibits the estradiol-dependent proliferation of human mammary tumor cells [9]. Thus, RLX is used to prevent and treat osteoporosis and decrease the risk of developing invasive breast cancer in postmenopausal women. Although RLX is rapidly absorbed following oral administration (Approximately 60%), yet its absolute bioavailability is only 2% [10]. This is mostly because RLX undergoes extensive first-pass metabolism to the glucuronide conjugates raloxifen-49-glucuronide, RLX 6-glucuronide, and RLX-6, 49-diglucuronide [10]. As well, RLX belongs to class II drugs according to biopharmaceutical classification system (BCS), that is, low solubility and high permeability.

For bioavailability enhancement, many drug delivery systems have been adopted with the aim of efficient drug delivery to the plasma. As a rule of thumb, the right drug delivery system prominently participates in reducing side effects, improving patient's compliance, and enhancing the bioavailability where controlling drug dissolution and absorption leads to improved therapeutic efficacy [11]. Nanotechnology, for example lipid microspheres, polymeric micelles, or liposomes, among many other manipulative techniques, tailor drugs at the nanosized range for nanomedicines development. However, there have been some worriment related to the potential harm these nanomaterials present [12]. The green synthesis performs simple, cost-effective, eco-friendly, with no organic solvents, and non-toxic final dosage forms [13]. Thus, our study aimed at formulating a buccal nanofiber (NF) film utilizing green nanotechnology as a tool to enhance RLX oral bioavailability with reduced potential hazards to the patient and the environment.

Aiming to avoid first pass metabolism, various formulation techniques and routes of administrations have been adopted, however, buccal drug delivery is known for its high patient acceptability compared to other non-oral transmucosal routes of drug administration. The direct access to the systemic circulation through the internal jugular vein avoids acidic hydrolysis in the gastrointestinal tract (GIT) and bypasses the hepatic first pass metabolism enhancing the oral bioavailability. An added advantage is the rapid cellular recovery of the buccal mucosa after administration [14]. NFs offer the advantageous physical properties of the traditional polymeric films while maintaining the drug in the form of fibers in the nanometric size range, with a large surface area to volume ratio. Thus, it is being presented as a potent drug delivery vehicle with high permeability through mucosal membranes [15,16], flexibility in surface functionalities and superior mechanical performance (e.g., stiffness and tensile strength) [17]. Moreover, NFs has the benefit of preparing solid dispersions of a drug in a crystalline carrier which in turn increases its solubility and absorption [18]. NFs were prepared using the emulsion electrospinning method. It involves emulsification of active agents within the solution which is then dissolved in the appropriate solvents. Emulsion electrospinning has gained various levels of success in the production of drugs in the form of electrospun NFs for treatment of different diseases [19–21].

Different strategies have been made to improve RLX bioavailability: Murthy et al. [22] developed Self-assembled lecithin-chitosan RLX nanoparticles, Shah et al. [23] developed RLX nanostructured lipid carriers, Ağardan et al. [24] prepared RLX-loaded liposomes, Pandya et al. [25] prepared an RLX nanoemulsion, and Saini et al. [26] developed RLX-loaded chitosan nanoparticles.

Thus, the aim of this study was to formulate a novel solid dispersed buccal core-sheath RLX-loaded NFs through electrospinning of the oil in water raloxifene nanoemulsion (O/W) RLX-NE with polymer solution to improve RLX solubility, dissolution and, hence, bioavailability and use the benefits of buccal delivery to bypass the first-pass metabolism. An in vivo pharmacokinetic study of the buccal RLX-NF films compared to RLX oral dispersion was conducted in rabbits.

2. Materials and Methods

2.1. Materials

Raloxifene HCl (RLX), USP 99.7% purity, batch number CFR018007 was purchased from BDR Pharmaceuticals International (Mumbai, India). Cremophor® RH40 (polyoxyl 40 hydrogenated castor oil) was kindly provided by BASF SE (Ludwigshafen, Germany). Transcutol® HP was a kind gift from Gattefosse (Lyon, France). Miglyol® 812 (caprylic/capric triglyceride) was a kind gift from EIPICO for pharmaceutical industries (Cairo, Egypt). Hydroxypropyl methyl cellulose (HPMC), methocel™ K100LV (100 cP at a 2% addition rate in water), was purchased from Chempoint (Bellevue, WA, USA). Poly vinyl alcohol (PVA) (average Mw 146,000–186,000), chitosan (CS) high molecular weight (800–2000 cP), and acetonitrile were purchased from Sigma–Aldrich, Co. (St. Louis, MO, USA). Normal saline was purchased from Otsuka Pharmaceutical Co., S.A.E (Cairo, Egypt). Xylazine (XYLA-JECT®) was purchased from ADWIA Pharmaceuticals Co. (Cairo, Egypt). Ethyl acetate and formic acid were bought from El-Nasr Pharmaceutical Chemicals Co. (Cairo, Egypt). Olmesartan obtained from AstraZeneca (Cairo, Egypt). Potassium dihydrogen phosphate and disodium hydrogen phosphate were purchased from Sisco Research Laboratories (Mumbai, India). HPLC-grade methanol was purchased from Research-lab Fine Chem Industries (Mumbai, India). All other chemicals and solvents were of analytical grade and used as received.

2.2. Preparation of RLX-Loaded Nanofibers

2.2.1. Preparation of RLX-Nanoemulsion

RLX-nanoemulsion was prepared using the Abd-Elrasheed et al. method [27]. Briefly, 5 mg RLX was vortexed first with 0.1 mL of Miglyol® 812 (oil phase) for 1–2 min, then 0.4 mL of Cremophor® RH40 (surfactant) and 0.4 mL of Transcutol® HP (co-surfactant) were added and vortexed for 5 min, 0.1 mL of water was added dropwise while vortexing. Finally, homogenization was performed for 1 min at 10,000 rpm. RLX-NE was then sonicated for 10 min to eliminate any air bubbles.

2.2.2. Preparation of Polymer Solutions

The choice of the polymers used was based on a preliminary study to choose the optimum type which could be electrospun in a continuous manner into NFs. PVA and HPMC were dissolved in double-distilled water to obtain 10% w/v and 1% w/v, respectively. For CS solution, CS was dissolved in 0.5% acetic acid to obtain a 1.5% w/v CS solution. All polymer solutions were stirred on a magnetic stirrer at 150 rpm and 80 °C for 1 h until complete dissolution, then stirring was continued at room temperature for 12 h.

2.2.3. Preparation of RLX-Loaded Nanofibers

Nine RLX-loaded NFs were prepared by the electrospinning technique comprising RLX-NE and different types of polymers with varied ratio. The composition of the developed RLX-NFs is shown in Table 1. The investigated variables were A: type of polymer (PVA, PVA + HPMC, PVA + CS), B: RLX-NE:polymer ratio (1:9, 2:8, or 4:6). For polymer solutions, fixed concentrations were used for each type and in case of polymer mixtures (i.e., PVA and HPMC, and PVA and CS) a 2:8 ratio were kept constant throughout the study as shown in Table 1.

Table 1. The composition and in vitro characterization results (mean ± S.D., n = 3) of the developed raloxifene nanofibers.

System	Composition (mL)				In Vitro Characterization Data of RLX-NFs				
	RLX-NE (5 mg/mL)	PVA (10% w)	HPMC (1% w)	Chitosan (1.5% w/v)	Drug Content (%)	Mucoadhesion Time (h)	Fiber Size (nm)	Q60 * (%)	Q240 * (%)
E1	1	9	-	-	99.90 ± 5.52	24 ± 0.00	555.82 ± 29.88	61.14 ± 7.5	91.10 ± 3.63
E2	2	8	-	-	96.56 ± 0.62	24 ± 0.00	594.67 ± 26.63	56.04 ± 16.02	94.61 ± 8.16
E3	4	6	-	-	102.45 ± 4.98	0.38 ± 0.01	1005.93 ± 3.52	83.36 ± 0.14	102.11 ± 2.61
E4	1	7.2	-	1.8	105.33 ± 7.10	24 ± 0.00	374.54 ± 20.74	19.26 ± 0.65	45.20 ± 0.31
E5	2	6.4	-	1.6	98.20 ± 14.62	24 ± 0.00	391.10 ± 15.42	33.91 ± 4.93	60.06 ± 2.60
E6	4	4.8	-	1.2	97.31 ± 0.80	0.5 ± 0.07	1513.61 ± 46.23	68.87 ± 9.29	76.09 ± 15.59
E7	1	7.2	1.8	-	102.33 ± 6.59	24 ± 0.00	321.20 ± 26.14	44.47 ± 7.53	87.76 ± 2.10
E8	2	6.4	1.6	-	93.38 ± 7.03	24 ± 0.00	529.27 ± 79.49	66.92 ± 16.26	91.43 ± 1.77
E9	4	4.8	1.2	-	92.38 ± 2.14	0.42 ± 0.05	1899 ± 2.19	80.08 ± 9.59	96.62 ± 0.57

* Q60 and Q240 represent the % drug released after 60 min and 240 min. RLX-NE; raloxifene nanoemulsion, PVA; polyvinyl alcohol and HPMC; hydroxypropyl methylcellulose.

RLX-NE was stirred with polymer solutions at varied volume ratios as presented in Table 1 on a magnetic stirrer at 700 rpm for 30 min to obtain homogenous solutions followed by sonication for 90 min to get rid of any air bubbles before electrospinning. The homogenous solutions were then electrospun into NFs by electrospinner (NANON-01A, MECC, Fukuoka, Japan). The polymer solution was placed into 5 mL syringe (22 G). NFs were fabricated through electrospinning under optimized parameters set after preliminary experiments, and optimal spinning conditions were set to be: Applied voltage (25 KV), flow rate (0.9 mL/h), 15 cm distance between the nozzle and the aluminum foil collector and width (20 mm) [28]. All NFs samples were electrospun at ambient conditions with humidity of 48–50%.

2.3. *Characterization of RLX-Nanoemulsion*

2.3.1. Determination of Particle Size (PS), Polydispersity Index (PDI), and Zeta Potential (ZP)

The mean PS (Z-Average) and PDI of RLX-NE were determined at 25 ± 1 °C, in triplicates, via dynamic light scattering (DLS) analysis (Malvern Zetasizer Nano ZS; Worcestershire, UK). A 0.1 mL of the prepared RLX-NE formulation was diluted to 100 times with double distilled water. Following dilution, the glass tube was vortexed to ensure complete dispersion of the formulation. DLS analyzes the fluctuation of intensity of a scattered laser beam (caused by the Brownian random motion of droplets), at an angle of 90°, as a function of time [29]. The derived data were needed to estimate the diffusion coefficient and hence, the Z-average of droplets according to the Stoke–Einstein equation [30].

A laser doppler anemometer, coupled with the Zetasizer Nano ZS, was utilized to determine the electrophoretic mobility and hence, the zeta potential of the charged droplets.

2.3.2. Drug Content

A total of 0.1 mL of the NE formulation was diluted with 140 mL of PBS solution, pH 6.8, and the absorbance of this solution was measured spectrophotometrically at 286 nm against a blank (0.1 mL plain NE mixed with 140 mL PBS solution pH 6.8) and the concentration was calculated.

2.3.3. Spectroscopic Characterization of Percentage Transmission

A total of 0.1 mL of the NE formulation was diluted to 100 times with double-distilled water. The percentage transmission of the NE formulation after dilution was measured at 400–800 nm to determine the degree of clarity of the NE formulation.

2.4. Characterization of RLX-Loaded Nanofibers

2.4.1. Solid State Characterization of RLX-Loaded Nanofibers

All RLX-NFs formulae were morphologically investigated using a scanning electron microscope (SEM) and subjected to thermal analysis, differential scanning calorimetry DSC, Fourier-transform Infrared Spectra (IR), and powder x-ray diffraction PXRD

Scanning Electron Microscope (SEM)

The fiber surface morphologies of the RLX-NFs were assessed via a SEM (JEOL GSM6610LV, JEOL Ltd., Tokyo, Japan) operating at acceleration voltage of 30 kV. NFs samples were placed on the carbon tape. The fiber diameter was measured using Image J software from the SEM pictures in original magnification. At least 15 isolated NFs were randomly selected, and their diameters and diameter distributions were measured and averaged.

Differential Scanning Calorimetry (DSC)

The physical state of all RLX-NFs formulae and their individual components, as well as the pure drug, were performed using a Shimadzu differential scanning calorimeter (DSC-50, Shimadzu, Japan). Samples were placed in a standard aluminum pan and then heated at a constant rate of 5 °C/min under the atmosphere of nitrogen gas carrier in a temperature range from 20 to 350 °C.

Fourier-Transform Infrared Spectra (FT-IR)

FTIR of RLX-NFs was recorded by a Fourier transform infrared spectrometer (FTIR, Shimadzu FTIR-8400 S, Kyoto, Japan). The experiments were carried out in the range of 4000 to 500 cm^{-1}. The infrared spectra were recorded in the transmission mode using thick mats of RLX-NFs.

Powder X-ray Diffraction (PXRD)

PXRD studies of RLX and all RLX-NFs were assessed at room temperature (PANalytical X'Pert PRO diffractometer; Almelo, Netherlands). The samples were prepared using nickel filtered Cu Kα radiation (λ = 1.542 Å, 45 kV, and 35 mA). The results were presented as intensity versus 2θ (5–80°) [31].

2.4.2. Determination of Drug Content and Homogeneity of RLX-Loaded Nanofibers

Accurate weights from different sites of RLX-PVA-NFs, RLX-PVA-HPMC NFs were dispersed in double-distilled water and RLX-PVA-CS NFs was dissolved in 1% acetic acid by homogenization at 10,000 rpm. The drug content and the homogeneity of NF films were then analyzed using UV spectrophotometer. Measurements were performed, in triplicate, at room temperature.

2.4.3. In Vitro Release Studies

After drug content determination of all RLX-NFs formulae, samples of RLX-NFs equivalent to 1 mg RLX were accurately weighed and then, the RLX-NFs were immersed in 25 mL release medium (PBS, pH 6.8 of 0.1% tween) [32] in accordance with optimum sink conditions and gently stirred at 37 $\pm$ 1 °C with a speed of 50 rpm in a shaking water bath. At certain time intervals (10, 20, 30, 40, 50, 60, 120, 180, and 240 min), a certain volume of the solution (1 mL) was removed and replaced with the same volume of a fresh medium. Withdrawn samples were suitably diluted and analyzed using UV spectrophotometer and the mean drug released percentages ($\pm$ S.D.) were plotted versus time [28].

2.4.4. Bioadhesion Potential of RLX-Loaded Nanofibers

Adhesion study was performed to RLX-loaded NF films by adapting a method of Nakamura et al. [33]. 100 g of a hot agar/mucin solution (1 and 2% w/w, respectively, in phosphate buffer pH 6.8) were casted on a glass plate (10 cm $\times$ 10 cm) and left to gel at

4–8 °C for 3 h. The gel was then equilibrated for one hour to the test conditions of 25 °C and 75% relative humidity (saturated sodium chloride solution) in a chamber. NF films were placed on top of the gel, moved downward due to gravity after the glass plate was turned into a vertical position. The time taken to move from the top to the bottom of the plate was recorded (n = 3).

2.5. *Optimization of RLX-Loaded Nanofibers via D-Optimal Response Surface Methodology*

A response surface methodology was used for optimization using D-optimal design to study the effect of the different formulation variables on the characteristics of the developed RLX-loaded NFs using Design-Expert® software (Stat-Ease, Inc., Minneapolis, MN, USA). The independent variables were, A: Type of polymer (PVA, PVA + HPMC, PVA + CS), B: RLX-NE:polymer ratio (1:9, 2:8, or 4:6), while the % drug released at time 60 min, Q60 (Y1), % drug released at time 240 min, Q240 (Y2), fiber size (Y3), and mucoadhesion time (Y4) were chosen as the dependent variables (responses).

The desirability formula was adjusted to minimize fiber size and maximize Q60, Q240 and mucoadhesion time. One-way ANOVA and LSD post hoc analysis were attempted to compare the results of the selected responses using SPSS software 27.0 (SPSS Inc., Chicago, IL, USA). A p-value < 0.05 was considered statistically significant.

Numerical optimization technique was chosen for optimization of the responses. This method is based on the utilization of desirability functions and the optimum RLX-loaded NFs was selected.

2.6. *Characterization of the Optimum RLX-Loaded Nanofibers*

2.6.1. High-Resolution Transmission Electron Microscope (HRTEM) of the Selected RLX-Loaded NFs

The morphology of the selected RLX-NFs formula was evaluated by high-resolution transmission electron microscopy (HRTEM, JEM-2010, JEOL) with an accelerating voltage of 200 KV. TEM was operated at 15 kV, and the fiber samples were prepared by making a suspension of the NFs in ethyl alcohol and directly depositing the as-spun ultrafine fiber suspension onto copper grids.

2.6.2. Ex Vivo Drug Permeation Studies

The protocol of the ex vivo permeation studies was approved by the institutional review board, Research Ethics Committee Faculty of Pharmacy, Cairo University (approval no. PI 2762). Cow buccal mucosa, without any treatment, was freshly obtained from a local slaughterhouse immediately after the animal was killed and stored on normal saline in an ice box until it was transferred to our laboratory [34]. According to the results of the in vitro characterization studies, freshly prepared RLX-suspension (2 mg/2 mL PBS pH 6.8 of 0.1% tween 80) and the optimum RLX-loaded NFs formula equivalent to 2 mg RLX, were exposed to permeation testing of the drug through cow buccal membrane using the method described by Tayel et al. [35]. The apparatus used to test the permeation consisted of a glass tube (1.3 cm diameter) opened from both ends. The membrane was stretched over one open end of the glass tube and made water-tight by a rubber band then loaded with the formula forming the donor chamber. Two milliliters of phosphate buffer, pH 6.8, of 0.1% tween was transferred to the donor chamber to simulate the conditions inside the buccal cavity. The tube was attached to the shaft of the USP dissolution apparatus. The tube was then immersed in phosphate buffer, pH 6.8, of 0.1% tween 80 contained in the USP dissolution apparatus flask so that the membrane was just below the surface of the recipient solution. The temperature was maintained at 37 ± 0.5 °C, and the apparatus was run at 100 rpm for 6 h. Samples of 0.5 mL were withdrawn at 0.5, 1, 2, 3, 4, and 6 h, and were compensated for by equal volume of fresh buffer. The withdrawn samples were analyzed using a validated LC/MS/MS method. The drug permeation studies were conducted in triplicate and the results were presented as mean ± SD. The cumulative drug amounts permeated through the buccal mucosa per unit area (μg/cm^2) were plotted against time (h). The flux (Jss) at steady state was calculated. The enhancement ratio (ER) was calculated via dividing the flux value of the RLX-NFs formula by that of the aqueous dispersion, according to the following equations [36]:

$$J_{ss} = \text{Cumulative amount of drug permeated/time} \times \text{Diffusion area} \quad (1)$$

$$\text{ER} = \text{Flux of RLX-NFs/Flux of aqueous RLX dispersion} \quad (2)$$

2.7. In Vivo Estimation of RLX Pharmacokinetic Parameters in Rabbits

2.7.1. Study Design

The in vivo study was conducted, to compare the pharmacokinetics parameters of RLX-NFs following the buccal application of formula E2 (test treatment) and the oral administration of RLX aqueous dispersion (reference treatment). The study was performed using a non-blind, two-treatment, two-period, randomized, crossover design. The protocol of the in vivo study was approved (PI–2762) by the Research Ethics Committee at the Faculty of Pharmacy at Cairo University, Cairo, Egypt.

2.7.2. Animals

Adult male New Zealand albino rabbits (weighing 3.0 ± 0.1 kg each) were derived from the animal house (Faculty of Pharmacy, Cairo University, Egypt). They were housed under optimum environmental conditions with respect to room temperature (25 °C), humidity (50%), ventilation (15–20 AC/h), and adoption of alternating 12 h light/dark cycles. The animals were accommodated as one rabbit per cage, fed with standard dry food, and had free access to water.

2.7.3. Administration of Treatments and Blood Collection

A crossover design was conducted where the animals were randomly divided into two groups and each group contained three rabbits. The rabbits were anaesthetized using subcutaneous injection of xylazine HCL (20 mg/kg body weight) [37]. Then, a second dose of xylazine HCL was given after four hours to keep them sedated for 8 h to ensure proper application of the formulations and protect them till complete absorption of formulations.

Group I administered the optimum RLX-NFs films (formula E2) through the buccal mucosa in a dose level equivalent to 2.64 mg/kg body weight [38], Group II administered RLX oral suspension [32] in the same dose level. Following a wash out period of eight days, the reverse of randomization of the treatments took place in phase II. Blood samples of (1–2 mL) were collected into the EDTA-added centrifuge tubes at the time points (0, 0.25, 0.5, 1, 1.5, 2, 4, 6, 8, 12, 24, and 48 h). Regular check was done on the NFs in the rabbits' buccal cavity in order to check on their degradation pattern induced by the salivary flow. Blood samples were centrifuged at 4000 rpm for 15 min at room temperature. The plasma was separated and stored at −20 °C. Later RLX concentration in plasma was determined using a validated LC/MS/MS method.

2.7.4. Estimation of RLX Concentration in Rabbit Plasma

The concentration of RLX in the thawed plasma samples were analyzed using a triple quadrupole LC/MS/MS mass spectrometer (Micromass, Manchester, UK), according to the method developed by Said et al. [39]. Briefly, 200 μL of Olmesartan solution (internal standard; 100 ng/mL) was vortexed with 0.5 mL of plasma samples. The drug and the (internal standard) were extracted with ethyl acetate (4 mL), and vortexed for 1 min. The organic layer was separated by centrifugation (Eppendorf centrifuge 5804 R, Hamburg, Germany) at 4000 rpm for 10 min at 4 °C, and finally dried using a vacuum concentrator (Eppendorf 5301, Hamburg, Germany). The dried sample was reconstituted in 200 μL of the mobile phase, prior to analysis. For analysis, the samples (20 μL) were injected into the mass spectrometer. The elutes were separated on Agilent TMZORBAX Eclipse Plus (4.6 mm × 50 mm, 100 Å, 5 μm) using an isocratic mobile phase consisting of a mixture of 80% acetonitrile and 20% of 0.1% formic acid in water, at a flow rate 1 mL/min. The

LC/MS/MS mass spectrometer was equipped with an electrospray ionization (ESI) source that was adjusted to operate in the positive ion mode. The multiple reaction monitoring mode was employed to detect the transitions of the m/z 244.0 and 315.96 precursor ions to the m/z 185.30 and 270.00 product ions for RLX and IS, respectively. A linear calibration curve (r2 = 0.9961) was constructed between RLX plasma concentrations (ng/mL) and RLX/IS peak area ratios over RLX concentration range of 1–300 ng/mL.

2.7.5. Pharmacokinetics and Statistical Analyses

The mean RLX plasma concentrations (± S.D) of both treatments were plotted versus time. The pharmacokinetic parameters; peak RLX plasma concentration (C_{max}, ng/mL), the time to reach C_{max} (T_{max}, h), the mean residence time ($MRT_{0-\infty}$, h), the elimination half-life ($t_{1/2}$), the area under the curve from zero to the last sampling point ($AUC_{0-48\ h}$, ng.h/mL) and from zero to infinity ($AUC_{0-\infty}$ ng·h/mL) were estimated by application of non-compartmental analysis using WinNonlin software Ver. 8.3 (New Jersey, NJ, USA), and multivariate ANOVA using general linear model in SPSS software 27.0 (SPSS Inc., Chicago, IL, USA) and a non-parametric Wilcoxon signed rank test for Tmax. The $AUC_{(0-48\ h)}$ and $AUC_{(0-\infty)}$ values were used for calculating the relative bioavailability of the test treatment. Results were expressed as the mean (± S.D) except for T_{max} which was expressed as the median value.

3. Results and Discussion

3.1. *Preparation of RLX-Loaded Nanofibers*

Herein, A novel buccal RLX-NFs were formulated via electrospinning of RLX-NE with polymer solutions that yielded a core-sheath NFs without the use of organic solvents, which paves the road to "green" drug delivery systems. Nine RLX-loaded NFs were successfully prepared via the electrospinning of RLX-NE with polymer solution (at NE:polymer volume ratios 1:9, 2:8, and 4:6) using (PVA, PVA + CS, or PVA + HPMC) polymer solutions as presented in Table 1. The incorporation of the drug in the form of O/W NE presents the drug into the NFs in a more soluble form. Hence, the drug diffusion is enhanced with subsequent efficient penetration, increased rate of drug absorption, and enhanced bioavailability [19,40,41].

3.2. *Characterization of RLX-Loaded Nanoemulsion*

3.2.1. Determination of Particle Size (PS), Polydispersity Index (PDI), and Zeta Potential (ZP)

It was observed that the prepared NE system has a globule size 62.60 ± 6.82 nm. The small mean droplet size (MDS) might be attributed to the use of the proper surfactant/Co-surfactant mixture which can be adsorbed around oil-water interface reducing the free energy of the system, giving a small globule size [42]. Such surfactant/co-surfactant mixture composition and ratio was designated based on thorough preliminary experiments.

Poly-dispersity index (PDI) values ranged from 0.122 to 0.127, these small values of PDI indicates a monodispersed population. RLX-NE showed a ZP value of (4.33 ± 0.31 mV). The significance of such low ZP was minimum because the RLX-NE system will be further incorporated into nanofiber formulations.

3.2.2. Drug Content

The mean RLX content in the formulated NE system was found to be 100.34 ± 1.17% confirming the absence of drug loss during the formulating procedures.

3.2.3. Spectroscopic Characterization of Percentage Transmission

Percentage transmission (%T) of diluted RLX-NE was measured spectrophotometrically in the visible range (400–800 nm) and found to be (99.86% T), indicating highly clear and stable system obtained by aqueous dilution of the prepared NE formula. This might be attributed to the fact that the oil droplets are in a state of tremendously fine dispersions as presented above by the miniature globule size [43].

3.3. *Characterization of RLX-Loaded Nanofibers*

D-optimal Response Surface methodology was employed to estimate the significance of the direct and/or interactive contribution of the formulation variables on the characteristics of the developed RLX-loaded NFs. One categorical factor (polymer type, A) and one numerical factor (NE: polymer ratio, B) were investigated. The results were statistically analyzed and the final equations for the four responses (Q60 min, Q240 min, fiber size, and mucoadhesion time) were derived in terms of coded factors, as follow:

$$\begin{gathered} \text{Y1 (Q60)} = 59.125 + 9.1113 \times \text{A [1]} - 15.66 \times \text{A[2]} + 18.042 \times \text{B} - 5.594 \times \text{A[1]B} + 6.9625 \times \text{A[2]B} \\ \text{Y2 (Q240)} = 83.701 + 12.859 \times \text{A[1]} - 21.58 \times \text{A[2]} + 8.2778 \times \text{B} - 2.753 \times \text{A[1]B} + 6.6794 \times \text{A[2]B} \\ \text{Y3 (Fiber Size)} = 647.96 - 115 \times \text{A[1]} - 32.79 \times \text{[2]} + 527.83 \times \text{B} - 319.2 \times \text{A[1]B} + 52.321 \times \text{A[2]B} + 297.06 \times \text{B}^2 \\ \text{Y4 (Mucoadhesion Time)} = 21.0542 - 0.02095 \times \text{A[1]} + 0.02619 \times \text{A[2]} - 11.7833 \times \text{B} - 0.02857 \times \text{A[1]B} + \\ 0.03571 \times \text{A[2]B} - 8.8375 \times \text{B}^2 \end{gathered} \tag{3}$$

where, A[1], A[2], and B are the coefficients of multi-level categorical and numerical factors.

3.3.1. Solid State Characterization of RLX-Loaded Nanofibers

Scanning Electron Microscope (SEM) and Fiber Size

The SEM photographs and distribution of the average diameter are shown in Figure 1. SEM images of NFs films formulae E1-E8 revealed that the nanofibers were randomly aligned, interconnected and continuous. It can also be distinctly seen that smooth, bead-free, and relatively uniform fibrous diameter fibers were obtained. Since the drug was solubilized in a NE system containing a mixture of surfactant and co-surfactant, consequently it can be suggested that the NE played a role in decreasing the surface tension which enhances the electrospinnability of NFs. This comes in agreement with Wang et al. who observed that using non-ionic surfactant produced a smaller NFs that dramatically reduced surface tension than those obtained without surfactants [44].

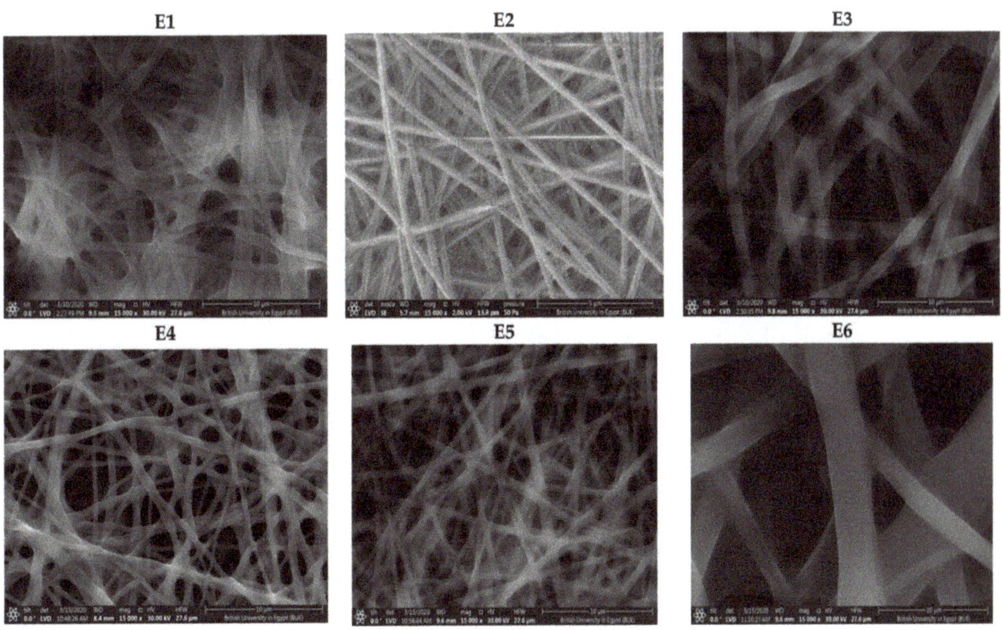

Figure 1. *Cont.*

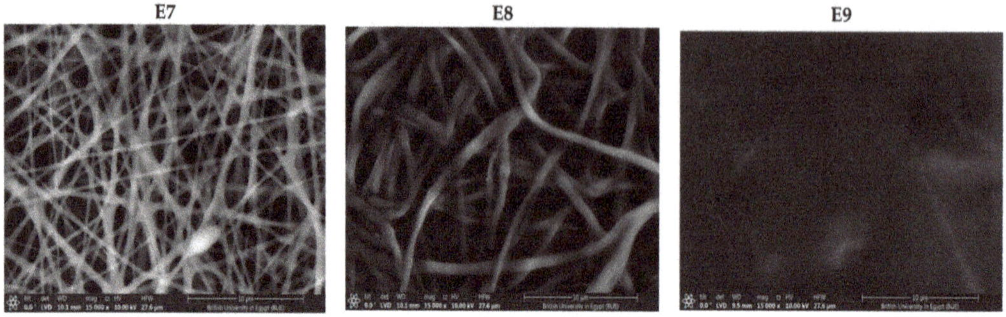

Figure 1. SEM of NFs formulae (**E1–E9**).

Moreover, it was found that as the polymer to NE volume ratio increased, consequently the polymer concentration increased and resulted in enhanced electrospinnability of NFs and a uniform bead-free NFs is produced. Therefore, it could be observed that formulae with highest polymer ratio (E1, E4, and E7) produced smoother, bead-free, uniform diameter fibers. On the other hand, SEM images of E6 (least polymer:NE ratio) revealed less uniform, wider diameter NFs. Additionally, SEM images of E9 revealed that NFs were fused together which emerged as thin sheet in spite of a fibrous mat-like structure. Formulations E6 and E9 comprised least NE:polymer volume ratio (4:6) than other formulae.

Similar trends were observed by Jia et al. [45], who concluded that as the concentration of the polymer (CS/PVA) solution increased, beaded NFs transformed into uniform bead-free structure and also by Geng et al. [46], Who observed that the morphology of the nanofibers changed gradually from the highly beaded structure to uniform fiber structure with increasing the polymer concentration.

CS and HPMC pure solutions have poor electrospinning properties. Blending these biopolymers with PVA resulted in enhanced electrospinning properties which requires a higher PVA concentration in order to overcome the repulsive forces of the biopolymer molecules which prevent chains to be adequately entangled. On a study by Sedghi et al. [47] it was found that blending CS with PVA considerably enhanced its electrospinnability and subsequently produced NFs of uniform and smooth morphology [48,49].

Furthermore, Sedghi et al. found that as PVA concentration increased, the beads existed in NFs were completely resolved and simultaneously junction points or fused NFs disappeared. Similar results were observed by Aydogdu et al. and Lu et al. who found that non-homogeneous beaded NFs were obtained when pure HPMC was electrospun. Hence, polyethylene oxide was added to the HPMC solutions at different ratios. Due to the strong repulsive force between biopolymer molecules, sufficient chain entanglement cannot occur. Adding polyethylene oxide reduced the repelling force between molecules, promoting NFs formation [50,51].

As revealed in Table 1, the average diameters of the NFs ranged between 321.207 $\pm$ 26.14 nm (E7) to 1899.00 $\pm$ 2.9 nm (E9) associated with narrow size distribution and small standard deviations of NFs. Such results indicate that uniform NFs were obtained by electrospinning process [51]. The fiber size was significantly affected by NE:polymer ratio (p = 0.006). It was found that as the NE:polymer ratio increase, a significant increase in diameter of NFs was obtained.

Although the polymer concentration decreased as the NE:polymer volume ratio increased, the concentration of NE components (Miglyol® 812 as an oil, Cremophor® RH40 as surfactant and Transcutol® HP as co-surfactant) increased which resulted in an increase in the solution viscosity. The diameter of fibers is directly proportional to the viscosity of the solutions. This was explained by Neo et al. [52] who proved that having high viscosity, prevented the instability of bending and reduced jet path. Due to the low stretching of the solutions, reducing the jet path resulted in high diameters. In a previous study, it was reported that increasing oil concentration during preparation of intranasal zaleplon NE from 10% to 20% resulted in a significantly larger globule size [27].

A direct correlation was observed between the fiber diameter and the type of polymer. PVA NFs have larger diameter than PVA/CS and PVA/HPMC-loaded NFs at the same NE:polymer volume ratio. This could be due to blending of polymers leads to smaller diameter than using one polymer. Similar observations have been made by Lin et al. [53] and Ignatova et al. [54] who investigated a series of blended NFs at different weight ratios and found a decrease in the average diameter of the NFs with increasing the blended content.

For PVA/CS NFs, the presence of amino groups in CS can be protonated in the acidic media. Therefore, the density of electrical charges on the surface of the jet increases in presence of CS than in PVA alone leading to higher charge density in the polymeric jet. Increasing the surface charge imposed greater elongation force on the jet under the electrical field, leading to decrease NFs size [49]. Similarly, Bahrami et al. found that finer fibers were produced with increasing CS content in the blend solution; on the other hand, fibers became more fragile [55].

Differential Scanning Calorimetry (DSC)

The DSC thermograms of RLX, PVA, CS, HPMC, and RLX-loaded NF films are represented in Figure 2A. The DSC thermogram of RLX alone shows sharp endothermic peak at 262.71 °C which came in agreement with Patil et al. [56]. Such sharp peak is indicative of strong crystal lattice energy which can be considered as one of the factors responsible for lower RLX aqueous solubility [57].

Regarding the polymers, DSC thermogram of PVA showed two broad endothermic peaks at 194.24 °C and 325.19 °C corresponding to its melting points, which agrees with the finding of Salah et al. [58]. DSC thermogram of CS also presented broad endothermic peak at 79.22 °C that can be ascribed to the loss of water and a second exothermic peak at 305.69 °C that might be related to the decomposition of its amine unit according to Ferrero et al. [59]. Further, HPMC DSC thermograms showed a broad endothermic transition corresponding to its melting point. The endothermic transition of HPMC starts from 54.15 °C to 122.03 °C, with a broad peak value at 87.39 °C which is in agreement with Demappa et al. [60].

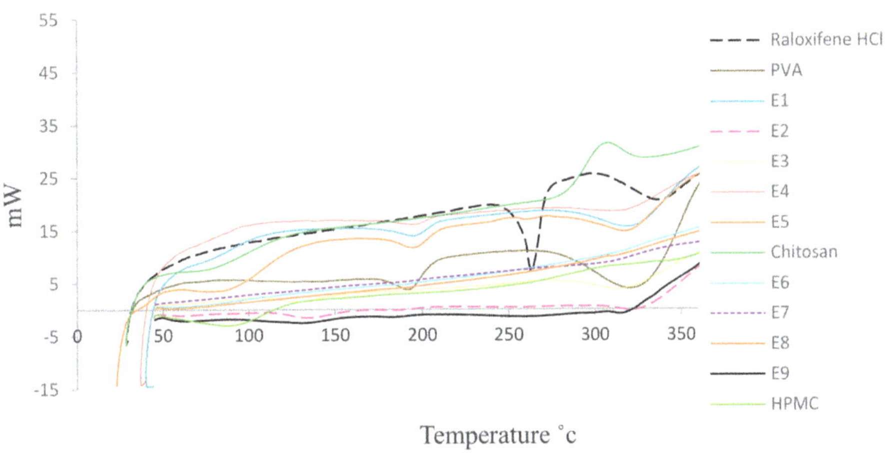

(A)

Figure 2. *Cont.*

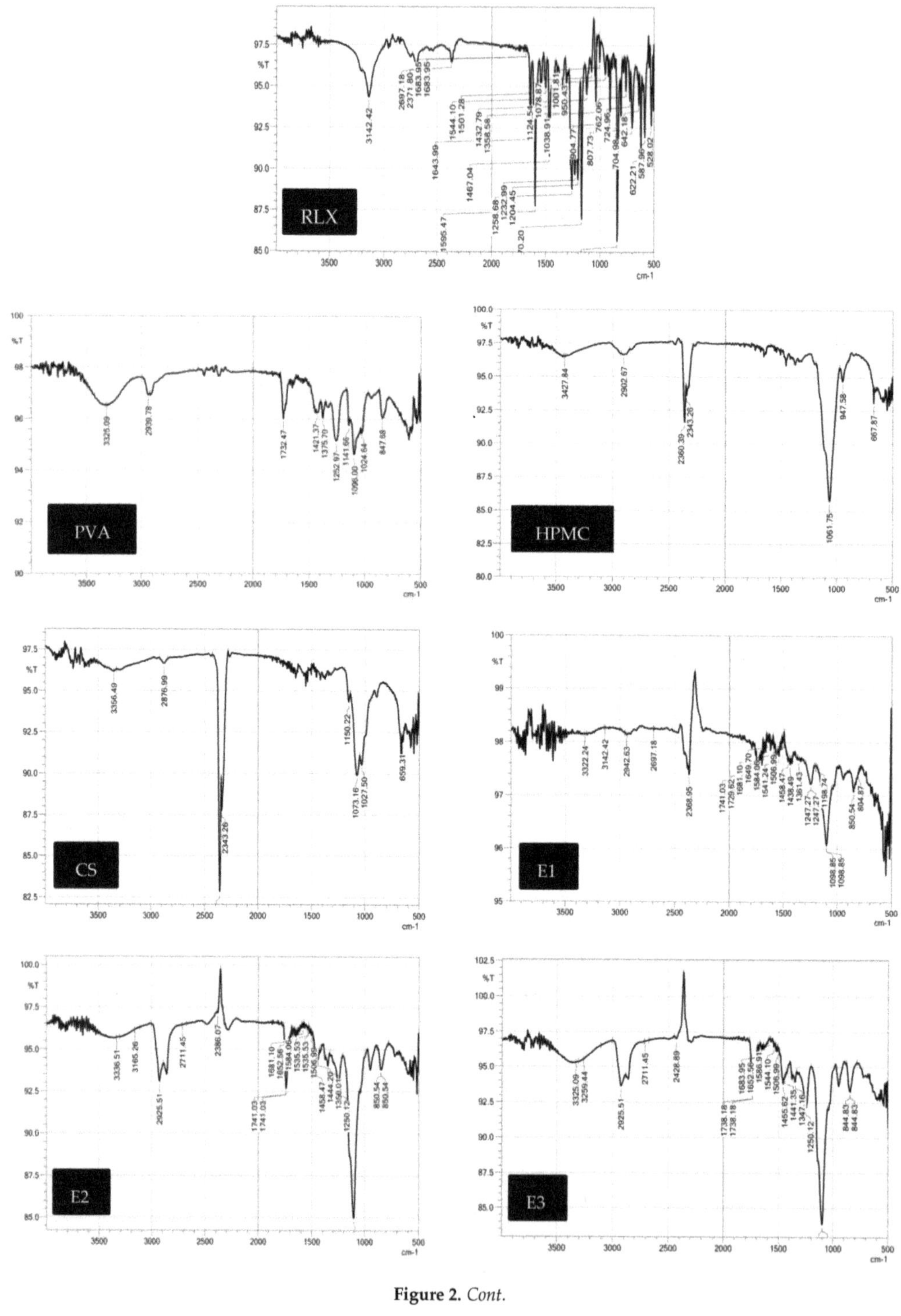

Figure 2. *Cont.*

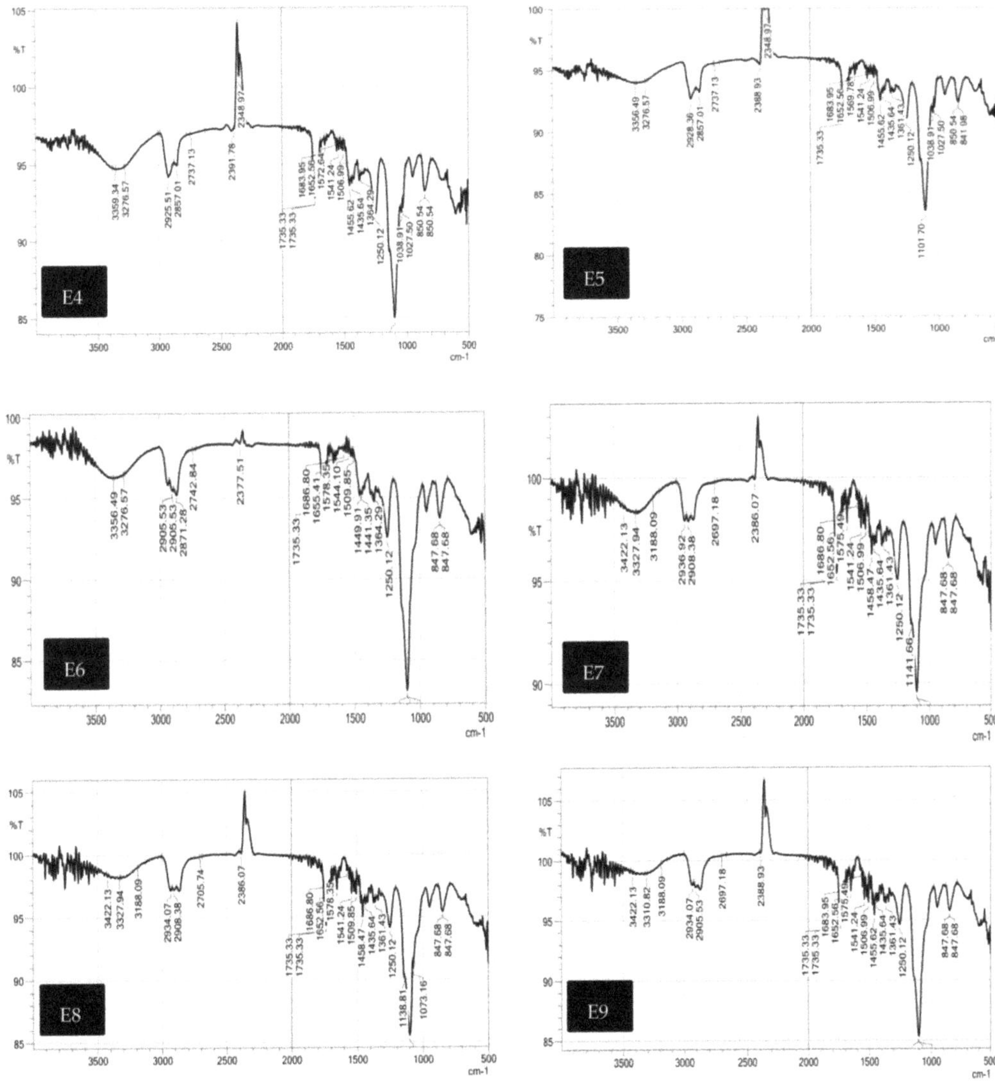

(B)

Figure 2. (**A**) DSC thermogram of RLX-NFs formulae (E1–E9) in comparison to that of the pure RLX, PVA, CS, and HPMC, (**B**) IR spectrum of pure (RLX, PVA, HPMC, CS, E1–E9).

On the other hand, the sharp melting point peak of the drug completely disappeared in all RLX-NFs formulae, except for E5 where it was greatly decreased. Such disappearance of RLX characteristic peak could indicate the change of the crystalline drug into the amorphous state which might be molecularly dispersed throughout the developed NFs formulation. such changes in the drug state could contribute to the increase in RLX solubility and thus the enhanced dissolution profile [61].

Fourier-Transform Infra-Red Spectra (FT-IR)

The infrared spectrum of RLX, as shown in Figure 2B, exhibited the characteristic bands at 3142.42 cm^{-1} due to aromatic OH group, 2959.76 cm^{-1} due to CH aliphatic, 1680 cm^{-1}, 1595.47 and 1432.79 cm^{-1} due to C=C, and 1258.68 cm^{-1} due to C–O stretching. 1170.20 cm^{-1} due to C–N and 850 cm^{-1} due to benzene ring para-substitution. For PVA, IR spectrum showed characteristic broad band at 3325.09 cm^{-1} corresponding to O–H alcoholic, band at 2939.78 cm^{-1} attributed to C–H SP3 and a band at 1089.38 corresponding to C–O–H. Additionally, HPMC IR spectrum showed characteristic broad band at 3427.84 cm^{-1} corresponding to O–H alcoholic, band at 2902.67 cm^{-1} attributed to C–H SP3 and a band at 1050 cm^{-1} corresponding to C–O. As for CS, IR spectrum showed characteristic biforked band at 3356.49 cm^{-1} corresponding to NH2 and a band at 2876.99 cm^{-1} attributed to O–H.

Alternatively, RLX-loaded NFs films IR spectrum showed decreased intensity in the characteristic bands of RLX. Most of RLX-peaks were found to be smoothened suggesting strong physical interaction with the polymers used [62] and the absence of formation of new peaks confirming the absence of any chemical interaction [63].

Powder X-ray Diffraction (PXRD) Studies

The X-ray diffraction pattern for the pure drug, PVA, CS, HPMC, and RLX-loaded NF films are shown in Figure 3. The crystallinity of the phases influences the dissolution of the dosage form; an amorphous form dissolves at a faster rate relative to crystalline form because of its higher internal energy and greater molecular motion, which enhances the thermodynamic properties [64,65].

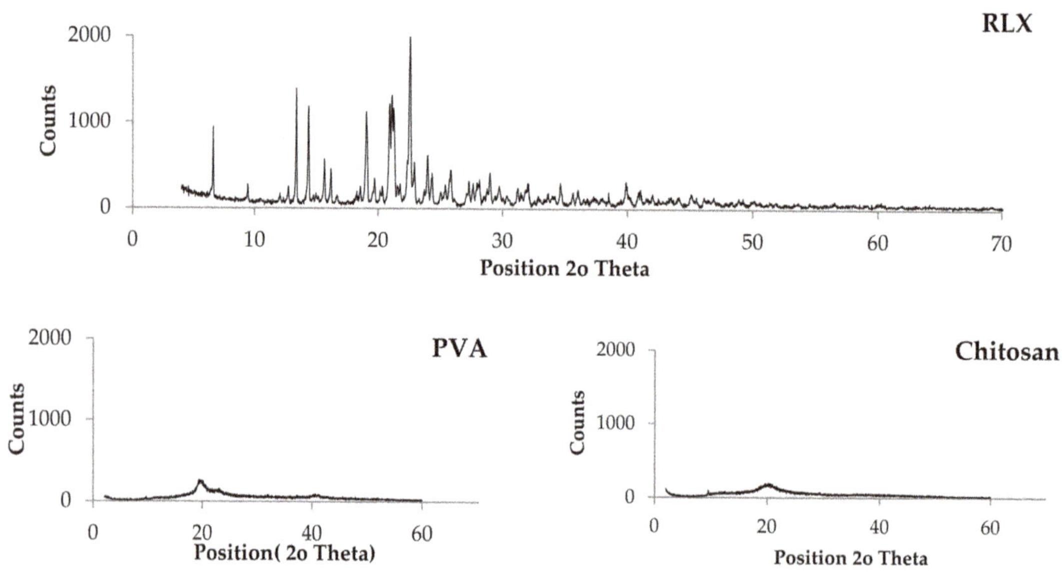

Figure 3. *Cont.*

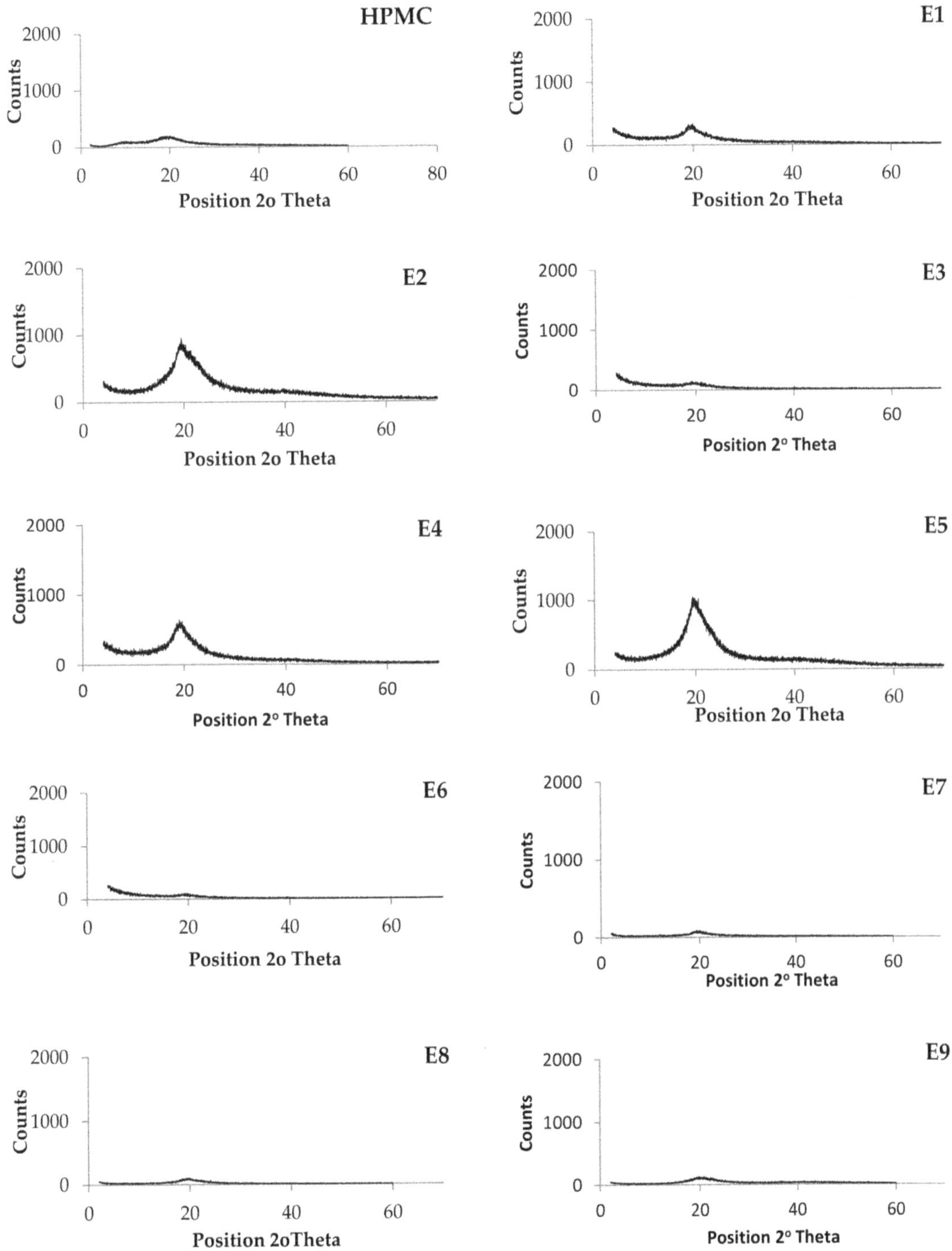

Figure 3. Powder X-ray diffraction of NFs formulae (E1–E9) in comparison to that of the pure (RLX, PVA, CS, HPMC).

The PXRD pattern of the RLX exhibited sharp and intense peaks which is indicative of its strong crystalline nature with seven prominent peaks of high intensity at 2θ = 13.41°, 14.41°, 19.06°, 20.94°, 21.13°, 21.28°, and 22.62°. The other four less prominent peaks are at 2θ = 6.68°, 15.70°, 22.98°, and 24.01°.

Regarding the polymers used, all the polymer PXRD showed less sharp and intense peaks indicating lower crystallinity. The PXRD of PVA showed one prominent peak at 2θ = 19.65°. The PXRD of HPMC showed two prominent peaks at 2θ = 20° and 10.69°. The PXRD of CS showed one prominent peak at 2θ = 20.27°.

A comparison of PXRD of pure drug with that of RLX-loaded NFs films showed disappearance of all the drug peaks in all the formulae which could be attributed to the destruction of the drug crystal lattice because of progressive amorphization [66], which indicate that the drug might be molecularly dispersed inside the NF films which might lead to enhanced RLX solubility associated with augmented dissolution profile and bioavailability. It is worthy to note that such PXRD results are in great agreement with the DSC results.

3.3.2. Determination of Drug Content and Homogeneity of RLX-Loaded Nanofibers

The mean values of RLX content in different RLX-loaded NFs films are presented in Table 1 and ranged from 92.38% to 105.33%. These results confirmed the absence of drug loss during the electrospinning procedures and uniform distribution of the drug through RLX-loaded NFs films.

3.3.3. In Vitro Release Studies

The in vitro drug release profiles from RLX-loaded PVA, RLX-loaded PVA/HPMC and RLX-loaded PVA/CS core-sheath NFs in PBS pH 6.8 with 0.1% tween at 37 °C in comparison to RLX aqueous dispersion, are displayed in Figure 4. Practically, the burst drug release is required in order to rapidly achieve the minimum effective drug concentration, while the prolonged drug release is required to maintain the therapeutic drug response for an extended period [67].

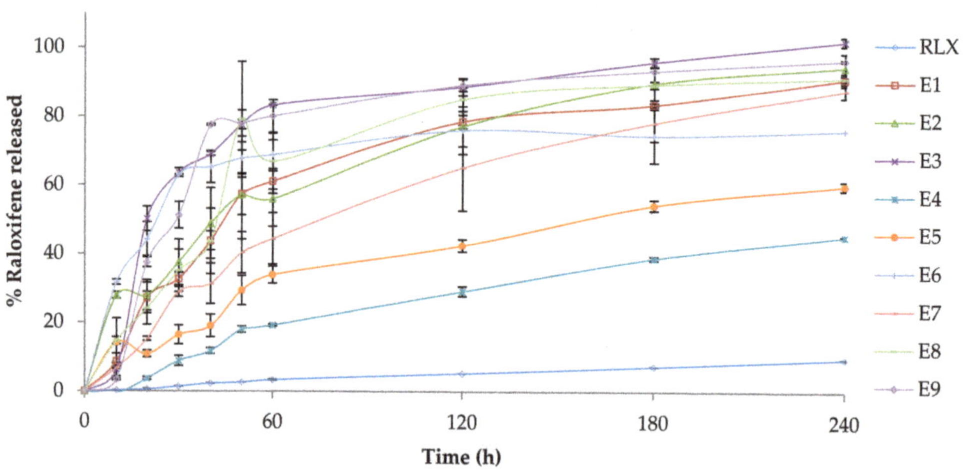

Figure 4. In vitro drug release profiles from RLX-loaded nanofibers (E1–E9) in comparison to RLX aqueous dispersion, in phosphate buffered saline (pH 6.8 of 0.1% *w*/*v* tween 80) at 37 ± 0.5 °C (mean ± S.D., *n* = 3).

A fast release of the drug was observed with an initial burst release of formulae E3, E6, E9, E2, E8, E1, and E7. This burst release could be attributed to the use of higher volume ratios of NE to polymer solution. As a result, core-sheath structure with an incomplete inward movement of the NE droplets during electrospinning process occurred. This allowed NE droplets to be located close to or loosely associated with the fiber surface or adsorbed on the fiber surface. Therefore, after 20 min, the initial burst release of the drug from NFs formulae with higher polymers ratio (E3(50.13%), E6(44.24%), and E9(37.43%)] was higher than NFs formulae with lower polymers ratio [E2(27.81%), E8(24.05), E1(27.07%), and E7(15.21%)). E4 and E5 showed (3.69%) and (10.83%) release of the drug respectively at time 20 min which could be attributed to the complete inward movement of NE inside the NFs. This comes in agreement with Yang, Y., Li et al. [68] who studied preparing proteins NFs using W/O emulsion-based electrospinning. They verified that the correlation between the volume ratio of aqueous to organic phase and burst release may be attributed to the completeness of core-sheath structure and unencapsulated proteins located close to, or loosely associated, with the fiber surface [68].

As well, O'Donnell et al. performed a study for evaluation of acetaminophen release from biodegradable PVA and nanocellulose films. It revealed that the fraction of burst release increased with increasing drug concentration in the initial formulations and this accounts for drug attached to the surface of the film and release, rather than being entrapped deeper in the polymer matrix [69].

Table 1 also presents the drug release profile discriminators (Q60 and Q240; response variables Y1 and Y2) of RLX-loaded NFs. Statistical analysis of the response variables confirmed that the type of polymer (A) and the NE:polymer ratio (B) had significant impacts on both Q60 and Q240.

Q60 at (p = 0.0097) and Q240 at (p = 0.0024) of RLX-loaded NFs significantly increased by increasing the NE:polymer ratio. As well, it was observed that for all the formulated NFs systems, increasing the volume ratio of NE to polymer solution resulted in a significant faster release from the NFs. This could be attributed to the fact that by increasing the ratio of NE to polymer solution, more drug will be in the NF films and also the polymer concentration will decrease, which results in a thinner shell phase and a faster release of the drug is observed. Yang et al. found that fibers prepared from a high volume ratio of aqueous to organic phase led to a reduction in the effective release lifetime, which may be due to the thinner shell phase of the matrix [68].

Additionally, the polymer type had significant effect on Q60 at (p = 0.0439) and Q240 at (p = 0.0005). As PVA, HPMC, and CS are hydrophilic biodegradable swellable polymers [70,71], when NFs are introduced in aqueous media, the possible mechanism of drug release is mostly through solvent penetration into the polymer matrix causing burst release, degradation (relaxation of the network) that creates more free volume for drug dissolution and drug diffusion to the surrounding medium. [72]. The free volume of a polymer matrix is dependent on its composition. Homogenous polymer solution exhibits increased free volume than heterogenous ones (i.e., polymeric mixture of PVA and HPMC or CS). This theory is based on the fact that there is an increased difference between ambient temperature and glass transition temperature in homogenous polymer solution. Consequently, drug diffusion and mobility are increased [73]. Additionally, as the viscosity of the polymer increases, the molecular mobility decreases and the diffusion of the drug through the NFs film decreases [72]. PVA has a viscosity of 55–65 cp, HPMC K100 LV has a viscosity of 100 cp, and CS has a viscosity of 800–2000 cp, therefore, PVA-loaded NFs show a faster release profile than PVA/HPMC and PVA/CS NFs due to higher drug mobility and diffusion through the NFs. PVA/HPMC-loaded NFs also showed faster release profiles than PVA/CS NFs. On top of that, the in vitro performance at PH 6.8 decreased CS solubility and the release of the drug decreased due to prolonged-diffusional release. That explains why the release of PVA-loaded NFs and PVA/HPMC-loaded NFs reported faster release than CS-loaded NFs at all NE-to-polymer ratios 1:9, 2:8 and 4:6.

Comparable results were introduced by O'Donnell et al. who found that control (PVA) films had the highest release, followed by other cellulose blended NF films due to the added resistance to the matrix [68].

3.3.4. Bioadhesion Potential of RLX-Loaded NFs

Mucoadhesion strength of NFs films were measured by the mucoadhesion time that NFs films took to move from the top to the bottom of the agar/mucin gel plate. ANOVA results pointed out that NE:polymer ratio had significant effect on mucoadhesion time (Y4) with respective p-values of ($p = 0.0001$). An inverse correlation was established between NE:polymer ratio and mucoadhesion time, as the NE:polymer volume ratio decrease, (polymer concentration increase), a significant increase in the mucoadhesion time of NFs films was observed. Therefore, for the NFs formulae of high polymer concentrations, E1, E2, E4, E5, E7, and E8, the NFs mucoadhesion lasted till the end of the experiment (24 h). These results are on good agreement with Mishra et al. who developed a mucoadhesive buccal patches using 10% w/v PVA and 6% w/v HPMC and reported a mucoadhesion residence time more than 18 h [74].

On the other hand, for NFs E3, E6, and E9, the volume ratio of polymer:NE is 6:4, which has a lower polymer concentration than other NFs, consequently the mucoadhesion time of E3, E6, and E9 were 0.38, 0.5, and 0.42 h, respectively.

The concentration of the polymer is the mainstay factor affecting the mucoadhesion. It affects the development of a strong adhesive force with the mucus and it can be explained by the polymer chain length available for penetration into the mucus layer. When the concentration of the polymer is too low, the number of penetrating polymer chains per unit volume of the mucus is small, and the interaction between the polymer and mucus is unstable [75]. As the concentration of the polymer increase, it would result in a longer penetration chain length and better adhesion.

A direct correlation between the type of polymer and mucoadhesion time was observed. The mucoadhesion time of NFs E6 > E9 > E3. CS NFs (E6) has –OH and –NH2 groups leading to the formation of both hydrogen and covalent bonding. In addition, due to its positive charge, ionic interaction occurred with the negative charge of the sialic acid residues of mucus as well as epithelial surfaces [76]. HPMC NFs (E9) is proposed to form hydrogen bonding with the mucin present in the mucosal layer [77]. Nafee et al. found that mucoadhesive force of polyacrylic acid derivatives (PAA) showed the highest mucoadhesion, prolonged residence time. CS ensured promising mucoadhesive characteristics, whilst HPMC and exhibited weaker mucoadhesion [78].

3.4. Elucidation of Optimum RLX-NFs Films

Statistical optimization of the results was performed setting the optimization criteria for RLX-loaded NFs to minimize fiber size and maximize Q60, Q240, and mucoadhesion time. Formula E2 was chosen with the highest desirability value of (0.859). Hence, it was promoted for further investigations.

3.5. Characterization of the Optimum RLX-Loaded NFs

3.5.1. HRTEM of the Selected RLX-Loaded NFs Film

The morphology of the NFs E2 was evaluated by high resolution TEM (HRTEM; JEM-2010, JEOL) with an accelerating voltage of 200 Kv. HRTEM was operated at 15 kV, and the fiber sample were prepared by making a suspension of the NFs in ethyl alcohol and directly depositing the as-spun ultrafine fibers suspension onto copper grids.

HRTEM observation of the selected optimized NFs formula E2 was conducted to obtain an evidence that RLX-NE was indeed encapsulated within the shell material. It was observed that RLX-NE was wrapped into the center of the fibers (Figure 5). The sharp boundaries in the HRTEM images are indicative of the difference of electron transmission ability between the core and sheath materials [79]. As the NE emulsion exits the capillary of the electrospinner, stretching of the polymer and bending with the dispersed phase occurred, causing concentration of NE in the middle during incipient fiber elongation. During electrospinning, the emulsion droplets achieve their enrichment in the axial region and stretching into an oviform in the direction of the fiber path. The inward movement of emulsion droplets is due to the rapid elongation and evaporation of the solvents during the electrospinning. As a result of water evaporation, a rapid increase of the viscosity of the polymer solution was achieved. This viscosity gradient between the outer and the inner layer resulted in the inward movement of the NE droplets and, furthermore, the evaporation of water may cause their coalescence to some extent [80].

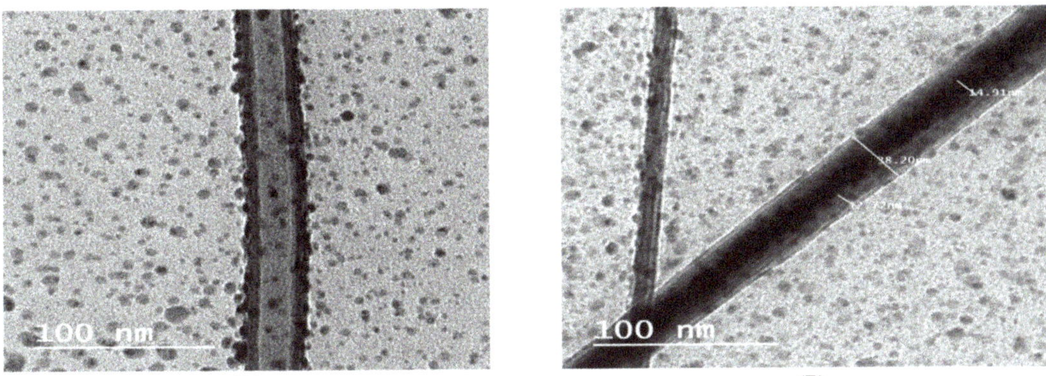

(A) **(B)**

Figure 5. High resolution transmission electron micrographs (HRTEM) of the optimized formula (E2) showing; (**A**) core-sheath structure where a clear visual of the incorporated nanoemulsion is observed. Additionally, demonstrating the adsorption of some of the NE on the fiber surface. (**B**) NFs with clearly discriminated core and sheath with defined sizes.

3.5.2. Ex Vivo Drug Permeation Studies

The ex vivo drug permeation from the selected RLX-loaded NFs formula (E2) was studied and compared to its permeation from RLX aqueous dispersion and presented in Figure 6A. The permeation parameters studied were the cumulative amount permeated per unit area (Q6h), drug flux (Jss), and enhancement ratio (ER). It was clear that all the permeation parameters of RLX-NFs (E2) showed superiority over RLX aqueous dispersion. One way ANOVA revealed that NFs had significantly (p = 0.006) higher Q6h (33.06 ± 2.75 μg/cm^2), significantly (p = 0.013) higher Jss (5.51 ± 0.45 μg/cm^2/h) and significantly (p = 0.000) higher ER (2.12 ± 0.10) than RLX aqueous dispersion; Q6h (15.12 ± 1.12 μg/cm^2), and Jss (2.52 ± 0.18 μg/cm^2/h).

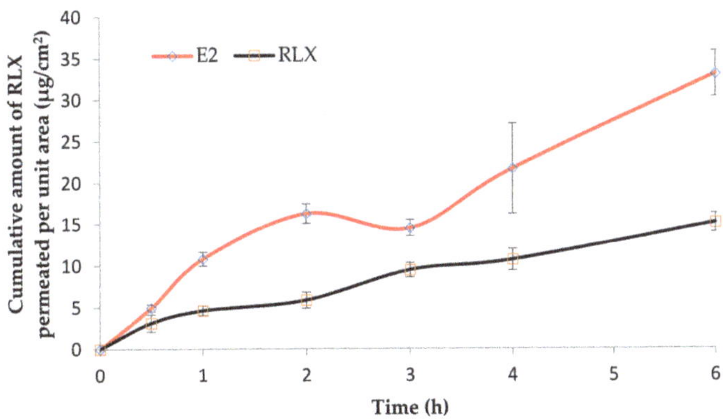

(A)

Figure 6. *Cont.*

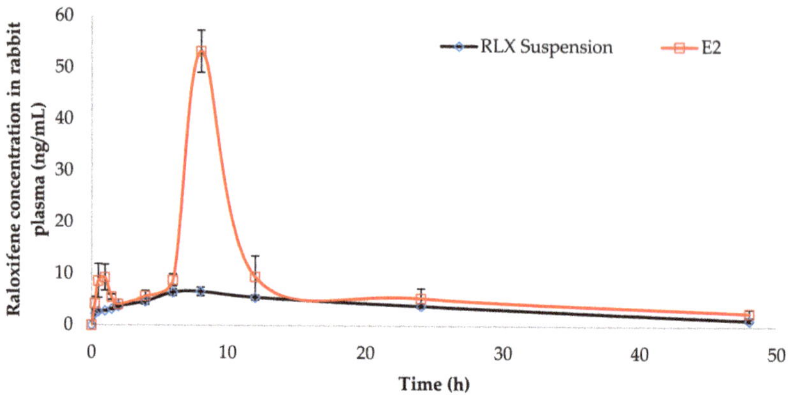

(B)

Figure 6. (**A**) Ex vivo drug permeation from RLX-loaded NFs formula (E2), in comparison to RLX aqueous dispersion, in phosphate buffer saline (pH 6.8 of 0.1% tween 80) at 37 ± 0.5 °C (mean ± SD; $n = 3$). (**B**) RLX plasma concentration-time profiles following the buccal application of RLX-loaded nanofibers formula (E2) and the oral administration of RLX aqueous dispersion in rabbits (mean ± S.D., $n = 6$).

This could be explained by the extremely large surface area, unique surface topology, and porosity of the electrospun NFs, which significantly intensifies the intimate contact between the NF-based formula and mucosal surface, as well as the high drug concentration at the site of administration [17]. The influence of the NE system on the permeation of the loaded drug should be also kept into consideration; the NE's nano-size and penetration-enhancing capability related to its component (surfactant; Cremophor RH 40® and Transcutol HP®) greatly support and augment the drug permeation through the buccal mucosa [81,82].

3.6. *In Vivo Estimation of RLX Pharmacokinetics in Rabbits*

The RLX plasma concentration-time profiles following the buccal application of RLX-loaded NF films (E2; test treatment) and the oral administration of RLX aqueous suspension (reference treatment) in rabbits are displayed in Figure 6B and the observed RLX pharmacokinetic parameters were summarized in Table 2. Regular checking on the formulation in the buccal cavity of the rabbits was done to monitor the degradation progress. NFs kept their integrity up to 8 h with a decrease in the dimension as time elapsed.

The plasma concentration-time profiles of the test NFs (E2) formula showed a small peak after one hour which could be due to the burst release and rapid permeation by passive diffusion of the adsorbed, unloaded NE present on the surface of NFs [83]. Such results agree with the in vitro release results discussed earlier.

Statistical analysis of the pharmacokinetic parameters using multivariate ANOVA proved significant difference ($p = 0.0001$) between the C_{max} values of RLX following the oral administration of the reference treatment (7.04 ± 0.26 ng/mL) and the NFs formula E2 (53.18 ± 4.56 ng/mL).

The other hand, non-parametric Wilcoxon signed rank test revealed that both treatments had a non-significantly different median T_{max} of 8 h at ($p = 1.00$), which means that we retain the null hypothesis. These findings were in line with Shah et al. who reported T_{max} of RLX oral suspension to be 8 h [23]. This might be attributed to the gradual degradation of the multilayers of the NFs in order to release RLX-NE to be permeated through the buccal mucosa and the presence of the solid dispersed multilayered core-sheath NFs (Figures S1 and S2 in Supplementary Material). After layer-by-layer degradation of the NFs, release of NE occurred followed by passive diffusion of the NE.

Table 2. The pharmacokinetic parameters and relative bioavailability of RLX following buccal application of RLX-loaded NF (E2) films and oral administration of RLX aqueous dispersion in rabbits (mean ± S.D., n = 6)

Treatments	Oral RLX Aqueous Dispersion	RLX-Loaded NF (E2) Buccal Film
C_{max} (ng/mL)	7.04 ± 0.26	53.18 ± 4.56
T_{max} (h) *	8	8
$MRT_{0-\infty}$ (h)	17.81 ± 0.80	15.11 ± 1.70
$t_{1/2}$ (h)	16.98 ± 3.31	24.93 ± 11.51
AUC_{0-48} (ng·h/mL)	177.92 ± 11.51	408.74 ± 59.21
$AUC_{0-\infty}$ (ng·h/mL)	209.61 ± 22.20	509.84 ± 71.72
% relative bioavailability based on AUC $_{(0-48)}$		229.73
% relative bioavailability based on AUC $_{(0-\infty)}$		243.22

* Median (range).

Additionally, an insignificant difference was observed between the mean residence time ($MRT_{0-\infty}$) at (p = 0.580) and the elimination half-life ($T_{1/2}$) at (p = 0.279) of RLX following the oral administration of the reference treatment (17.81 h ± 0.80 and 16.98 ± 3.31 h, respectively) and the buccal application of the test treatment; E2 (15.11 ± 1.70 h and 24.93 ± 11.51 h, respectively).

Taking into consideration the low oral bioavailability of RLX (2%) a significant improvement in the bioavailability of RLX was noted following the buccal application of the test treatment. The relative bioavailability of RLX was 2.29-fold higher (with respect to the mean AUC_{0-48} h values, (p = 0.001)) and 2.43-fold higher (with respect to the mean $AUC_{0-\infty}$ values, (p = 0.002)) (Table 2). The improved RLX bioavailability when formulated into NF buccal films could be attributed to several factors including, (i) enhancing the solubility of RLX via its formulation in the form of solid dispersed core-sheath electrospun NFs, (ii) the extremely large surface of NFs which promotes the intimate drug contact with the buccal mucosa promoting high drug concentration at the site of administration, (iii) the use Cremophor RH 40 and Transcutol HP in the formulation of the NE, which increases the permeation through the buccal mucosa, and finally (iv) the avoidance of the first pass metabolism due to buccal administration of the NFs films.

For the purpose of enhancing RLX bioavailability, Wang et al. [84] reported the formulation of RLX inclusion complex using sulfo-butyl-ether-β-cyclodextrin followed by conjugation to chitosan to form RLX-loaded cyclodextrin/chitosan nanoparticles (RLX-ccNPs). The enhancement of oral bioavailability of RLX-ccNPs relative to RLX suspensions was about 3.60-fold. Saini et al. [26] developed intranasal RLX-loaded chitosan nanoparticles, which enhance RLX bioavailability by ~1.7-fold. Murthy et al. [22] developed soy lecithin-chitosan complex nanoparticles (LCNPs) for oral delivery of RLX. After loading RLX into LCNPs, its bioavailability increased significantly (p < 0.05) by over ~4.2-fold compared to the free drug suspension. The observed 2.43-fold higher $AUC_{0-\infty}$ values of the developed NFs demonstrated how this technique established itself among other nanocarrier systems. Additionally, NFs films present the advantages of solid dosage forms over nanocarrier dispersion systems: Stability, easy handling, increased patient compliance, and accurate dosing.

4. Conclusions

Solid dispersed multilayered core-sheath RLX-loaded NFs films were successfully developed utilizing the nanoemulsion electrospinning technique without the use of organic solvents, which paves the road to "green" drug delivery systems. This would vastly extend their potential in therapy. According to the D-optimal response surface methodology, the investigated variables included the NE:polymer ratio and polymer type. Considering the Q60, Q240, fiber size, and mucoadhesion time, the highest desirability values were achieved with E2, comprising RLX-NE and NE:polymer (PVA) at a ratio of 2:8. The superiority of E2 over RLX aqueous dispersion was revealed with respect to the ex vivo drug permeation parameters, Q6h, Jss, and ER and improved bioavailability of RLX by ≈2.29-fold. RLX-loaded NFs (E2) would be expected to rapidly initiate and maintain the therapeutic response of RLX for a sufficient period, thus enhancing RLX bioavailability. Further clinical studies can be conducted to confirm these findings.

Supplementary Materials: The following are available online at https://www.mdpi.com/article/10.3390/pharmaceutics13040474/s1, Figure S1: Nanofiber film (E2) showing multilayered layout, Figure S2: A photo of nanofibers film E2, Table S1: Non-parametric test summary, Table S2: Related-samples Wilcoxon signed rank test summary, Table S3: Between-subjects factors, Table S4: Multivariate tests, Table S5: Tests of between-subjects effects.

Author Contributions: Data curation: E.A.-E., M.M.E.-A.; formal analysis: E.A.-E., M.M.E.-A.; investigation: E.A.-E., S.N.E.-H.; methodology: E.A.-E., S.A.S.; statistical analysis: S.N.E.-H.; project administration: S.S., R.H.F.; resources: E.A.-E., S.N.E.-H.; validation: R.H.F.; visualization: S.S.; writing—original draft: E.A.-E.; writing—review and editing: S.N.E.-H., R.H.F.; final approval: E.A.-E., M.M.E.-A., S.N.E.-H., R.H.F., S.S., S.A.S. All authors have read and agreed to the published version of the manuscript.

Funding: The APC was partially funded by Faculty of Pharmacy, Cairo University, Cairo, 11562, Egypt.

Institutional Review Board Statement: The animal procedures were conducted at the animal house (Faculty of Pharmacy, Cairo University, Egypt) and approved by the institutional review board, Research Ethics Committee Faculty of Pharmacy, Cairo University (approval no. PI 2762) according to the guidelines of Council for international organizations of Medical Sciences (CIOMS) and the international Council for Laboratory Animal Science (ICLAS).

Informed Consent Statement: Not applicable.

Data Availability Statement: All data generated or analyzed during this study are included in this published article (and its Supplementary material file).

Conflicts of Interest: The authors declare no conflict of interest.

References

1. Kannel, W.B.; Hjortland, M.C.; McNamara, P.M.; Gordon, T. Menopause and risk of cardiovascular disease. The framingham study. *Obstet. Gynecol. Surv.* **1977**, *32*, 239–242. [CrossRef]
2. Parrish, H.M.; Carr, C.A.; Hall, D.G.; King, T.M. Time interval from castration in premenopausal women to development of excessive coronary atherosclerosis. *Am. J. Obstet. Gynecol.* **1967**, *99*, 155–162. [CrossRef]
3. Aitken, J.M.; Hart, D.M.; Anderson, J.B.; Lindsay, R.; Smith, D.A.; Speirs, C.F. Osteoporosis after Oophorectomy for Non-malignant Disease in Premenopausal Women. *BMJ* **1973**, *2*, 325–328. [CrossRef] [PubMed]
4. Richelson, L.S.; Wahner, H.W.; Melton, L., III; Riggs, B.L. Relative contributions of aging and estrogen deficiency to post-menopausal bone loss. *N. Engl. J. Med.* **1984**, *311*, 1273–1275. [CrossRef] [PubMed]
5. Delmas, P.D.; Bjarnason, N.H.; Mitlak, B.H.; Ravoux, A.-C.; Shah, A.S.; Huster, W.J.; Draper, M.; Christiansen, C. Effects of Raloxifene on Bone Mineral Density, Serum Cholesterol Concentrations, and Uterine Endometrium in Postmenopausal Women. *N. Engl. J. Med.* **1997**, *337*, 1641–1647. [CrossRef]
6. Walsh, B.W.; Kuller, L.H.; Wild, R.A.; Paul, S.; Farmer, M.; Lawrence, J.B.; Shah, A.S.; Anderson, P.W. Effects of raloxifene on serum lipids and coagulation factors in healthy postmenopausal women. *JAMA* **1998**, *279*, 1445–1451. [CrossRef]
7. Jordan, V.; Glusman, J.; Eckert, S.; Lippman, M.; Powles, T.; Costa, A.; Morrow, M.; Norton, L. Incident primary breast cancers are reduced by raloxifene: Integrated data from multicenter, double-blind, randomized trials in~ 12,000 postmenopausal women. *Proc. Am. Soc. Clin. Oncol.* **1998**, *17*, 122a.
8. Cummings, S.R.; Eckert, S.; Krueger, K.A.; Grady, D.; Powles, T.J.; Cauley, J.A.; Norton, L.; Nickelsen, T.; Bjarnason, N.H.; Morrow, M.; et al. The Effect of Raloxifene on Risk of Breast Cancer in Postmenopausal Women: Results from the MORE Randomized Trial. *Obstet. Gynecol. Surv.* **2000**, *55*, 100. [CrossRef]
9. Tu, Z.; Li, H.; Ma, Y.; Tang, B.; Tian, J.; Akers, W.; Achilefu, S.; Gu, Y. The enhanced antiproliferative response to combined treatment of trichostatin A with raloxifene in MCF-7 breast cancer cells and its relevance to estrogen receptor β expression. *Mol. Cell. Biochem.* **2012**, *366*, 111–122. [CrossRef]
10. Hochner-Celnikier, D. Pharmacokinetics of raloxifene and its clinical application. *Eur. J. Obstet. Gynecol. Reprod. Biol.* **1999**, *85*, 23–29. [CrossRef]

11. Vasconcelos, T.; Sarmento, B.; Costa, P. Solid dispersions as strategy to improve oral bioavailability of poor water soluble drugs. *Drug Discov. Today* **2007**, *12*, 1068–1075. [CrossRef]
12. Lam, P.-L.; Wong, W.-Y.; Bian, Z.; Chui, C.-H.; Gambari, R. Recent advances in green nanoparticulate systems for drug delivery: Efficient delivery and safety concern. *Nanomedicine* **2017**, *12*, 357–385. [CrossRef]
13. Korany, M.A.; Mahgoub, H.; Haggag, R.S.; Ragab, M.A.A.; Elmallah, O.A. Green chemistry: Analytical and chromatography. *J. Liq. Chromatogr. Relat. Technol.* **2017**, *40*, 839–852. [CrossRef]
14. Shojaei, A.H.; Chang, R.K.; Guo, X.; Burnside, B.A.; Couch, R.A. Systemic drug delivery via the buccal mucosal route. *Pharm. Technol.* **2001**, *25*, 70–81.
15. Modgill, V.; Garg, T.; Goyal, A.K.; Rath, G. Permeability study of ciprofloxacin from ultra-thin nanofibrous film through various mucosal membranes. *Artif. Cells Nanomed. Biotechnol.* **2014**, *44*, 122–127. [CrossRef]
16. Weng, L.; Xie, J. Smart electrospun nanofibers for controlled drug release: Recent advances and new perspectives. *Curr. Pharm. Des.* **2015**, *21*, 1944–1959. [CrossRef]
17. Huang, Z.-M.; Zhang, Y.-Z.; Kotaki, M.; Ramakrishna, S. A review on polymer nanofibers by electrospinning and their applications in nanocomposites. *Compos. Sci. Technol.* **2003**, *63*, 2223–2253. [CrossRef]
18. Mašek, J.; Mašková, E.; Lubasová, D.; Špánek, R.; Raška, M.; Turánek, J. Nanofibers in Mucosal Drug and Vaccine Delivery. In *Nanofibers-from Preparation to Applications, Simona Clichici, Adriana Filip, Gutavo M. do Nascimento*; IntechOpen: London, UK, 2018; pp. 171–198.
19. Nikmaram, N.; Roohinejad, S.; Hashemi, S.; Koubaa, M.; Barba, F.J.; Abbaspourrad, A.; Greiner, R. Emulsion-based systems for fabrication of electrospun nanofibers: Food, pharmaceutical and biomedical applications. *RSC Adv.* **2017**, *7*, 28951–28964. [CrossRef]
20. Qi, H.; Hu, P.; Xu, A.J.; Wang, A. Encapsulation of Drug Reservoirs in Fibers by Emulsion Electrospinning: Morphology Characterization and Preliminary Release Assessment. *Biomacromolecules* **2006**, *7*, 2327–2330. [CrossRef]
21. Tian, L.; Prabhakaran, M.P.; Ding, X.; Ramakrishna, S. Biocompatibility evaluation of emulsion electrospun nanofibers using osteoblasts for bone tissue engineering. *J. Biomater. Sci. Polym. Ed.* **2013**, *24*, 1952–1968. [CrossRef]
22. Murthy, A.; Ravi, P.R.; Kathuria, H.; Vats, R. Self-assembled lecithin-chitosan nanoparticles improve the oral bioavailability and alter the pharmacokinetics of raloxifene. *Int. J. Pharm.* **2020**, *588*, 119731. [CrossRef] [PubMed]
23. Shah, N.V.; Seth, A.K.; Balaraman, R.; Aundhia, C.J.; Maheshwari, R.A.; Parmar, G.R. Nanostructured lipid carriers for oral bioavailability enhancement of raloxifene: Design and in vivo study. *J. Adv. Res.* **2016**, *7*, 423–434. [CrossRef]
24. Ağardan, N.B.M.; DeĞim, Z.; Yılmaz, Şükran; Altıntaş, L.; Topal, T. The Effectiveness of Raloxifene-Loaded Liposomes and Cochleates in Breast Cancer Therapy. *AAPS PharmSciTech* **2015**, *17*, 968–977. [CrossRef] [PubMed]
25. Pandya, B.R.; Chotai, N.P.; Suthar, R.M.; Patel, H.K. Formulation, In-vitro Evaluation and Optimization of Nanoemulsion of Raloxifene Hydrochloride. *Int. J. Pharm. Sci. Res.* **2017**, *9*, 38–48.
26. Saini, D.; Fazil, M.; Ali, M.M.; Baboota, S.; Ali, J. Formulation, development and optimization of raloxifene-loaded chitosan nanoparticles for treatment of osteoporosis. *Drug Deliv.* **2014**, *22*, 1–14. [CrossRef]
27. Abd-Elrasheed, E.; El-Helaly, S.N.; El-Ashmoony, M.M.; Salah, S. Brain Targeted Intranasal Zaleplon Nano-emulsion: In-Vitro Characterization and Assessment of Gamma Aminobutyric Acid Levels in Rabbits' Brain and Plasma at Low and High Doses. *Curr. Drug Deliv.* **2018**, *15*, 898–906. [CrossRef] [PubMed]
28. Liu, D.-Q.; Cheng, Z.-Q.; Feng, Q.-J.; Li, H.-J.; Ye, S.-F.; Teng, B. Polycaprolactone nanofibres loaded with 20 (S)-protopanaxadiol for in vitro and in vivo anti-tumour activity study. *R. Soc. Open Sci.* **2018**, *5*, 180137. [CrossRef]
29. Abd-Elbary, A.; Makky, A.M.; Tadros, M.I.; Alaa-Eldin, A.A. Laminated sponges as challenging solid hydrophilic matrices for the buccal delivery of carvedilol microemulsion systems: Development and proof of concept via mucoadhesion and pharmacokinetic assessments in healthy human volunteers. *Eur. J. Pharm. Sci.* **2016**, *82*, 31–44. [CrossRef]
30. Yeap, S.P.; Lim, J.; Ngang, H.P.; Ooi, B.S.; Ahmad, A.L. Role of Particle–Particle Interaction Towards Effective Interpretation of Z-Average and Particle Size Distributions from Dynamic Light Scattering (DLS) Analysis. *J. Nanosci. Nanotechnol.* **2018**, *18*, 6957–6964. [CrossRef]
31. Ammar, H.O.; Mohamed, M.I.; Tadros, M.I.; Fouly, A.A. Transdermal Delivery of Ondansetron Hydrochloride via Bilosomal Systems: In Vitro, Ex Vivo, and In Vivo Characterization Studies. *AAPS PharmSciTech* **2018**, *19*, 2276–2287. [CrossRef] [PubMed]
32. Elnaggar, Y.S.R.; Abdallah, O.Y.; Gohar, E.Y.; Elsheikh, M.A. Nanoemulsion liquid preconcentrates for raloxifene hydrochloride: Optimization and in vivo appraisal. *Int. J. Nanomed.* **2012**, *7*, 3787–3802. [CrossRef]
33. Nakamura, F.; Ohta, R.; Machida, Y.; Nagai, T. In vitro and in vivo nasal mucoadhesion of some water-soluble polymers. *Int. J. Pharm.* **1996**, *134*, 173–181. [CrossRef]
34. Ahmed, T.A.; Bawazir, A.O.; Alharbi, W.S.; Safo, M.K. Enhancement of Simvastatin ex vivo Permeation from Mucoadhesive Buccal Films Loaded with Dual Drug Release Carriers. *Int. J. Nanomed.* **2020**, *15*, 4001–4020. [CrossRef] [PubMed]
35. Tayel, S.A.; Soliman, I.I.; Louis, D. Formulation of Ketotifen Fumarate Fast-Melt Granulation Sublingual Tablet. *AAPS PharmSciTech* **2010**, *11*, 679–685. [CrossRef] [PubMed]
36. Tawfik, M.A.; Tadros, M.I.; Mohamed, M.I.; El-Helaly, S.N. Low-Frequency versus High-Frequency Ultrasound-Mediated Transdermal Delivery of Agomelatine-Loaded Invasomes: Development, Optimization and in-vivo Pharmacokinetic Assessment. *Int. J. Nanomed.* **2020**, *15*, 8893–8910. [CrossRef] [PubMed]
37. Mohammed, A.; Sayed, M.; Abdelnabi, M. A new protocol of anesthesia using thiopental, diazepam and xylazine in white New Zealand rabbits. *Aust. J. Basic Appl. Sci.* **2011**, *5*, 1296–1300.

38. Shin, J.-W.; Seol, I.-C.; Son, C.-G. Interpretation of animal dose and human equivalent dose for drug development. *J. Korean Med.* **2010**, *31*, 1–7.
39. Said, M.; Elsayed, I.; Aboelwafa, A.A.; Elshafeey, A.H. Transdermal agomelatine microemulsion gel: Pyramidal screening, statistical optimization and in vivo bioavailability. *Drug Deliv.* **2017**, *24*, 1159–1169. [CrossRef]
40. Bhatt, P.; Madhav, S. A detailed review on nanoemulsion drug delivery system. *Int. J. Pharm. Sci. Res.* **2011**, *2*, 2482–2489.
41. Reza, K.H. Nanoemulsion as a novel transdermal drug delivery system. *Int. J. Pharm. Sci. Res.* **2011**, *2*, 1938–1946.
42. Nepal, P.R.; Han, H.-K.; Choi, H.-K. Preparation and in vitro–in vivo evaluation of Witepsol® H35 based self-nanoemulsifying drug delivery systems (SNEDDS) of coenzyme Q10. *Eur. J. Pharm. Sci.* **2010**, *39*, 224–232. [CrossRef]
43. Singh, S.K.; Verma, P.R.P.; Razdan, B. Glibenclamide-loaded self-nanoemulsifying drug delivery system: Development and characterization. *Drug Dev. Ind. Pharm.* **2010**, *36*, 933–945. [CrossRef] [PubMed]
44. Wang, S.-Q.; He, J.-H.; Xu, L. Non-ionic surfactants for enhancing electrospinability and for the preparation of electrospun nanofibers. *Polym. Int.* **2008**, *57*, 1079–1082. [CrossRef]
45. Jia, Y.-T.; Gong, J.; Gu, X.-H.; Kim, H.-Y.; Dong, J.; Shen, X.-Y. Fabrication and characterization of poly (vinyl alcohol)/chitosan blend nanofibers produced by electrospinning method. *Carbohydr. Polym.* **2007**, *67*, 403–409. [CrossRef]
46. Geng, X.; Kwon, O.-H.; Jang, J. Electrospinning of chitosan dissolved in concentrated acetic acid solution. *Biomaterial* **2005**, *26*, 5427–5432. [CrossRef]
47. Sedghi, R.; Shaabani, A.; Mohammadi, Z.; Samadi, F.Y.; Isaei, E. Biocompatible electrospinning chitosan nanofibers: A novel delivery system with superior local cancer therapy. *Carbohydr. Polym.* **2017**, *159*, 1–10. [CrossRef]
48. Desai, K.; Kit, K.; Li, J.; Zivanovic, S. Morphological and Surface Properties of Electrospun Chitosan Nanofibers. *Biomacromolecules* **2008**, *9*, 1000–1006. [CrossRef]
49. Duan, B.; Dong, C.; Yuan, X.; Yao, K. Electrospinning of chitosan solutions in acetic acid with poly (ethylene oxide). *J. Biomater. Sci. Polym. Ed.* **2004**, *15*, 797–811. [CrossRef]
50. Lu, J.-W.; Zhu, Y.-L.; Guo, Z.-X.; Hu, P.; Yu, J. Electrospinning of sodium alginate with poly (ethylene oxide). *Polymer* **2006**, *47*, 8026–8031. [CrossRef]
51. Aydogdu, A.; Sumnu, G.; Sahin, S. A novel electrospun hydroxypropyl methylcellulose/polyethylene oxide blend nanofibers: Morphology and physicochemical properties. *Carbohydr. Polym.* **2018**, *181*, 234–246. [CrossRef]
52. Neo, Y.P.; Ray, S.; Easteal, A.J.; Nikolaidis, M.G.; Quek, S.Y. Influence of solution and processing parameters towards the fabrication of electrospun zein fibers with sub-micron diameter. *J. Food Eng.* **2012**, *109*, 645–651. [CrossRef]
53. Lin, T.; Fang, J.; Wang, H.; Cheng, T.; Wang, X. Using chitosan as a thickener for electrospinning dilute PVA solutions to improve fibre uniformity. *Nanotechnology* **2006**, *17*, 3718–3723. [CrossRef]
54. Ignatova, M.; Starbova, K.; Markova, N.; Manolova, N.; Rashkov, I. Electrospun nano-fibre mats with antibacterial properties from quaternised chitosan and poly (vinyl alcohol). *Carbohydr. Res.* **2006**, *341*, 2098–2107. [CrossRef] [PubMed]
55. Bahrami, S.H.; Nouri, M. Chitosan-poly (vinyl alcohol) blend nanofibers: Morphology, biological and antimicrobial properties. *e-Polymers* **2009**, *9*, 1580–1591.
56. Patil, P.H.; Belgamwar, V.S.; Patil, P.R.; Surana, S.J. Solubility enhancement of raloxifene using inclusion complexes and cogrinding method. *J. Pharm.* **2013**, *2013*. [CrossRef]
57. Yalkowsky, S.H. Solubility and Partitioning V: Dependence of Solubility on Melting Point. *J. Pharm. Sci.* **1981**, *70*, 971–973. [CrossRef]
58. Salah, S.; Mahmoud, A.A.; Kamel, A.O. Etodolac transdermal cubosomes for the treatment of rheumatoid arthritis: Ex vivo permeation and in vivo pharmacokinetic studies. *Drug Deliv.* **2017**, *24*, 846–856. [CrossRef] [PubMed]
59. Ferrero, F.; Periolatto, M. Antimicrobial finish of textiles by chitosan UV-curing. *J. Nanosci. Nanotechnol.* **2012**, *12*, 4803–4810. [CrossRef] [PubMed]
60. Demappa, T.; Ganesha, S.; Divakara, S.; Pattabi, M.; Somashekar, R. Physical and thermal properties of 8MeV electron beam irradiated HPMC polymer films. *Nucl. Instrum. Methods Phys. Res. Sect. B Beam Interact. Mater. At.* **2008**, *266*, 3975–3980.
61. Tadros, M.I.; Fahmy, R.H. Controlled-release triple anti-inflammatory therapy based on novel gastroretentive sponges: Characterization and magnetic resonance imaging in healthy volunteers. *Int. J. Pharm.* **2014**, *472*, 27–39. [CrossRef] [PubMed]
62. Kenechukwu, F.C.; Ofokansi, K.C.; Ezugwu, R.O.; Attama, A.A. Improved dissolution and anti-inflammatory activity of ibuprofen-polyethylene glycol 8000 solid dispersion systems. *Int. J. Pharm. Investig.* **2016**, *6*, 139–147. [CrossRef]
63. Jain, A. Solubilization of indomethacin using hydrotropes for aqueous injection. *Eur. J. Pharm. Biopharm.* **2008**, *68*, 701–714. [CrossRef]
64. Damian, F.; Blaton, N.; Naesens, L.; Balzarini, J.; Kinget, R.; Augustijns, P.; Van den Mooter, G. Physicochemical characterization of solid dispersions of the antiviral agent UC-781 with polyethylene glycol 6000 and Gelucire 44/14. *Eur. J. Pharm. Sci.* **2000**, *10*, 311–322. [CrossRef]
65. Hancock, B.C.; Zografi, G. Characteristics and Significance of the Amorphous State in Pharmaceutical Systems. *J. Pharm. Sci.* **1997**, *86*, 1–12. [CrossRef]
66. Fini, A.; Moyano, J.R.; Ginés, J.M.; Perez-Martinez, J.I.; Rabasco, A.M. Diclofenac salts, II. Solid dispersions in PEG6000 and Gelucire 50/13. *Eur. J. Pharm. Biopharm.* **2005**, *60*, 99–111. [CrossRef] [PubMed]

67. Ghoneim, A.M.; Tadros, M.I.; Alaa-Eldin, A.A. Spray-Dried Silica Xerogel Nanoparticles as a Promising Gastroretentive Carrier System for the Management of Chemotherapy-Induced Nausea and Vomiting. *Int. J. Nanomed.* **2019**, *14*, 9619–9630. [CrossRef] [PubMed]
68. Yang, Y.; Li, X.; Cui, W.; Zhou, S.; Tan, R.; Wang, C. Structural stability and release profiles of proteins from core-shell poly (DL-lactide) ultrafine fibers prepared by emulsion electrospinning. *J. Biomed. Mater. Res. Part A* **2008**, *86*, 374–385. [CrossRef]
69. O'Donnell, K.L.; Oporto-Velásquez, G.S.; Comolli, N. Evaluation of Acetaminophen Release from Biodegradable Poly (Vinyl Alcohol) (PVA) and Nanocellulose Films Using a Multiphase Release Mechanism. *Nanomaterials* **2020**, *10*, 301. [CrossRef] [PubMed]
70. Parameswara, P.; Demappa, T.; Mahadevaiah Prakash, Y.; Somashekarappa, H.; Byrappa, K.; Somashekar, R. Polymeric degradation of water soluble chitosan/HPMC films using WAXS data. *Mater. Res. Innov.* **2012**, *16*, 126–129. [CrossRef]
71. Hassan, C.M.; Trakampan, P.; Peppas, N.A. Water Solubility Characteristics of Poly (vinyl alcohol) and Gels Prepared by Freezing/Thawing Processes. In *Water Soluble Polymers*; Metzler, J.B., Ed.; Springer: Boston, MA, USA, 2005; pp. 31–40.
72. Varma, M.V.S.; Kaushal, A.M.; Garg, A.; Garg, S. Factors Affecting Mechanism and Kinetics of Drug Release from Matrix-Based Oral Controlled Drug Delivery Systems. *Am. J. Drug Deliv.* **2004**, 2, 43–57. [CrossRef]
73. Siepmann, J.; Siegel, R.A.; Rathbone, M.J. *Fundamentals and Applications of Controlled Release Drug Delivery*; Springer: New York, NY, USA, 2012; Volume 3, pp. 33–34.
74. Mishra, S.K.; Garud, N.; Singh, R. Development and evaluation of mucoadhesive buccal patches of flurbiprofen. *Acta Pol. Pharm. Drug Res.* **2011**, *68*, 955–964.
75. Peppas, N.A.; Buri, P.A. Surface, interfacial and molecular aspects of polymer bioadhesion on soft tissues. *J. Control. Release* **1985**, *2*, 257–275. [CrossRef]
76. Ways, T.M.; Lau, W.M.; Khutoryanskiy, V.V. Chitosan and its derivatives for application in mucoadhesive drug delivery systems. *Polymers* **2018**, *10*, 267. [CrossRef]
77. Patel, H.; Srinatha, A.; Sridhar, B.K. External Cross-linked Mucoadhesive Microbeads for Prolonged Drug Release: Development and In vitro Characterization. *Indian J. Pharm. Sci.* **2014**, *76*, 437–444.
78. Nafee, N.A.; Ismail, F.A.; Boraie, N.A.; Mortada, L.M. Mucoadhesive Delivery Systems. I. Evaluation of Mucoadhesive Polymers for Buccal Tablet Formulation. *Drug Dev. Ind. Pharm.* **2004**, *30*, 985–993. [CrossRef]
79. Yu, J.H.; Fridrikh, S.V.; Rutledge, G.C. Production of Submicrometer Diameter Fibers by Two-Fluid Electrospinning. *Adv. Mater.* **2004**, *16*, 1562–1566. [CrossRef]
80. Xu, X.; Zhuang, X.; Chen, X.; Wang, X.; Yang, L.; Jing, X. Preparation of Core-Sheath Composite Nanofibers by Emulsion Electrospinning. *Macromol. Rapid Commun.* **2006**, *27*, 1637–1642. [CrossRef]
81. Helmy, S.A.; El-Bedaiwy, H.M.; El-Masry, S.M. Applying Biopharmaceutical Classification System criteria to predict the potential effect of Cremophor® RH 40 on fexofenadine bioavailability at higher doses. *Ther. Deliv.* **2020**, *11*, 447–464. [CrossRef] [PubMed]
82. Javadzadeh, Y.; Adibkia, K.; Hamishekar, H. Transcutol®(Diethylene Glycol Monoethyl Ether): A Potential Penetration Enhancer. In *Percutaneous Penetration Enhancers Chemical Methods in Penetration Enhancement*; Metzler, J.B., Ed.; Springer: Berlin/Heidelberg, Germany, 2015; pp. 195–205.
83. Bhati, R.; Nagrajan, R.K. A detailed review on oral mucosal drug delivery system. *Int. J. Pharm. Sci. Res.* **2012**, *3*, 659–681.
84. Wang, Z.; Li, Y. Raloxifene/SBE-β-CD Inclusion Complexes Formulated into Nanoparticles with Chitosan to Overcome the Absorption Barrier for Bioavailability Enhancement. *Pharmaceutics* **2018**, *10*, 76. [CrossRef]

pharmaceutics

MDPI

Article

Improved Dissolution Rate and Intestinal Absorption of Fexofenadine Hydrochloride by the Preparation of Solid Dispersions: In Vitro and In Situ Evaluation

Basanth Babu Eedara *,†, **Dinesh Nyavanandi** ‡, **Sagar Narala** ‡, **Prabhakar Reddy Veerareddy** § and **Suresh Bandari** ‡

Department of Pharmaceutics, St. Peter's Institute of Pharmaceutical Sciences, Vidyanagar, Warangal 506001, India; dnyavana@go.olemiss.edu (D.N.); snarala@go.olemiss.edu (S.N.); vpreddyidia@gmail.com (P.R.V.); sbandari@olemiss.edu (S.B.)

* Correspondence: basantheedara@arizona.edu or babubasanth@gmail.com

† Current affiliation: Skaggs Pharmaceutical Sciences Center, The University of Arizona, College of Pharmacy, 1703 E. Mabel St, Skaggs Pharmaceutical Sciences Center, Tucson, AZ 85721-0207, USA.

‡ Current affiliation: Department of Pharmaceutics and Drug Delivery, School of Pharmacy, The University of Mississippi, Oxford, MS 38677, USA.

§ Current affiliation: College of Pharmacy, Palamuru University, Mahabubnagar, Telangana 509001, India.

Abstract: The objective of this study was to enhance dissolution and permeation of a low soluble, absorbable fexofenadine hydrochloride (FFH) by preparing solid dispersions using polyethylene glycol 20,000 (PEG 20,000) and poloxamer 188 as carriers. The phase solubility measurement for the supplied FFH revealed a linear increase in the solubility of fexofenadine with increasing carrier concentration in water (1.45 mg/mL to 11.78 mg/mL with 0% *w/v* to 30% *w/v* PEG 20,000; 1.45 mg/mL to 12.27 mg/mL with 0% *w/v* to 30% *w/v* poloxamer 188). To select the appropriate drug carrier concentration, a series of solid dispersions were prepared in the drug carrier weight ratios of 1:1, 1:2 and 1:4 by fusion method. The solid dispersions composed of drug carrier at 1:4 weight ratio showed highest dissolution with the time required for the release of 50% of the drug <15 min compared to the supplied FFH (>120 min). The intestinal absorption study presented a significant improvement in the absorption of drug from the solid dispersions composed of poloxamer 188 than PEG 20,000. In summary, the solid dispersions of FFH prepared using PEG 20,000 and poloxamer 188 demonstrated improved dissolution and absorption than supplied FFH and could be used to improve the oral bioavailability of fexofenadine.

Keywords: fexofenadine hydrochloride; PEG 20,000; poloxamer188; dissolution; absorption

Citation: Eedara, B.B.; Nyavanandi, D.; Narala, S.; Veerareddy, P.R.; Bandari, S. Improved Dissolution Rate and Intestinal Absorption of Fexofenadine Hydrochloride by the Preparation of Solid Dispersions: In Vitro and In Situ Evaluation. *Pharmaceutics* **2021**, *13*, 310. https://doi.org/:10.3390/pharmaceutics13030310

Academic Editor: Vitaliy Khutoryanskiy

Received: 29 January 2021
Accepted: 23 February 2021
Published: 27 February 2021

1. Introduction

Allergic rhinitis (AR), an immunoglobulin E-mediated inflammation of the nasal mucosa, is estimated to affect approximately 10% to 25% of the worldwide population [1]. The pathophysiology of AR involves the release of several proinflammatory mediators. Histamine is one of the key proinflammatory mediators released by the mast cells, which stimulates H1- receptors and leads to the characteristic symptoms of rhinorrhea, nasal itching, sneezing, and nasal congestion [2]. Oral antihistamines have been the first-line treatment for AR over decades and are effective at treating the symptoms caused by allergen exposure [2].

Fexofenadine hydrochloride (FFH) is a safe, long acting, non-sedating anti-histamine with selective peripheral H1- receptor antagonist activity used in allergic conditions like seasonal allergic rhinitis and chronic idiopathic urticaria [3]. It is a white to off-white crystalline powder and its oral bioavailability in humans is about 33% [4]. The low oral bioavailability of fexofenadine is attributed to the slightly soluble nature in water, low intestinal permeability, and active efflux by intestinal P-glycoprotein (P-gp) [5,6].

Several formulation strategies such as crystal modification (salts, polymorphs, and co-crystals), surfactant systems, particle size reduction, amorphization, cyclodextrin complexation, lipid formulations and pH modification have been followed to enhance the oral bioavailability of poorly absorbed drugs [7–16]. Among these formulations, the solid dispersions are easy to prepare, and extensively used approach to enhance the solubility and dissolution behavior of low soluble drugs as it can produce a solid dosage formulation of a drug in an amorphous or mono-molecularly dispersed state surrounded by an inert carrier [7]. Moreover, solid dispersion approach can also be used to enhance the intestinal permeability of poorly absorbable drugs by using a surface active or lipid carrier [17].

In the literature, several formulation approaches such as disintegrating tablets [18], water-in-oil micro emulsions [19], and phospholipid complexation [20] have been reported to improve the oral bioavailability of fexofenadine. In our previous study, Gelucire 44/14 and d-α-tocopheryl polyethylene glycol 1000 succinate (TPGS) have been used as carriers for the preparation of dispersions and the results showed a significant improvement in the dissolution as well as intestinal permeability of FFH [21]. However, the developed dispersions were sticky due to the semisolid nature of the carriers at room temperature. The current study aimed to develop non-sticky solid dispersions of FFH using polyethylene glycol 20,000 (PEG 20,000) and poloxamer 188 as carriers.

Polyethylene glycol (PEG) is a water-soluble, biocompatible, nonpolar polymer of ethylene oxide widely used in a variety of pharmaceutical formulations. They are available over a wide range of molecular weights from 200 Da to 35,000 Da. High molecular weight PEGs shows greater amphiphilic character [22]. Previous reports demonstrated the P-gp inhibitory activity of PEG and increased the absorption of P-gp substrate drugs [23,24]. Shen et al. demonstrated that the intestinal absorption of quinidine was significantly improved in the presence of PEG 20,000 by inhibiting P-gp activity [25].

Poloxamers are nonionic amphiphilic triblock copolymers composed of a central hydrophobic poly(propylene oxide) (PPO) chain linked with two hydrophilic poly(ethylene oxide) (PEO) chains [26]. Poloxamers have been used as a surfactant, solubilizing agent, wetting agent, and stabilizing agent in various pharmaceutical formulations [27]. Several studies reported the use of poloxamer 188 (molecular weight 8400 Da) as a carrier for the preparation of solid dispersions and has been improved the solubility and dissolution of low water-soluble drugs [28,29]. Furthermore, it is known to inhibit P-gp-mediated efflux activity and enhance the permeability of several P-gp substrate drugs across the intestine [30].

The low melting points of PEG 20,000 (60 to 63 °C) and poloxamer 188 (~50 °C) are also beneficial in the preparation of solid dispersions by the fusion method. The solid dispersions were produced by varying the FFH to carrier (PEG 20,000/poloxamer 188) weight ratios (1:1, 1:2 and 1:4) following the fusion method. Furthermore, this study investigated the influence of PEG 20,000 and poloxamer 188 on the solid-state nature of FFH and their ability to improve the drug solubility, dissolution behavior in water and intestinal absorption rate in rats.

2. Materials and Methods

2.1. Materials

Fexofenadine hydrochloride (FFH) was obtained as a generous gift sample from Ami Lifesciences Pvt. Ltd. (Vadodara, India). Polyethylene glycol 20,000 (PEG 20,000) was purchased from E. Merck (India) Ltd. (Mumbai, India). Poloxamer 188 was donated by BASF SE (Ludwigshafen, Germany). All other reagents and chemicals used were of analytical or high-performance liquid chromatography (HPLC) grade.

2.2. Animals

Healthy male Wistar rats (age: 10 weeks, body weight: 250–300 g) were purchased from Mahaveera Enterprises (146-CPCSEA no:199; Hyderabad, India), acclimatized for at least one week in separate cages in a clean room and maintained under the controlled conditions of temperature with free access to food and water. All the experimental protocols including animal care and handling were reviewed and approved (1003/SPIPS/Wgl/IAEC/2011; 30 April 2011) by the Institutional Animal Ethical Committee, St. Peter's Institute of Pharmaceutical Sciences (Hanamkonda, India).

2.3. Drug Quantification by HPLC

FFH was quantified using a validated HPLC method [21]. Briefly, a Shimadzu (Shimadzu Corporation, Kyoto, Japan) HPLC system equipped with LC-10 AT solvent delivery unit, SPD-10 AVP UV-spectrophotometric detector with Lichrospher C18 column (5 μm, 4.6 × 250 mm was used. The system was operated under isocratic flow at 1 mL/min using a mobile phase consisting of a mixture of 0.05M potassium dihydrogen orthophosphate buffer containing 0.5% *v/v* triethylamine (pH 4.2 adjusted with orthophosphoric acid), acetonitrile, and methanol (50:38:12 *v/v/v*). Samples of 20 μL were injected using an auto sampler and analyzed at a wavelength of 220 nm. The calibration curve was constructed over the concentration range of 0.25–16 μg/mL and was linear with a good correlation coefficient ($R^2 > 0.9996$). The limit of detection (LOD) and limit of quantification (LOQ) of FFH were 0.125 μg/mL and 0.25 μg/mL, respectively.

2.4. Phase Solubility Study

Solubility of FFH in aqueous solution containing PEG 20,000 and poloxamer 188 was determined according to the method reported by Higuchi and Connors [31]. An excess amount (300 mg) of FFH was mixed with 10 mL of aqueous solutions containing carrier (PEG 20,000 and poloxamer 188) at 5%, 10%, 15%, 20%, 25%, and 30% *w/v* in screw-capped glass vials. These dispersions were maintained at 37 °C in a thermostatically controlled water bath with continuous shaking for 48 h. The contents of each vial were filtered through 0.45 μm membrane filter and the filtrate obtained was analyzed by HPLC after appropriate dilution using methanol.

The apparent stability constant (K_s), between drug–carrier combination was calculated from the slope of the phase solubility diagrams i.e., drug solubility change with increasing carrier concentration, using the following equation (Equation (1)).

$$K_s = \frac{\text{Slope}}{S_0}(1 - \text{Slope}) \tag{1}$$

where S_0 is the solubility of FFH in water, in the absence of carrier.

The Gibbs-free energy of transfer (ΔG°_{tr}) of FFH from plain water without carrier to the aqueous solution of carrier was calculated using the following equation (Equation (2)).

$$\Delta G^\circ_{tr} = -2.303RT\ \text{Log}\frac{S_S}{S_0} \tag{2}$$

where S_S and S_0 are the molar solubility of FFH in an aqueous solution of the carrier and the plain water without carrier, respectively. R is the general gas constant ($R = 8.31\ J\ K^{-1}\ mol^{-1}$) while T is the temperature in degree Kelvin.

2.5. Preparation of Solid Dispersions and Physical Mixtures

Solid dispersions of FFH were prepared using PEG 20,000 and poloxamer 188 as carriers in 1:1, 1:2 and 1:4 weight ratios by fusion method. Briefly, an accurately weighed amount of carrier was melted in a porcelain dish using a water bath maintained at a temperature of 65 °C. The pre-weighed amount of FFH was then added to the molten carrier and mixed thoroughly using a glass stirrer for 5 min to get a uniform dispersion. The resultant mixture was cooled overnight at room temperature. The hardened solid masses were ground in a glass mortar, passes through 250 μm sieve (mesh size #60), transferred into a screw-capped glass vial and stored at room temperature (22 ± 1 °C) in a desiccator with silica gel (30 ± 2% relative humidity).

Physical mixtures of FFH and carriers (PEG 20,000/poloxamer 188) at 1:1, 1:2 and 1:4 weight ratios were gently mixed in a glass mortar with a spatula for 5 min, transferred into a screw-capped glass vial and stored at room temperature (22 ± 1 °C) in a desiccator with silica gel (30 ± 2% relative humidity).

2.6. Drug Content Estimation

Drug content in the produced solid dispersions and physical mixtures was estimated by dissolving the powder equivalent to 10 mg of FFH in 100 mL of methanol. An aliquot of sample was centrifuged at 10,000 rpm for 15 min. The supernatant samples were filtered through 0.45 μm membrane filter and the filtrates obtained were analyzed by HPLC after appropriate dilution into the validated range with methanol.

2.7. In Vitro Dissolution Studies

The dissolution studies of supplied FFH, solid dispersions and physical mixtures (equivalent to 30 mg of FFH) were conducted using United States Pharmacopeia type II (paddle) dissolution apparatus (Electrolab, Mumbai, India). The dissolution study was conducted using 900 mL of distilled water in a standard, hemispherical bottomed dissolution vessel at the paddle rotation speed of 50 rpm. Aliquots of 5 mL were withdrawn at different time intervals from 15 min to 120 min with fresh medium replacement, immediately filtered using a 0.45 μm filter and quantified by HPLC after appropriate dilution with the mobile phase.

Various parameters such as Q_{15} and Q_{120} i.e., percent drug released at 15 min and 120 min, dissolution efficiency at 15 min and 120 min (DE_{15} and DE_{120}), and time taken for 50% release ($t_{50\%}$) were calculated as reported in our previous publication [21] to assess the dissolution behavior of solid dispersions in relation to the supplied drug and physical mixture.

2.8. Characterization of the Physicochemical Properties

2.8.1. Scanning Electron Microscopy (SEM)

The surface morphologies of supplied FFH, PEG 20,000, poloxamer 188, optimized solid dispersions and their respective physical mixtures were examined using a scanning electron microscope (SEM, Jeol Corporation, Tokyo, Japan) operating at 15 Kv. The powder sample was directly sprinkled onto the double-sided adhesive tape which was affixed to aluminum stub and made electrically conductive by coating them with platinum (approximately 5 nm) in vacuum for 100 s at 30 W. Samples were observed under SEM and micrographs were recorded at different magnifications to study the surface characteristics.

2.8.2. Differential Scanning Calorimetry (DSC)

DSC thermograms of supplied FFH, PEG 20,000, poloxamer 188, optimized solid dispersions, and their respective physical mixtures were obtained using a differential scanning calorimeter unit (TA-60WSI, Shimadzu, Kyoto, Japan) calibrated with sapphire (heat capacity) and indium (heat flow) standards. Powder sample (~4 mg) was weighed in a flat-bottomed aluminum pan (Shimadzu DSC-60, Kyoto, Japan) and press sealed with a standard aluminum lid. All the samples were scanned at a heating rate of 10 °C/min from 30 °C to 300 °C using an empty pan as a reference. Nitrogen gas was purged at a flow rate of 80 mL/min. The heat of fusion of samples was calculated from the peak area of the melting endotherm. 2.8.3. X-ray powder diffraction (XRPD).

X-ray powder diffractograms of supplied FFH, PEG 20,000, poloxamer 188, optimized solid dispersions and their respective physical mixtures were collected using an X'Pert PRO MPD diffractometer (PANalytical, Almelo, The Netherlands) with a copper anode (Cu Kα radiation, λ = 0.15406 nm, 45 kV, 40 mA). Powder samples were gently pressed on an aluminum sample holder and scanned over the range of $3° < 2\theta < 50°$ 2θ at a scanning speed of 5° 2θ/min ambient temperature.

2.8.3. Fourier Transform Infrared Spectroscopy (FTIR)

FTIR spectra of supplied FFH, PEG 20,000, poloxamer 188, optimized solid dispersions and their respective physical mixtures were recorded using FTIR spectrophotometer (Spectrum GX-FT-IR, PerkinElmer Inc., Waltham, MA, USA). Powder samples (~2 mg) were gently mixed with IR grade dry potassium bromide (200 mg) and compressed in a hydraulic press to form disks. The spectrum was recorded as an average of 64 scans in the frequency range of 4000 cm^{-1} to 400 cm^{-1} with the resolution of 4 cm^{-1} at room temperature.

2.9. In Situ Intestinal Absorption Study

The in situ intestinal absorption of fexofenadine from the prepared solid dispersions was investigated following the single pass intestinal absorption (SPIP) study as reported previously [32]. Healthy male Wistar rats (age: 10 weeks, body weight: 250 g to 300 g) were fasted overnight before the experiment with free access to water. After anesthesia via intraperitoneal injection of thiopental sodium (50 mg/kg), rats were placed in a supine position on a heating pad to maintain normal body temperature. The abdomen was cut open along the midline and a 10 cm segment of the proximal rat jejunum was selected. Small "V" shape incisions were made at both ends of the selected segment, rinsed with the physiological saline (37 °C), cannulated with perfusion tubing, ligated with a cotton thread, and connected to a syringe pump (Harvard Apparatus PHD 2000 pump, Holliston, MA, USA). The surgical area was kept hydrated by covering with a cotton gauze wetted with physiological saline. Using syringe pump, the intestinal segments were perfused with blank perfusion buffer (PBS, pH 7.4 at 37 °C) for 15 min at 0.5 mL/min flow rate to clear out the intestinal content. The perfusion buffer containing known concentrations (5 mg/mL) of drug (supplied FFH and solid dispersions dispersed in perfusion buffer) was then perfused at a flow rate of 0.2 mL/min for 20 min to reach an initial steady state according to the previous reports [33]. After equilibration, samples were collected at every 15 min for a 90 min perfusion period and frozen at −20 °C until analysis by HPLC.

At the end of the study, animals were euthanized with a cardiac injection of saturated potassium chloride solution. The intestinal segments were excised, and their length and perimeter were measured. Then, the intestinal segments were cut open, washed with blank perfusate buffer, placed in a formaldehyde saline solution (10% *v/v*) for 24 h and embedded in paraffin wax. The paraffin-embedded sections were stained with hematoxylin/eosin and observed under microscope for histological evaluation.

Various permeability parameters such as effective permeability coefficient in rats ($P_{eff(rat)}$, cm/s), predicted effective permeability coefficient in human ($P_{eff(human)}$, cm/s), absorption rate constant (K_a), and enhancement ratio (ER) were calculated as reported in our previous publication [21] to assess the absorption behavior of fexofenadine from solid dispersions in relation to the supplied drug.

2.10. Statistical Analysis

All the experimental data were expressed as mean ± SD (n = 3). Statistical analyses were performed by one-way analysis of variance (ANOVA) with Student-Newman-Keuls posthoc testing using GraphPad Prism 5 software (GraphPad Software, San Diego, CA, USA) with $p \leq 0.05$ as the level of significance.

3. Results

3.1. Phase Solubility Study

The phase solubility of FFH in water alone and with PEG 20,000/poloxamer 188 (0% to 30% *w/v*) at 37 °C is shown in Table 1. Also, a phase solubility diagram of solubility of FFH (mg/mL) against carrier concentration (% *w/v* of PEG 20,000/poloxamer 188) was plotted (Figure S1, supplementary information) [31]. The solubility of supplied FFH in water at 37 °C was found to be 1.45 ± 0.15 mg/mL. A linear increase in FFH solubility was observed with increasing carrier (PEG 20,000/poloxamer 188) concentration and the solubility curves represented an A_L type with the slope less than unity (Figure S1, supplementary information). The solubility of FFH increased by almost seven times in the presence of PEG 20,000/poloxamer 188 in water at a concentration of 30% *w/v* with an R^2 value > 0.99. This indicated that both the carriers could efficiently increase the solubility of FFH owing to the wettability effect as well as hydrophilic or surface active-like properties [26,28,34–36].

Table 1. Phase solubility and Gibbs-free energy of transfer ($\Delta G°_{tr}$) of fexofenadine hydrochloride in different concentrations of PEG 20,000 and poloxamer 188 water systems at 37 ± 0.5 °C (Mean ± SD, *n* = 3).

Concentration of Carrier (% *w/v*)	PEG 20,000		Poloxamer 188	
	Drug Solubility (mg/mL)	**$\Delta G°_{tr}$ (J/mole)**	**Drug Solubility (mg/mL)**	**$\Delta G°_{tr}$ (J/mole)**
0	1.45 ± 0.15	-	1.45 ± 0.15	-
5	3.14 ± 0.19	−1.99	4.06 ± 0.21	−2.66
10	5.06 ± 0.23	−3.22	5.73 ± 0.24	−3.55
15	7.16 ± 0.18	−4.12	6.90± 0.19	−4.03
20	7.99 ± 0.26	−4.40	9.32 ± 0.14	−4.80
25	10.02 ± 0.17	−4.99	10.65 ± 0.28	−5.14
30	11.78 ± 0.18	−5.40	12.27 ± 0.22	−5.51

Gibbs-free energy is the function of energy formation after the phase transformation of insoluble drug into soluble form. The $\Delta G°_{tr}$ value indicates the favorable or unfavorable nature of carrier for drug solubilization in an aqueous medium. The $\Delta G°_{tr}$ values (Table 1) of FFH were negative and decreased with increasing carrier concentration which indicated the spontaneous nature of the drug solubilization process in the presence of PEG 20,000 and poloxamer 188 [37]. The values of apparent stability constant (K_s) calculated using Equation (1) for FFH-PEG 20,000 and FFH- poloxamer 188 combinations were 145.1 mL/g and 118.5 mL/g, respectively which indicated a strong binding affinity between fexofenadine and the carrier [38].

3.2. Drug Content

The drug content of all prepared physical mixtures (1:4 *w/w*) and the solid dispersions (1:1 *w/w*, 1:2 *w/w* and 1:4 *w/w*) of FFH with PEG 20,000/poloxamer 188 was found to be 95.2 ± 3% to 99.3 ± 2%. All the solid dispersions showed the presence of high drug content which indicated better suitability of the fusion method used for the preparation of solid dispersions with high content uniformity.

3.3. *In Vitro* Dissolution Study

Figure 1 shows the dissolution profiles of supplied FFH, the solid dispersions (SD-$FP20K_{(1:1)}$, SD-$FP20K_{(1:2)}$, SD-$FP20K_{(1:4)}$, SD-$FP188_{(1:1)}$, SD-$FP188_{(1:2)}$, and SD-$FP188_{(1:4)}$) and the physical mixtures (PM-$FP20K_{(1:4)}$ and PM-$FP188_{(1:4)}$) prepared using PEG 20,000 and poloxamer 188 in distilled water as dissolution media. The supplied FFH has the lowest dissolution (38.9 ± 2.1%) compared to the solid dispersions and physical mixtures prepared. It was noted that the solid dispersions showed improved dissolution with increasing amounts of PEG 20,000/poloxamer 188 (with PEG 20,000: SD-$FP20K_{(1:1)}$—70.0 ± 1.9% < SD-$FP20K_{(1:2)}$—78.3 ± 2.6% < SD-$FP20K_{(1:4)}$—80.6 ± 1.8%; with poloxamer 188: SD-$FP188_{(1:1)}$—78.2 ± 2.3% < SD-$FP188_{(1:2)}$—84.7 ± 2.3% < SD-$FP188_{(1:4)}$—94.7 ± 1.9%).

Dissolution parameters, represented as the percentage of the drug released (Q_{15} and Q_{120}), the percentage dissolution efficiency (DE_{15} and DE_{120}), as well as the time required for the release of 50% of the drug ($t_{50\%}$) were calculated and presented in Table 2. Supplied FFH showed poor dissolution with Q_{15} of only 16.9%. The overall amount of the drug dissolved from supplied FFH after 120 min was 38.9%, while DE_{15} and DE_{120} were 8.5% and 24.3%, respectively. All the solid dispersions and physical mixtures prepared using

PEG 20,000/poloxamer 188 showed significant ($p < 0.05$) differences in FFH dissolution parameters (Table 2).

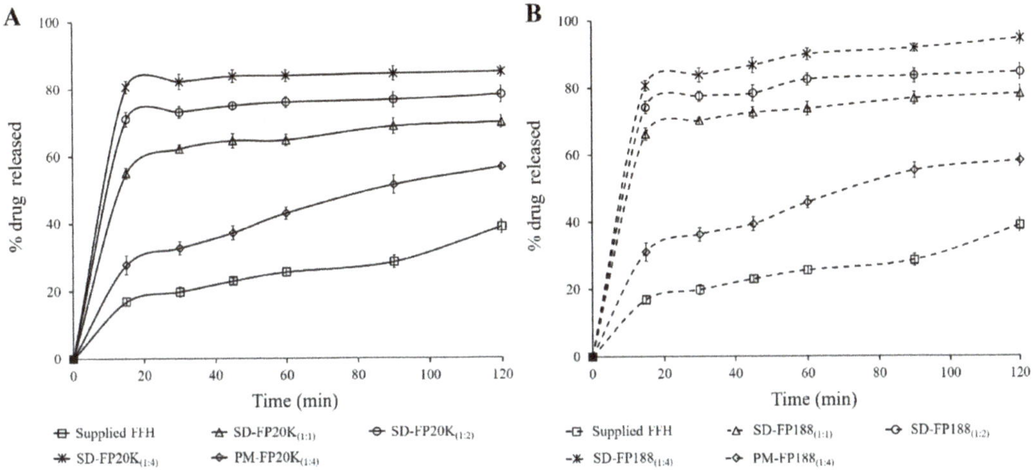

Figure 1. (**A**) In vitro dissolution profiles of supplied FFH, solid dispersions of FFH with PEG 20,000 at 1:1 (SD-FP20K$_{(1:1)}$), 1:2 (SD-FP20K$_{(1:2)}$) and 1:4 (SD-FP20K$_{(1:4)}$) weight ratios, and physical mixture of FFH with PEG 20,000 at 1:4 weight ratio (PM-FP20K$_{(1:4)}$). (**B**) In vitro dissolution profiles of supplied FFH, solid dispersions of FFH with poloxamer 188 at 1:1 (SD-FP188$_{(1:1)}$), 1:2 (SD-FP188$_{(1:2)}$) and 1:4 (SD-FP188$_{(1:4)}$) weight ratios, and physical mixture of FFH with PEG 20,000 at 1:4 weight ratio (PM-FP188$_{(1:4)}$) Each point represents the mean ± SD. (n = 3).

Table 2. Summary of dissolution parameters for supplied fexofenadine hydrochloride, solid dispersions and physical mixtures prepared using PEG 20,000 and poloxamer 188.

Parameters	Q_{15}	Q_{120}	DE_{15}	DE_{120}	$t_{50\%}$
Supplied FFH	16.9 ± 1.2	38.9 ± 2.1	8.5 ± 0.6	24.3 ± 1.4	>120
SD-FP20K$_{(1:1)}$ *	55.1 ± 1.6	70.0 ± 1.9	27.5 ± 0.8	60.9 ± 1.8	>120
SD-FP20K$_{(1:2)}$ *	71.1 ± 2.2	78.3 ± 2.6	35.5 ± 1.1	70.7 ± 1.8	60
SD-FP20K$_{(1:4)}$ *	80.6 ± 1.8	85.1 ± 1.5	40.3 ± 0.9	78.4 ± 1.8	15
PM-FP20K$_{(1:4)}$ *	27.8 ± 2.8	56.7 ± 1.1	13.9 ± 1.4	40.3 ± 2.0	>120
SD-FP188$_{(1:1)}$ *	66.3 ± 1.8	78.2 ± 2.3	27.5 ± 0.9	68.9 ± 1.7	>120
SD-FP188$_{(1:2)}$ *	74.2 ± 1.9	84.7 ± 2.3	35.5 ±1.0	75.7 ± 1.9	60
SD-FP188$_{(1:4)}$ *	80.7 ± 1.6	94.7 ± 1.9	40.3 ± 0.8	83.1 ± 1.6	15
PM-FP188$_{(1:4)}$ *	31.1 ± 2.8	58.2 ± 1.8	13.9 ± 0.6	43.0 ± 1.4	>120

Q_{15} and Q_{120}—percent drug released at 15 and 60 min; DE_{15} and DE_{120}: dissolution efficiency at 15 and 120 min; $t_{50\%}$—time (min) taken to release 50% of the drug, FFH, fexofenadine hydrochloride; SD-FP20K—solid dispersion of FFH with PEG 20,000, PM- FP20K—physical mixture of FFH with PEG 20,000, SD-FP188—solid dispersion of FFH with poloxamer 188, PM-FP188—physical mixture of FFH with poloxamer 188, respectively. * Numbers represent the weight/weight ratios of FFH/carriers.

Generally, the solid dispersions composed of hydrophilic or surface-active carriers improve the solubility and dissolution rate of poorly water-soluble drugs by reduction of the drug particle size, increase in surface area, by changing the crystalline form of the drug to amorphous form, and/or by enhancement of the drug wettability [39]. Poloxamer 188 is an amphiphilic block co-polymer consisting of ethylene oxide and propylene oxide blocks, which can self-aggregates to form micelles in aqueous solution and helps in solubilization of a hydrophobic drug [26]. Polyethylene glycols (PEGs) are polymers of ethylene oxide which improved the solubility and dissolution of FFH in solid dispersion may be due to the formation of a hydrophilic film of PEG around the drug particles which reduces the hydrophobic interaction of drug and improves the wettability [40].

Similarly, the enhanced drug dissolution obtained for physical mixtures (PM-FP20K$_{(1:4)}$ and PM-FP188$_{(1:4)}$) could be attributed to the wetting effect, and solubilization of poloxamer 188 and PEG 20,000. In the physical mixture, drug particles were intermixed or adhered on to the surface of the carrier particles (Figure 2). During dissolution, the carrier particles might have hydrated rapidly into a carrier solution, due to their hydrophilic/surface active nature, which improves the wetting and solubilization of the adjacent drug particles into the medium [34].

Figure 2. Scanning electron microscopic images of (**A**) Supplied FFH, (**B**) Supplied PEG 20,000, (**C**) Physical mixture of FFH with PEG 20,000 at 1:4 ratio (PM-FP20K$_{(1:4)}$), (**D**) Solid dispersion of FFH with PEG 20,000 at 1:4 ratio (SD-FP20K$_{(1:4)}$), (**E**) Supplied poloxamer 188, (**F**) Physical mixture of FFH with poloxamer 188 at 1:4 ratio (PM-FP188$_{(1:4)}$), (**G**) Solid dispersion of FFH with poloxamer 188 at 1:4 ratio (SD-FP188$_{(1:4)}$).

3.4. Solid State Characterization

3.4.1. Scanning Electron Microscopy

Figure 2 shows the representative SEM images of supplied FFH, PEG 20,000, poloxamer 188, the physical mixtures (PM-FP20K$_{(1:4)}$ and PM-FP188$_{(1:4)}$) and the solid ispersions (SD-FP20K$_{(1:4)}$ and SD-FP188$_{(1:4)}$). Supplied FFH was appeared as smooth surfaced crystalline particles of irregular size and shape adsorbed with tiny particles. The PEG 20,000 and poloxamer 188 showed large particles of irregular size and shape with blunt edges. Physical mixtures (PM-FP20K$_{(1:4)}$ and PM-FP188$_{(1:4)}$) contained FFH crystalline particles adsorbed on the surface of the respective carrier particles (PEG 20,000/poloxamer 188). The solid dispersions (SD-FP20K$_{(1:4)}$ and SD-FP188$_{(1:4)}$) appeared to be agglomerated particles of irregular size and shape with rough surface, but the crystalline drug particles or separate phases were indistinguishable.

In our previous study, Gelucire 44/14 and TPGS 1000 have been used as carriers in the preparation of the lipid surfactant-based dispersions of FFH and were waxy in nature due to the semi-solid nature of the lipids [21]. However, the solid dispersions (SD-FP20K$_{(1:4)}$ and SD-FP188$_{(1:4)}$) prepared using PEG 20,000 and poloxamer 188 clearly showed solid particles which can be formulated into a tablet with by adding other excipients.

3.4.2. Differential Scanning Calorimetry (DSC)

Figure 3 shows the DSC thermograms of supplied FFH, PEG 20,000, poloxamer 188, the physical mixtures (PM-FP20K$_{(1:4)}$ and PM-FP188$_{(1:4)}$) and the solid dispersions (SD-FP20K$_{(1:4)}$ and SD-FP188$_{(1:4)}$). The DSC thermogram of supplied FFH showed a sharp endothermic melting peak at 199.3 °C with an enthalpy of fusion (ΔH) of 51.4 J/g indicating crystalline form I of FFH [41]. The DSC thermograms of supplied PEG 20,000 and poloxamer 188 showed an endothermic melting peak at 61.3 °C (ΔH-107.5 J/g) and 51.1 °C (ΔH-73.5 J/g), respectively, which are consistent with the literature [42,43]. The physical mixtures and solid dispersions produced using FFH and PEG 20,000/poloxamer at 1:4 weight ratio showed the endothermic peaks of the drug and respective carrier (PM-FP20K(1:4): FFH—195.5 °C, PEG 20,000—62.0 °C; PM-FP188$_{(1:4)}$: FFH—196.7 °C, poloxamer 188- 55.1 °C; SD-FP20K$_{(1:4)}$: FFH—194.9 °C, PEG 20,000—62.5 °C; and SD-FP188$_{(1:4)}$: FFH—193.3 °C, poloxamer 188—51.9 °C) with a slight shift in peak position. The presence of melting endotherms of FFH and carriers, PEG 20,000 and poloxamer 188 in PM-FP20K$_{(1:4)}$, PM-FP188$_{(1:4)}$, SD-FP20K$_{(1:4)}$, and SD-FP188$_{(1:4)}$) indicates the crystalline nature of drug and carriers in the prepared physical mixtures and solid dispersions. A slight shift in the melting peak position represents the degree of drug excipient miscibility during melting [44].

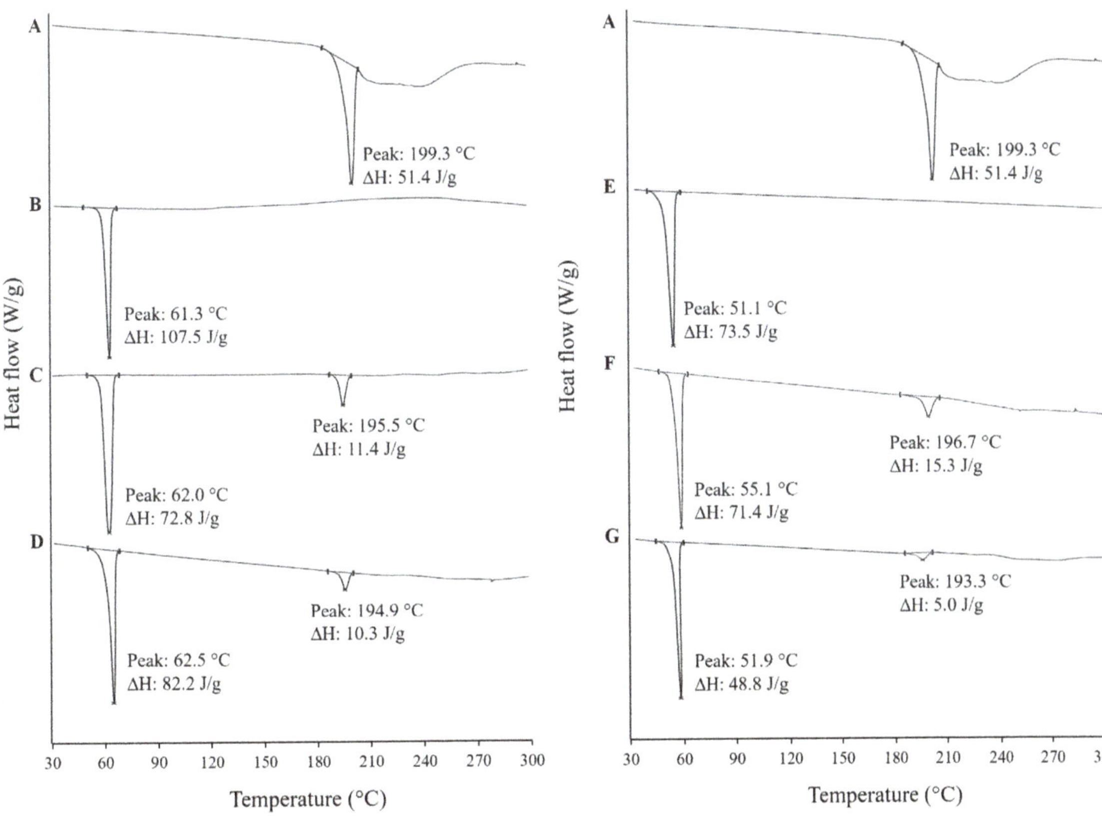

Figure 3. Differential scanning calorimetry (DSC) thermograms of (**A**) Supplied FFH, (**B**) Supplied PEG 20,000, (**C**) Physical mixture of FFH with PEG 20,000 at 1:4 ratio (PM-FP20K$_{(1:4)}$), (**D**) Solid dispersion of FFH with PEG 20,000 at 1:4 ratio (SD-FP20K$_{(1:4)}$), (**E**) Supplied poloxamer 188, (**F**) Physical mixture of FFH with poloxamer 188 at 1:4 ratio (PM-FP188$_{(1:4)}$), (**G**) Solid dispersion of FFH with poloxamer 188 at 1:4 ratio (SD-FP188$_{(1:4)}$).

3.4.3. X-Ray Powder Diffraction (XRPD)

Figure 4 shows the XRPD patterns of supplied FFH, PEG 20,000, poloxamer 188, the physical mixtures (PM-FP20K$_{(1:4)}$ and PM-FP188$_{(1:4)}$) and the solid dispersions (SD-FP20K$_{(1:4)}$ and SD-FP188$_{(1:4)}$) with sharp diffraction peaks indicating their crystalline nature. The diffractogram of supplied FFH showed characteristic peaks [2θ] at 14.1°, 16.0°, 17.9°, 18.3°, 19.5° and 19.9°, matching with the diffraction pattern of crystalline FFH reported by Kumar et al. (2009) [41]. PEG 20,000 and poloxamer 188 showed intense diffraction peaks [2θ] at 19.1° and 23.2°, matching with the diffractograms reported by Windbergs et al. (2009) [43] and Ige, Baria, and Gattani (2013) [45], respectively. The physical mixtures (PM-FP20K$_{(1:4)}$ and PM-FP188$_{(1:4)}$) and solid dispersions (SD-FP20K$_{(1:4)}$ and SD-FP188$_{(1:4)}$) produced a diffraction pattern that exactly matched with the superimposed diffractograms of the supplied FFH and respective carrier (PEG 20,000/poloxamer 188) indicating the crystalline nature of drug in its solid dispersion/physical mixture form. The absence of new peaks in the prepared solid dispersions and physical mixtures ruled out the formation of new crystalline phase [46].

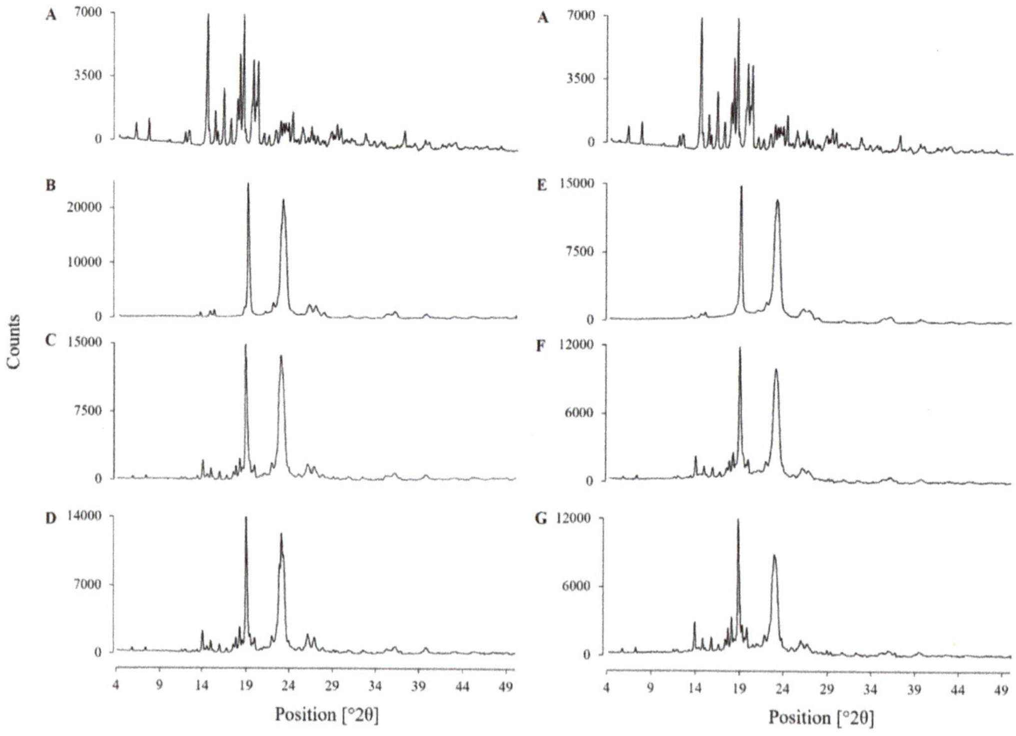

Figure 4. X-ray powder diffractograms (XRPD) of (**A**) Supplied FFH, (**B**) Supplied PEG 20,000, (**C**) Physical mixture of FFH with PEG 20,000 at 1:4 ratio (PM-FP20K$_{(1:4)}$), (**D**) Solid dispersion of FFH with PEG 20,000 at 1:4 ratio (SD-FP20K$_{(1:4)}$), (**E**) Supplied poloxamer 188, (**F**) Physical mixture of FFH with poloxamer 188 at 1:4 ratio (PM-FP188$_{(1:4)}$), (**G**) Solid dispersion of FFH with poloxamer 188 at 1:4 ratio (SD-FP188$_{(1:4)}$).

3.4.4. Fourier Transform Infrared Spectroscopy (FTIR)

Figure 5 shows the FTIR spectra of supplied FFH, PEG 20,000, poloxamer 188, the physical mixtures (PM-FP20K$_{(1:4)}$ and PM-FP188$_{(1:4)}$) and the solid dispersions (SD-FP20K$_{(1:4)}$ and SD-FP188$_{(1:4)}$). The FTIR spectrum of FFH showed characteristic peaks (cm^{-1}) at 3296 (ν O–H alcohol/phenol), 2935 (ν O–H carboxylic acid), 1705 (ν C=O carboxylic acid), 1278 (ν C–N amine), 744 and 702 (δ C–H aromatic); PEG 20,000 at 3300 to 3600 (ν O–H),

2800 to 2900 (ν C–H of OC2H5) and 1000 to 1200 (ν C–O); poloxamer 188 at 3502 (ν O–H), 2885 (ν C–H), and 1112 (ν C–O). All the characteristic peaks of FFH and respective carrier (PEG 20,000/poloxamer) were conserved at the same position in their physical mixture (PM-FP20K$_{(1:4)}$ and PM-FP188$_{(1:4)}$) and the solid dispersions (SD-FP20K$_{(1:4)}$ and SD-FP188$_{(1:4)}$) indicating the absence of interactions between FFH and carrier in the physical mixtures and solid dispersions.

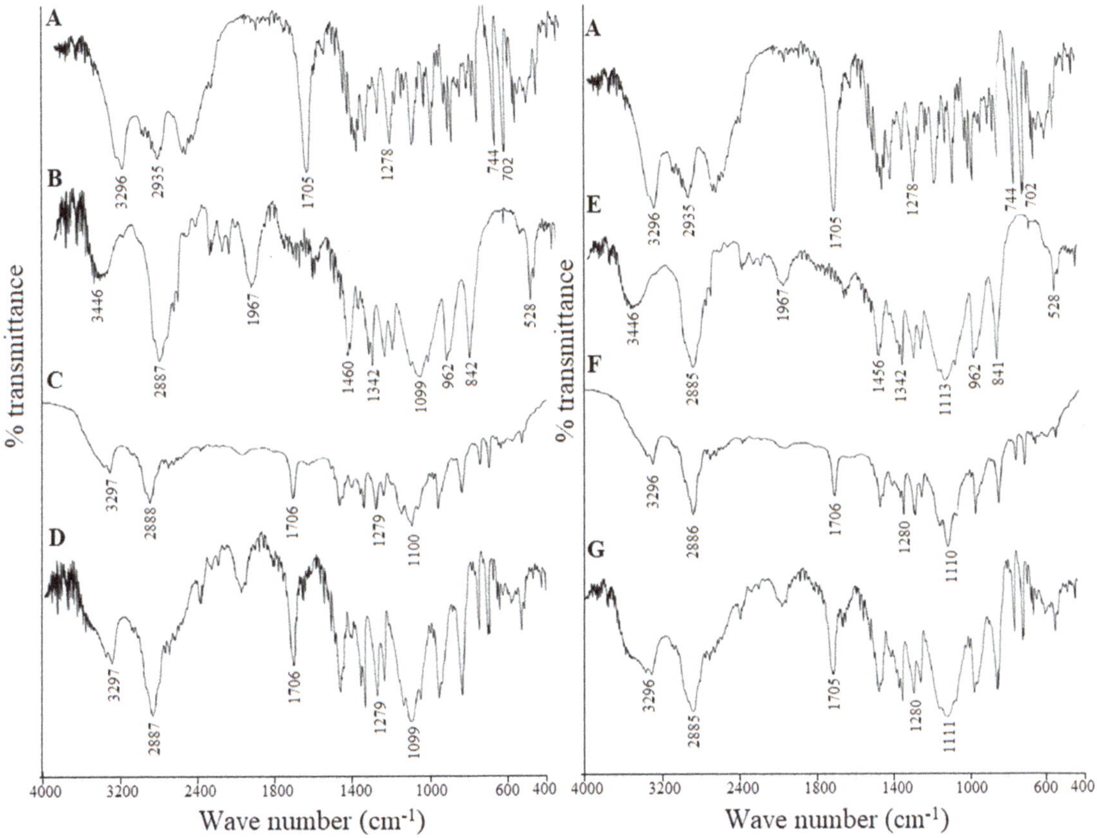

Figure 5. Fourier transform infrared (FTIR) spectra of (**A**) Supplied FFH, (**B**) Supplied PEG 20,000, (**C**) Physical mixture of FFH with PEG 20,000 at 1:4 ratio (PM-FP20K$_{(1:4)}$), (**D**) Solid dispersion of FFH with PEG 20,000 at 1:4 ratio (SD-FP20K$_{(1:4)}$), (**E**) Supplied poloxamer 188, (**F**) Physical mixture of FFH with poloxamer 188 at 1:4 ratio (PM-FP188$_{(1:4)}$), (**G**) Solid dispersion of FFH with poloxamer 188 at 1:4 ratio (SD-FP188$_{(1:4)}$).

3.5. *In Situ Intestinal Absorption Study*

The oral bioavailability of FFH in rats was reported to be about 4.2% [47] due to its low intestinal permeability, and the involvement of intestinal P-gp [19,48,49]. The phase solubility and in vitro dissolution studies have revealed the solubility and dissolution rate enhancement of FFH with PEG 20,000 and poloxamer 188. To understand the influence of the PEG 20,000 and poloxamer 188 on fexofenadine permeation, in situ single pass intestinal perfusion study was conducted in rats and the results are shown in Table 3. The effective permeability coefficient of fexofenadine across rat intestine ($P_{eff(rat)}$) from the dispersions of supplied FFH, solid dispersions SD-FP20K$_{(1:4)}$ and SD-FP188$_{(1:4)}$ was found to be 7.04 ± 0.56 ($\times 10^{-6}$) cm/s, 14.40 ± 0.27 ($\times 10^{-6}$) cm/s and 21.30 ± 0.45 ($\times 10^{-6}$) cm/s, respectively.

Table 3. In situ absorption parameters of fexofenadine hydrochloride from the prepared solid dispersions across rat intestine (Mean ± SD; $n = 3$).

Formulation	$P_{eff(rat)} \times 10^{-6}$ (cm/s)	$P_{eff\ (human)} \times 10^{-5}$ (cm/s)	k_a (min^{-1})	ER
Supplied FFH	7.04 ± 0.56	2.61 ± 0.18	0.016 ± 0.001	-
SD-FP20K$_{(1:4)}$	14.40 ± 0.27 †	5.00 ± 0.08 †	0.030 ± 0.001 †	2.04 †
SD-FP188$_{(1:4)}$	21.30 ± 0.45 †	7.26 ± 0.14 †	0.044 ± 0.001 †	3.03 †

FFH, fexofenadine hydrochloride; SD-FP20K$_{(1:4)}$—solid dispersion of FFH with PEG 20,000 at 1:4 weight ratio; SD-FP188$_{(1:4)}$—solid dispersion of FFH with poloxamer 188 at 1:4 weight ratio; $P_{eff(rat)}$—effective permeability coefficient in rat; $P_{eff(human)}$—predicted effective permeability coefficient in human; K_a—absorption rate constant; ER—enhancement ratio. † indicates significant difference at $p < 0.05$ against control.

The estimated human effective permeability coefficient ($P_{eff(human)}$) and absorption rate constant (K_a) values also indicate significantly ($p < 0.05$) higher rate of absorption of the drug from prepared solid dispersions compared to the supplied FFH. The enhancement ratio (ER) for SD-FP20K$_{(1:4)}$ and SD-FP188$_{(1:4)}$ was found to be 2.04 and 3.03, respectively which is greater than 1 indicating enhanced permeation of drug from the solid dispersions. Microscopic observation of the intestinal segments without and with treatment of supplied fexofenadine, SD-FP20K$_{(1:4)}$ and SD-FP188$_{(1:4)}$ (Figure S2, supplementary information) showed all layers of the intestine without any disruption to the epithelium indicating good biological acceptance of the carriers, PEG 20,000 and poloxamer 188 in the preparation of pharmaceutical formulations for oral administration.

Both SD-FP20K$_{(1:4)}$ and SD-FP188$_{(1:4)}$ were able to significantly enhance ($p < 0.05$) the intestinal permeability compared to that of supplied FFH dispersion. However, the extent of enhancement was significantly higher with non-ionic surfactant, poloxamer 188 than PEG 20,000 which could be attributed to the inhibition of P-gp efflux pump along with faster dissolution rate. Several studies have reported that poloxamer 188 inhibits P-gp mediated efflux activity and enhance the permeability of P-gp substrate drugs across rat intestinal segments [30,50]. Therefore, the present study suggests that poloxamer 188 could serve as a better carrier to improve the oral bioavailability of FFH by enhancing dissolution as well as intestinal permeability.

4. Conclusions

In this study, solid dispersions of fexofenadine hydrochloride were successfully prepared using PEG 20,000 and poloxamer 188 following fusion method. The phase solubility and in vitro dissolution studies revealed an improved solubility and dissolution rate of FFH with increased concentration of carriers compared to the supplied drug. The solid-state characterization using DSC and XRPD analysis suggested the crystalline nature of fexofenadine in the prepared solid dispersions. The FTIR observation supported the absence of the physicochemical interactions between drug carrier combinations studied. The in situ single pass intestinal perfusion study was carried out to evaluate the effect of the carriers in the intestinal absorption of fexofenadine from the prepared solid dispersions. Both the solid dispersions, SD-FP20K$_{(1:4)}$ and SD-FP188$_{(1:4)}$, showed 2–3 fold improvement in fexofenadine absorption compared to the control drug only dispersion. However, the extent of enhancement of intestinal absorption was significantly higher with poloxamer 188 than PEG 20,000 which could be attributed to the inhibition of P-gp efflux pump along with faster dissolution rate. The results of this study demonstrated that poloxamer 188 could serve as a better carrier to improve the oral bioavailability of FFH by enhancing dissolution as well as intestinal permeability.

Supplementary Materials: The following are available online at https://www.mdpi.com/1999-4923/13/3/310/s1, Figure S1: Phase solubility diagrams for fexofenadine hydrochloride in the presence of (**A**) PEG 20,000 and (**B**) poloxamer 188 in water at 37 ± 0.5 °C (mean ± SD, $n = 3$), Figure S2: Histological sections of rat intestine: (**A**) without treatment, after single pass intestinal perfusion studies with (**B**) supplied fexofenadine hydrochloride, solid dispersions (**C**) SD-FP20K$_{(1:4)}$ and (**D**) SD-FP188$_{(1:4)}$ showing intact layers of intestine (Hematoxylin/Eosin stain, 100X- magnification).

Author Contributions: Conceptualization, B.B.E., P.R.V. and S.B.; Methodology, B.B.E.; Formal analysis, B.B.E.; Writing-Original Draft Preparation, B.B.E., D.N. and S.N.; Writing-Review & Editing, B.B.E. and S.B.; Supervision, P.R.V. and S.B.; Project Administration, B.B.E. and P.R.V. All authors have read and agreed to the published version of the manuscript.

Funding: This research received no external funding.

Institutional Review Board Statement: All animal experiments were performed in accordance with the guidelines for animal experimentation of the Institutional Animal Ethical Committee (IAEC) at the St. Peter's Institute of Pharmaceutical Sciences (Approval No 1003/SPIPS/Wgl/IAEC/2011 and 30 April 2011)

Acknowledgments: The authors would like to thank Ami Life Sciences Pvt. Ltd., Gujarat, India for the generous gift sample of fexofenadine hydrochloride. The authors are also grateful to Gattefosse, France and Isochem, France for the donation of excipients. The financial assistance to Basanth Babu Eedara by All India Council of Technical Education (New Delhi, India) in the form of Junior Research Fellowship is duly acknowledged. The authors also thank T. Jayapal Reddy, Chairman, St. Peter's Institute of Pharmaceutical Sciences, Hanamkonda for providing the necessary facilities.

Conflicts of Interest: The authors declare no conflict of interest.

References

1. Pawankar, R.; Baena-Cagnani, C.E.; Bousquet, J.; Canonica, G.W.; Cruz, A.A.; Kaliner, M.A.; Lanier, B.Q.; Henley, K. State of world allergy report 2008: Allergy and chronic respiratory diseases. *World Allergy Organ. J.* **2008**, *1*, S4–S17. [CrossRef]
2. Howarth, P.; Wilson, S.; Lau, L.; Raj Akulasingam, K. The nasal mast cell and rhinitis. *Clin. Exp. Allergy* **1991**, *21*, 3–8. [CrossRef] [PubMed]
3. Axelrod, D.; Bielory, L. Fexofenadine hydrochloride in the treatment of allergic disease: A review. *J. Asthma Allergy* **2008**, *1*, 19.
4. Dresser, G.K.; Bailey, D.G.; Leake, B.F.; Schwarz, U.I.; Dawson, P.A.; Freeman, D.J.; Kim, R.B. Fruit juices inhibit organic anion transporting polypeptide–mediated drug uptake to decrease the oral availability of fexofenadine. *Clin. Pharm. Ther.* **2002**, *71*, 11–20. [CrossRef]
5. Drescher, S.; Schaeffeler, E.; Hitzl, M.; Hofmann, U.; Schwab, M.; Brinkmann, U.; Eichelbaum, M.; Fromm, M.F. MDR1 gene polymorphisms and disposition of the P-glycoprotein substrate fexofenadine. *Br. J. Clin. Pharm.* **2002**, *53*, 526–534. [CrossRef] [PubMed]
6. Clough, S.R. Fexofenadine. In *Encyclopedia of Toxicology*, 3rd ed.; Academic Press: Cambridge, MA, USA, 2014.
7. Carrier, R.L.; Miller, L.A.; Ahmed, I. The utility of cyclodextrins for enhancing oral bioavailability. *J. Control. Release* **2007**, *123*, 78–99. [CrossRef]
8. Bandari, S.; Jadav, S.; Eedara, B.B.; Jukanti, R.; Veerareddy, P.R. Physicochemical characterization and dissolution enhancement of loratadine by solid dispersion technique. *Korean J. Chem. Eng.* **2013**, *30*, 238–244. [CrossRef]
9. Eedara, B.B.; Kankane, M.; Jukanti, R.; Nagabandi, V.K.; Bandari, S. Enhanced solubility and permeability of exemestane solid dispersion powders for improved oral delivery. *J. Pharm. Investig.* **2013**, *43*, 229–242. [CrossRef]
10. Eedara, B.B.; Kallakunta, V.R.; Bandari, S. Self-nanoemulsifying powders for improved oral delivery of poorly water-soluble drugs. *Ther. Deliv.* **2015**, *6*, 899–901. [CrossRef]
11. Gangishetty, H.; Eedara, B.B.; Bandari, S. Development of ketoprofen loaded proliposomal powders for improved gastric absorption and gastric tolerance: In Vitro and in situ evaluation. *Pharm. Dev. Technol.* **2015**, *20*, 641–651. [CrossRef]
12. Sunkavalli, S.; Eedara, B.B.; Janga, K.Y.; Velpula, A.; Jukanti, R.; Bandari, S. Preparation and characterization of docetaxel self-nanoemulsifying powders (SNEPs): A strategy for improved oral delivery. *Korean J. Chem. Eng.* **2016**, *33*, 1115–1124. [CrossRef]
13. Bandari, S.; Dronam, V.R.; Eedara, B.B. Development and preliminary characterization of levofloxacin pharmaceutical cocrystals for dissolution rate enhancement. *J. Pharm. Investig.* **2017**, *47*, 583–591. [CrossRef]
14. Eedara, B.B.; Bandari, S. Lipid-based dispersions of exemestane for improved dissolution rate and intestinal permeability: In Vitro and ex vivo characterization. *Artif. Cells Nanomed. Biotechnol.* **2017**, *45*, 917–927. [CrossRef] [PubMed]
15. Nalla, P.; Bagam, S.; Eedara, B.B.; Dhurke, R. Formulation and evaluation of domperidone oral proliposomal powders. *Int. J. Pharm. Technol. Res.* **2015**, *7*, 108–118.
16. Ramasahayam, B.; Eedara, B.B.; Kandadi, P.; Jukanti, R.; Bandari, S. Development of isradipine loaded self-nano emulsifying powders for improved oral delivery: In Vitro and in vivo evaluation. *Drug Dev. Ind. Pharm.* **2015**, *41*, 753–763. [CrossRef]
17. Shimizu, M.; Uno, T.; Sugawara, K.; Tateishi, T. Effects of itraconazole and diltiazem on the pharmacokinetics of fexofenadine, a substrate of P-glycoprotein. *Br. J. Clin. Pharmacol.* **2006**, *61*, 538–544. [CrossRef] [PubMed]
18. Türkmen, Ö.; Ay Şenyiğit, Z.; Baloğlu, E. Formulation and evaluation of fexofenadine hydrochloride orally disintegrating tablets for pediatric use. *J. Drug Deliv. Sci. Technol.* **2018**, *43*, 201–210. [CrossRef]
19. Gundogdu, E.; Alvarez, I.G.; Karasulu, E. Improvement of effect of water-in-oil microemulsion as an oral delivery system for fexofenadine: In Vitro and in vivo studies. *Int. J. Nanomed.* **2011**, *6*, 1631. [CrossRef]

20. Mandeep; Kaur, S.; Samal, S.K.; Roy, S.; Sangamwar, A.T. Successful oral delivery of fexofenadine hydrochloride by improving permeability via phospholipid complexation. *Eur. J. Pharm. Sci.* **2020**, *149*, 105338. [CrossRef]
21. Eedara, B.B.; Veerareddy, P.R.; Jukanti, R.; Bandari, S. Improved oral bioavailability of fexofenadine hydrochloride using lipid surfactants: Ex Vivo, in situ and in vivo studies. *Drug Dev. Ind. Pharm.* **2014**, *40*, 1030–1043. [CrossRef]
22. Parray, Z.A.; Hassan, M.I.; Ahmad, F.; Islam, A. Amphiphilic nature of polyethylene glycols and their role in medical research. *Polym. Test.* **2020**, *82*, 106316. [CrossRef]
23. Hugger, E.D.; Audus, K.L.; Borchardt, R.T. Effects of poly (ethylene glycol) on efflux transporter activity in Caco-2 cell monolayers. *J. Pharm. Sci.* **2002**, *91*, 1980–1990. [CrossRef] [PubMed]
24. Johnson, B.M.; Charman, W.N.; Porter, C.J. An in vitro examination of the impact of polyethylene glycol 400, pluronic P85, and vitamin E da-tocopheryl polyethylene glycol 1000 succinate on P-glycoprotein efflux and enterocyte-based metabolism in excised rat intestine. *AAPS Pharmsci.* **2002**, *4*, 193–205. [CrossRef]
25. Shen, Q.; Li, W.; Lin, Y.; Katsumi, H.; Okada, N.; Sakane, T.; Fujita, T.; Yamamoto, A. Modulating effect of polyethylene glycol on the intestinal transport and absorption of prednisolone, methylprednisolone and quinidine in rats by in-vitro and in-situ absorption studies. *J. Pharm. Pharmacol.* **2008**, *60*, 1633–1641. [CrossRef]
26. Kabanov, A.V.; Batrakova, E.V.; Alakhov, V.Y. Pluronic® block copolymers as novel polymer therapeutics for drug and gene delivery. *J. Control. Release* **2002**, *82*, 189–212. [CrossRef]
27. Tambe, A.; Pandita, N. Enhanced solubility and drug release profile of boswellic acid using a poloxamer-based solid dispersion technique. *J. Drug Deliv. Sci. Technol.* **2018**, *44*, 172–180. [CrossRef]
28. Newa, M.; Bhandari, K.H.; Li, D.X.; Kwon, T.-H.; Kim, J.A.; Yoo, B.K.; Woo, J.S.; Lyoo, W.S.; Yong, C.S.; Choi, H.G. Preparation, characterization and in vivo evaluation of ibuprofen binary solid dispersions with poloxamer 188. *Int. J. Pharm.* **2007**, *343*, 228–237. [CrossRef] [PubMed]
29. Ghareeb, M.M.; Abdulrasool, A.A.; Hussein, A.A.; Noordin, M.I. Kneading technique for preparation of binary solid dispersion of meloxicam with poloxamer 188. *AAPS PharmSciTech* **2009**, *10*, 1206–1215. [CrossRef] [PubMed]
30. Huang, J.; Si, L.; Jiang, L.; Fan, Z.; Qiu, J.; Li, G. Effect of pluronic F68 block copolymer on P-glycoprotein transport and CYP3A4 metabolism. *Int. J. Pharm.* **2008**, *356*, 351–353. [CrossRef] [PubMed]
31. Higuchi, T. A phase solubility technique. *Adv. Anal. Chem. Instrum.* **1965**, *4*, 117–211.
32. Ni, J.; Tian, F.; Dahmani, F.Z.; Yang, H.; Yue, D.; He, S.; Zhou, J.; Yao, J. Curcumin-carboxymethyl chitosan (CNC) conjugate and CNC/LHR mixed polymeric micelles as new approaches to improve the oral absorption of P-gp substrate drugs. *Drug Deliv.* **2016**, *23*, 3424–3435. [CrossRef] [PubMed]
33. Varma, M.V.; Panchagnula, R. Enhanced oral paclitaxel absorption with vitamin E-TPGS: Effect on solubility and permeability in vitro, in situ and in vivo. *Eur. J. Pharm. Sci.* **2005**, *25*, 445–453. [CrossRef] [PubMed]
34. Chen, Y.; Zhang, G.; Neilly, J.; Marsh, K.; Mawhinney, D.; Sanzgiri, Y. Enhancing the bioavailability of ABT-963 using solid dispersion containing Pluronic F-68. *Int. J. Pharm.* **2004**, *286*, 69–80. [CrossRef]
35. Kolašinac, N.; Kachrimanis, K.; Homšek, I.; Grujić, B.; Đurić, Z.; Ibrić, S. Solubility enhancement of desloratadine by solid dispersion in poloxamers. *Int. J. Pharm.* **2012**, *436*, 161–170. [CrossRef]
36. Baghel, S.; Cathcart, H.; O'Reilly, N.J. Polymeric amorphous solid dispersions: A review of amorphization, crystallization, stabilization, solid-state characterization, and aqueous solubilization of biopharmaceutical classification system class II drugs. *J. Pharm. Sci.* **2016**, *105*, 2527–2544. [CrossRef] [PubMed]
37. Damian, F.; Blaton, N.; Naesens, L.; Balzarini, J.; Kinget, R.; Augustijns, P.; Van den Mooter, G. Physicochemical characterization of solid dispersions of the antiviral agent UC-781 with polyethylene glycol 6000 and Gelucire 44/14. *Eur. J. Pharm. Sci.* **2000**, *10*, 311–322. [CrossRef]
38. Mehanna, M.M.; Motawaa, A.M.; Samaha, M.W. In sight into tadalafil–block copolymer binary solid dispersion: Mechanistic investigation of dissolution enhancement. *Int. J. Pharm.* **2010**, *402*, 78–88. [CrossRef]
39. Vo, C.L.-N.; Park, C.; Lee, B.-J. Current trends and future perspectives of solid dispersions containing poorly water-soluble drugs. *Eur. J. Pharm. Biopharm.* **2013**, *85*, 799–813. [CrossRef]
40. Weerapol, Y.; Limmatvapirat, S.; Nunthanid, J.; Konthong, S.; Suttiruengwong, S.; Sriamornsak, P. Development and characterization of nifedipine-amino methacrylate copolymer solid dispersion powders with various adsorbents. *Asian J. Pharm. Sci.* **2017**, *12*, 335–343. [CrossRef]
41. Kumar, L.; Alam, M.S.; Meena, C.L.; Jain, R.; Bansal, A.K. Fexofenadine hydrochloride. In *Profiles of Drug Substances, Excipients and Related Methodology*; Elsevier: Amsterdam, The Netherlands, 2009; Volume 34, pp. 153–192.
42. Passerini, N.; Albertini, B.; González-Rodríguez, M.L.; Cavallari, C.; Rodriguez, L. Preparation and characterisation of ibuprofen–poloxamer 188 granules obtained by melt granulation. *Eur. J. Pharm. Sci.* **2002**, *15*, 71–78. [CrossRef]
43. Windbergs, M.; Strachan, C.J.; Kleinebudde, P. Influence of structural variations on drug release from lipid/polyethylene glycol matrices. *Eur. J. Pharm. Sci.* **2009**, *37*, 555–562. [CrossRef]
44. Pardhi, V.P.; Jain, K. Impact of binary/ternary solid dispersion utilizing poloxamer 188 and TPGS to improve pharmaceutical attributes of bedaquiline fumarate. *J. Drug Deliv. Sci. Technol.* **2021**, *62*, 102349. [CrossRef]
45. Ige, P.P.; Baria, R.K.; Gattani, S.G. Fabrication of fenofibrate nanocrystals by probe sonication method for enhancement of dissolution rate and oral bioavailability. *Colloids Surf. B Biointerfaces* **2013**, *108*, 366–373. [CrossRef]

46. Bandari, S.; Jadav, S.; Eedara, B.B.; Dhurke, R.; Jukanti, R. Enhancement of solubility and dissolution rate of Loratadine with Gelucire 50/13. *J. Pharm. Innov.* **2014**, *9*, 141–149. [CrossRef]
47. Strelevitz, T.J.; Foti, R.S.; Fisher, M.B. In Vivo use of the P450 inactivator 1-aminobenzotriazole in the rat: Varied dosing route to elucidate gut and liver contributions to first-pass and systemic clearance. *J. Pharm. Sci.* **2006**, *95*, 1334–1341. [CrossRef]
48. Lappin, G.; Shishikura, Y.; Jochemsen, R.; Weaver, R.J.; Gesson, C.; Houston, B.; Oosterhuis, B.; Bjerrum, O.J.; Rowland, M.; Garner, C. Pharmacokinetics of fexofenadine: Evaluation of a microdose and assessment of absolute oral bioavailability. *Eur. J. Pharm. Sci.* **2010**, *40*, 125–131. [CrossRef] [PubMed]
49. Piao, H.-M.; Balakrishnan, P.; Cho, H.-J.; Kim, H.; Kim, Y.-S.; Chung, S.-J.; Shim, C.-K.; Kim, D.-D. Preparation and evaluation of fexofenadine microemulsions for intranasal delivery. *Int. J. Pharm.* **2010**, *395*, 309–316. [CrossRef] [PubMed]
50. Bansal, T.; Akhtar, N.; Jaggi, M.; Khar, R.K.; Talegaonkar, S. Novel formulation approaches for optimising delivery of anticancer drugs based on P-glycoprotein modulation. *Drug Discov. Today* **2009**, *14*, 1067–1074. [CrossRef] [PubMed]

pharmaceutics

MDPI

Article

Novel Solid Dispersions of Naphthoquinone Using Different Polymers for Improvement of Antichagasic Activity

Verônica da Silva Oliveira [1], Elen Diana Dantas [1], Anna Thereza de Sousa Queiroz [1], Johny Wysllas de Freitas Oliveira [1], Marcelo de Sousa da Silva [1,2], Patricia Garcia Ferreira [3], Fernando de Carvalho da Siva [4], Vitor Francisco Ferreira [3] and Ádley Antonini Neves de Lima [1,*]

1 Department of Pharmacy, Health Sciences Center, Federal University of Rio Grande do Norte, Natal, Rio Grande do Norte 59012-570, Brazil; veronicasoliver47@gmail.com (V.d.S.O.); elendiana88@gmail.com (E.D.D.); anynhathereza@ufrn.edu.br (A.T.d.S.Q.); johnywysllas@gmail.com (J.W.d.F.O.); mssilva.ufrn@gmail.com (M.d.S.d.S.)

2 Global Health and Tropical Medicine, Institute of Hygiene and Tropical Medicine, NOVA University Lisbon, 1800-166 Lisbon, Portugal

3 Department of Pharmaceutical Technology, Faculty of Pharmacy, Federal Fluminense University, Niterói, Rio de Janeiro 24241-002, Brazil; patricia.pharma@yahoo.com.br (P.G.F.); vitorferreira@id.uff.br (V.F.F.)

4 Institute of Chemistry, Federal University Fluminense, Niterói, Rio de Janeiro 24020-150, Brazil; fcsilva@id.uff.br

* Correspondence: adleylima13@gmail.com; Tel.: +55-84-99928-8864

Received: 24 October 2020; Accepted: 20 November 2020; Published: 24 November 2020

check for updates

Abstract: IVS320 (3a,10b-dihydro-1*H*-cyclopenta[*b*]naphtho[2,3-*d*]furan-5,10-dione) is a naphthoquinone that has low solubility in aqueous medium, a physical behavior that limits its biological activities, considering that compounds from this class have several activities. In this work, solid dispersions (SDs) prepared between IVS320 and polymers hydroxypropyl methylcellulose (HPMC), polyethylene glycol (PEG), and polyvinylpyrrolidone (PVP) were developed using physical mixture (PM), kneading (KN), and rotary evaporation (RE) methods. Dispersions were investigated using Fourier transform infrared spectroscopy (FTIR), differential scanning calorimetry (DSC), thermogravimetry (TG), powder X-ray diffraction (PXRD), and scanning electron microscopy (SEM). In addition, in vitro antiparasitic activity in *Trypanosoma cruzi* Y strains was evaluated. Physical-chemical characterization demonstrated the formation of SDs through the interaction of IVS320 with polymeric matrices. SDs of IVS320-polymer presented a significant potentiation of antichagasic activity, with inhibitory growth around 62% (IVS320-HPMC/RE), 55% (IVS320-PEG/RE), and 85% (IVS320-PVP/RE), while pure IVS320 showed a value of 48% for the highest concentrations evaluated (50 μg/mL).

Keywords: solid dispersion; delivery system; polymeric carriers; antichagasic activity

1. Introduction

The parasitic infection caused by *Trypanosoma cruzi*, known as Chagas disease, is included among a list of 20 neglected tropical diseases, affecting around 6 to 8 million people and killing about 15 thousand per year [1]. The main form of transmission requires a vector, with *Triatoma infestans* being the most common insect [2,3]; this type of transmission corresponds to approximately 80% of infections. In addition to this route, *T. cruzi* can be transmitted by blood transfusion, transplant of solid organs, congenital infection, inoculation by accident at work, or orally, through the ingestion of contaminated substances [1].

In recent years, the interest in studying naphthoquinones has intensified, both for its relevance in vital biochemical processes, as well as for its variety of known pharmacological properties, such as anti-inflammatory [4,5], antitumor [6,7], antibacterial, antifungal [8], antiviral [7,9,10], antimalarial [11], leishmanicidal [12], and trypanocidal activities [13–15]. Among the naphthoquinone compounds included in the quinone family, there is the IVS320 molecule (Figure 1a), chemically named 3a,10b-dihydro-1*H*-cyclopenta[*b*]naphtho[2,3-*d*]furan-5,10-dione, produced synthetically and with proven antifungal [16] and antichagasic activity [17].

However, IVS320, like most new drug candidates, presents low solubility in aqueous medium [17], causing an impact on the dissolution rate and, consequently, on its bioavailability. There are several ways to get around this problem of solubility and stability, such as inclusion complexes [18] and solid dispersion systems.

In this case, for the elaboration of solid dispersions (SDs), polymers play a prominent role, as they can stabilize SDs due to their ability to retain the drug in an amorphous form in the polymeric matrix during storage [19], and can increase the wettability, dispersibility, and solubility of drugs [20].

Some examples of these polymers are polyvinylpyrrolidone (PVP) [21], polyethylene glycol (PEG), and various cellulose derivatives, such as hydroxypropyl methylcellulose (HPMC) [22,23], which are hydrophilic polymer carriers used as excipients in several pharmaceutical formulations, due to (among other characteristics) their low toxicity [24–26].

In view of the applications of naphthoquinones, pre-preliminary studies of the biological activities of compound IVS320, and the need to develop alternatives for the treatment of Chagas disease, this work aimed to use solutions centered on pharmaceutical technology to improve the physicochemical and biological properties of naphthoquinone IVS320, through the preparation of SDs in polymeric matrices.

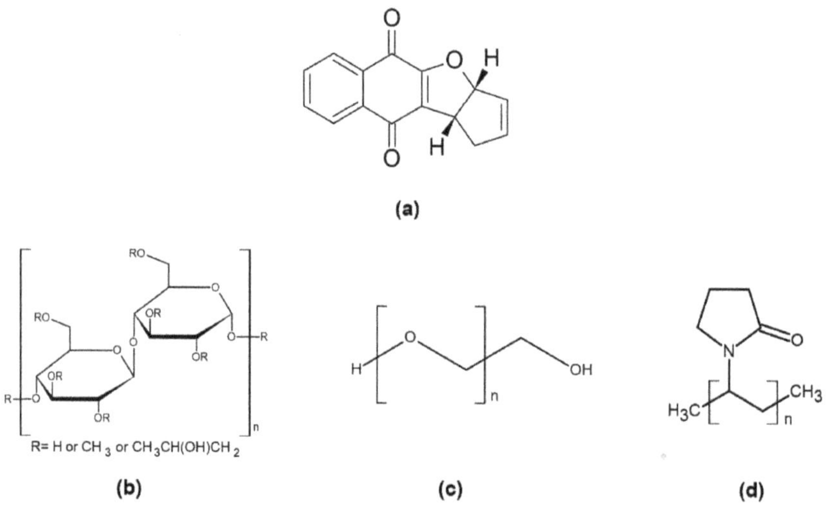

Figure 1. Chemical structure: (**a**) 3a,10b-dihydro-1*H*-cyclopenta[*b*]naphtho[2,3-*d*]furan-5,10-dione (IVS320), (**b**) hydroxypropyl methylcellulose (HPMC), (**c**) polyethylene glycol (PEG), and (**d**) polyvinylpyrrolidone (PVP).

In this article, SDs of IVS320 (Figure 1a) were prepared by kneading and rotary evaporation methods with polymers to improve the antichagasic activity of IVS320. Three polymers were selected as carriers: hydroxypropyl methylcellulose (HPMC) (Figure 1b), polyethylene glycol (PEG) (Figure 1c), and polyvinylpyrrolidone (PVP) (Figure 1d).

2. Materials and Methods

2.1. Material

The compound IVS320 (3a,10b-dihydro-1*H*-cyclopenta[*b*]naphtho[2,3-*d*]furan-5,10-dione) is a dihydrofuran fused 1,4-naphthoquinone that was synthesized by the LabSOA (Laboratory of Applied Organic Synthesis) at the Federal University Fluminense [17]. The polymers hydroxypropyl methylcellulose (HPMC), polyethylene glycol 6000 (PEG), and polyvinylpyrrolidone K-30 (PVP) were purchased from Sigma Aldrich Corporation (San Luis, MO, USA), and the solvents used were all analytical grade.

2.2. Preparation of the SDs

Solid dispersions (SDs) were prepared via physical mixture (PM), kneading (KN), and rotary evaporation (RE) methods using the ratio (1:1, *w*/*w*) IVS320:polymer for all methodologies.

The preparation of the physical mixture (PM) of IVS320 and the respective polymer (HPMC, PEG, and PVP) were carried out in a (1:1) ratio, subsequently homogenized with mortar and pestle, and then stored in the dissector.

In the kneading method (KN), the IVS320 and each polymer were weighed, homogenized with mortar and pestle, and then addition a mixture of water/acetone solvent (50:50, *v*/*v*) was added. Afterwards, this solution was dried in an oven at 60 °C for 24 h, and the solid obtained was stored in a desiccator.

The rotary evaporation method (RE) consisted of solubilizing the IVS320 and the polymers in a mixture of water/acetone solvents (50:50, *v*/*v*). Then, this solution was subjected to constant stirring at 25 °C for 72 h. Finally, the solution was evaporated under vacuum at 75 °C, and the solid obtained was stored in a desiccator.

2.3. Characterization of the SDs

2.3.1. Fourier Transform Infrared Spectroscopy (FTIR)

Infrared spectroscopic analysis was performed using IR Prestige-21 equipment (Shimadzu Corporation, Kyoto, Japan). The analysis was carried out in the 4000–600 cm^{-1} region with 15 scans.

2.3.2. Thermal Analysis

Differential scanning calorimetry (DSC) thermal analysis was carried out in a DSC-50 cell Shimadzu® (Tokyo, Japan) using approximately 2 mg of sample in aluminum crucibles under dynamic nitrogen atmosphere (50 $mL{\cdot}min^{-1}$) and with a heating rate of 10 $°C{\cdot}min^{-1}$ at a temperature range of 30–500 °C.

Thermogravimetry (TG) was obtained using a TGA-50 Shimadzu® (Tokyo, Japan) in the temperature range of 30–900 °C, using alumina crucibles with approximately 2 mg of samples under dynamic nitrogen atmosphere (50 $mL{\cdot}min^{-1}$) and with a heating rate of 10 $°C{\cdot}min^{-1}$.

2.3.3. Powder X-ray Diffraction (PXRD)

PXRD analysis was carried out using a Bruker D2 Phaser (Bruker Corporation, Billerica, MA, USA) with CuKα radiation (λ = 1.54 Å) at a voltage of 30 kV and a current of 15 mA, using a Lynxeye detector (Bruker Corporation, Billerica, MA, USA). Samples were scanned at room temperature for 2 h at a range of 5–40° at 0.05° s^{-1}.

2.3.4. Scanning Electronic Microscopy (SEM)

Samples were mounted on aluminum stubs using double-sided adhesive tape. Morphological analysis was performed on a Hitachi TM-3000 Tabletop Microscope (Hitachi Ltd., Tokyo, Japan) at a magnification of 500×. SEM images were obtained at an accelerating potential of 15 kV under reduced pressure.

2.4. *Antichagasic Activity*

2.4.1. Parasite

The *T. cruzi* Y strain was grown in LIT (liver infusion tryptose) medium supplemented with 10% SFB (serum fetal bovine) and 5% streptococcus/penicillin antibiotic (100 UI/mL), and kept at 27 °C in a BOD (Biochemical Oxygen Demand) oven to obtain the epimastigote forms of the parasite.

2.4.2. In Vitro Antichagasic Evaluation

The parasites were diluted and counted with the aid of a Neubauer improved camera (New optics, El Centro, CA, USA) and used at a concentration of 1×10^7 parasites/mL. The IVS320 compound and the IVS320-polymer systems were solubilized in dimethylsulfoxide (DMSO) with a maximum concentration of 1% in 96-well plates. Briefly, a stock solution was prepared for each of the evaluated systems, and subsequently diluted in the culture medium, obtaining concentrations in the range of 50 μg/mL to 2.5 μg/mL for the pure IVS320 and the concentration of the IVS320 in the SDs. Then cultures of the epimastigote form of *T. cruzi* were dispensed in 96-well plates and incubated for 24 h. The positive control contained only medium and strain; negative controls (medium and drug) and solvent control (contained 1% DMSO, medium and strain) were also obtained to ensure that the solvent did not influence the assay. The inhibition of *T. cruzi* was evaluated by resazurin reduction assay (Sigma Aldrich). After 24 h, a reading was performed at 570 nm and 600 nm using a microplate reader (Epoch, BioTek Instruments, Winooski, VT, USA). Experiments were performed in triplicate, and benznidazole was used as a trypanocidal reference drug. The inhibition percentage was based on the following formula:

$$\% \text{ Inhibition} = 100 - ((A_{570t} - (A_{600t} \times R_0))/(A_{570c} - (A_{600c} \times R_0))) \times 100 \quad (1)$$

A_{570t} = Absorption of the treatment at 570 nm wavelength.
A_{600t} = Absorption of the treatment at 600 nm wavelength.
A_{570c} = Absorption of control at 570 nm wavelength.
A_{600c} = Absorption of the control at 600 nm wavelength.
R_0 = Medium correction factor interacting with resazurin.

$$R_0 = Am_{570}/Am_{600} \quad (2)$$

Am_{570} = Absorbance of the medium at a wavelength of 570 nm.
Am_{600} = Absorbance of the medium at a wavelength of 600 nm.

2.4.3. Statistical Analysis

Data analysis was performed using GraphPad Prism software version 5.0 (La Jolla, CA, USA). The results are presented as mean ± standard error of the mean (S.E.M.). A one-way analysis of variance (ANOVA) was performed to determine the significant differences between the groups, followed by Dunnett's *t*-test for multiple comparisons. Values of $p < 0.05$ were considered significant.

3. Results and Discussion

3.1. *Fourier Transform Infrared Spectroscopy (FTIR)*

The infrared spectrum of IVS320 showed vibrational modes referring to νC=O asymmetrical and symmetrical, respectively, at 1678 and 1644 cm^{-1}, characteristic of the carbonyl group, corroborating with the stretching of 1,4-naphthoquinone, lawsone, and lapachol [27–29], as well as bands at 1614, 1591, and 1568 cm^{-1}, attributed to C=C stretching vibration present in the benzene ring and cyclopentene in the structure of the IVS320. Other bands were observed in the region from 1300 to 1020 cm^{-1}, corresponding to C-O-C and C-O (aromatic ether) stretching.

Analyzing the polymers spectra, HPMC exhibited bands at 3457 cm^{-1} attributed to the O-H stretching and at 1053 cm^{-1} corresponding to C-O-C stretching [20]. In the PEG, a broad band at 3477 cm^{-1} (νO-H) and a peak at 1096 cm^{-1} (νC-O) were observed, characteristic of ether. The PVP exhibited a broad band in the region of 3412 cm^{-1}, referring to the presence of water and corroborating with the DSC results of the polymer. Additionally, bands were also observed for PVP at 2933 cm^{-1}, 1648 cm^{-1}, and 1287 cm^{-1}, which were attributed to C-H, C=O (pyrrolidone group), and C-N stretching, respectively [30].

The SDs exhibited vibrational modes characteristic of the IVS320 molecule and isolated polymers, as well as a shift in the stretching vibration and changes in the profiles of some bands. Some changes were observed in the range of 1400–1170 cm^{-1}, corresponding to the vibrational modes of various C-C, C-O, and C-O-C bonds, and C-H and O-H deformations.

In the spectrum of SDs obtained with HPMC (Figure 2a), the presence of νO-H and νC-H was found to correspond with HPMC. In addition to a 1053 cm^{-1} band referring to C-O-C stretching, in the region from 1115 to 1016 cm^{-1} in the same spectrum, changes were observed in the profile of the three bands corresponding to the IVS320, located at 1102, 1069, and 1035 cm^{-1}, due to interactions between the IVS320 and the polymer, suggesting the formation of dispersions. By analyzing the spectra (Figure 2a), the physical mixture proved to be less effective at showing the IVS320-HPMC interaction, considering that the characteristic bands of the polymer were not very evident.

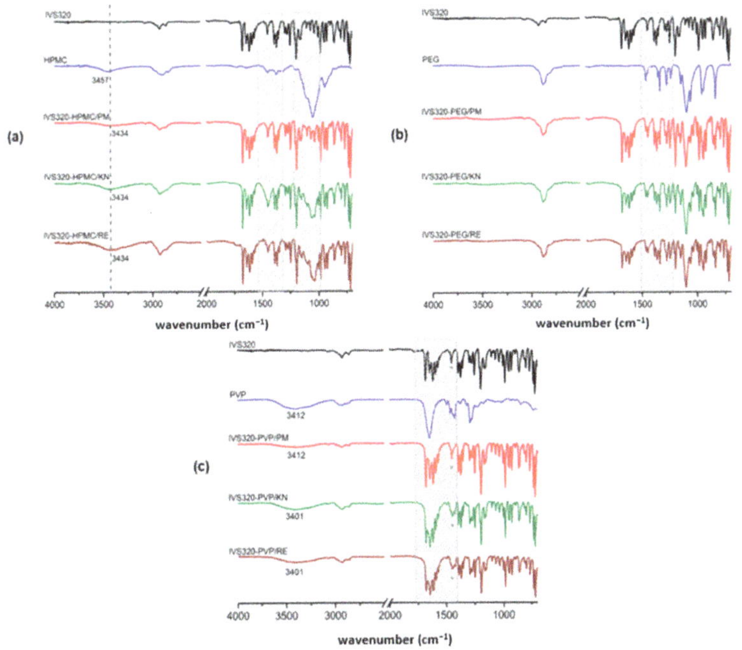

Figure 2. Infrared spectrum of systems: (**a**) IVS320-HPMC, (**b**) IVS320-PEG, and (**c**) IVS320-PVP.

IVS320-PEG SDs showed changes in the profile of the bands in the range of 1510 to 1213 cm^{-1}. In IVS320-PEG/KN and IVS320-PEG/RE SDs, the band displayed at 1248 cm^{-1} corresponding to the IVS320 underwent a small shift to 1250 cm^{-1} (C-O stretching). The presence of PEG in the dispersions was evidenced by the vibrational modes corresponding to this polymer, with small variations. In the SDs, bands were found at 1102 cm^{-1} referring to C-O deformation characteristic of ether, while in the PEG, it was at 1096 cm^{-1}.

In the range of 1678 to 1407 cm^{-1} in the spectrum of the IVS320-PVP SDs (Figure 2c), modifications were observed in the ratio of the intensity of the bands referring to C=O and C=C stretching, showing that such changes may be due to interactions between the IVS320 molecule and the polymeric chain through these functional groups, thus changing the profile of these bands. While in the spectrum of the physical mixture (IVS320-PVP/PM), the profile of this set of bands remained practically unchanged. Additionally, in the dispersions, it was possible to verify the O-H stretching referring to the presence of the polymer, showing small displacements when compared to the isolated polymer, with a value of 3412 cm^{-1}, and 3401 cm^{-1} in the PVP.

IVS320-HPMC SDs exhibited peak shifts for the O-H stretching (from 3457 cm^{-1} to 3434 cm^{-1}), C-O stretching (from 1248 cm^{-1} to 1250 cm^{-1}), and C-O-C stretching (from 1194 cm^{-1} to 1197 cm^{-1}). Similar shifts were observed for IVS320-PEG SDs (Figure 2b). In addition, in the IVS320-PEG SDs, a shift was observed regarding the C-O stretching of the PEG (from 1096 to 1102 cm^{-1}). These changes suggest the possibility of hydrogen bonding between the hydroxyl of HPMC and the carbonyl group of IVS320.

Hydrogen bonds between the carbonyl group (C=O) of the IVS320 and the hydroxyl of the HPMC or PEG polymers would be expected. However, spectral changes were observed, mainly in relation to the O-H, C-O, and C-O-C groups, which may suggest that interactions occur between the ether group of the IVS320 and the hydroxyl of the polymers. On the other hand, in the spectrum of the IVS320-PVP system, the main changes were found in the profile of the bands corresponding to C=O and C=C stretching, possibly attributed to the interactions between the C=O group of the IVS320 and polymer.

3.2. *Differential Scanning Calorimetry (DSC)*

The DSC curve of the IVS320 showed two events (Figure 3a): the first, an endothermic event peaking at 189 °C, corresponding to its melting point; the second, an exothermic event at 203 °C, attributed to possible changes in the crystalline form of the IVS320.

The thermogram of the HPMC and PVP polymers (Figure 3a) showed a broad endothermic peak, as a result of dehydration, in the range of 75 to 135 °C [30], while no peak corresponding to the melting point was found for these polymers, considering their amorphous structure [20]. However, PEG revealed an endothermic peak at a temperature around 63.9 °C, corresponding to its melting point, according to literature reports [20,31].

In the IVS320-HPMC/KN SD, an endothermic event at 184 °C and another exothermic at 200 °C were observed, both attributed to the IVS320 present in the polymeric matrix, while in IVS320-HPMC/RE SD, these same events were recorded at 181 °C and 195 °C. In the DSC thermograms (Figure 3b), characteristic fusion peaks of the IVS320 still existed, indicating the microcrystalline form of the IVS320 in the HPMC matrix, according to results reported in the 1:1 ratio (drug:polymer) [32].

In the DSC curve of the IVS320-PEG system, the process related to the melting point of the PEG showed variations that were observed at 65.2 °C in IVS320-PEG/KN SD and 62.7 °C in IVS320-PEG/RE SD. The endothermic events related to the IVS320 were not observed, while the exothermic events were observed with temperature variations, with values of 200 °C (IVS320-PEG/KN) and 198 °C (IVS320-PEG/RE) for SDs (Figure 3c).

There was also no endothermic event related to IVS320 in the IVS320-PVP SDs (Figure 3d); this indicates that the IVS320 was converted into an amorphous state, thus justifying its dispersion in the polymeric carrier. However, the exothermic event was found at 200 °C in the IVS320-PVP/KN SD and at 198 °C in the IVS320-PVP/RE SD [33].

Analyzing DSC data, it was found that in all dispersions, the exothermic event referring to the IVS320 underwent small temperature variations when compared to the isolated IVS320, implying that such changes occurred due to possible modifications in the crystalline form of the IVS320, thus suggesting its interaction in the polymeric matrices.

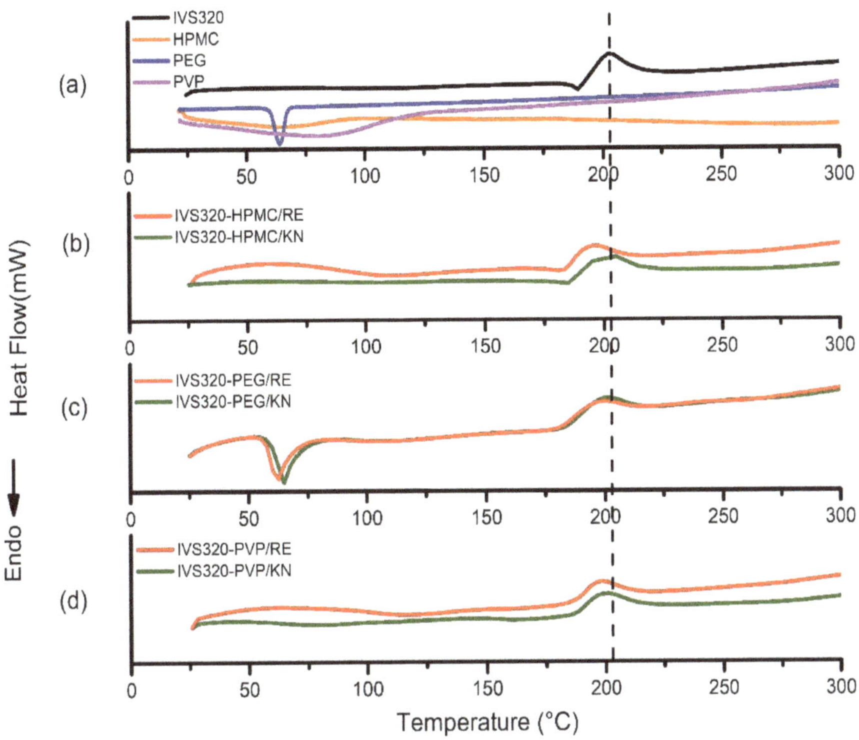

Figure 3. Differential scanning calorimetry (DSC) curve of (**a**) IVS320 and polymers, (**b**) solid dispersions (SDs) of IVS320-HPMC, (**c**) SDs of IVS320-PEG, and (**d**) SDs of IVS320-PVP. RE: rotary evaporation; KN: kneading.

3.3. Thermogravimetric Analysis (TG)

The TG curve of the IVS320 shows three mass loss events (Figure 4a). The first occurred in the range of 150 to 162 °C, with approximately 5% mass loss. The second variation occurred in the temperature range from 262 to 315 °C, with about 6.25% mass loss. Lastly, the third event was found in the range of 315 to 900 °C, with a reduction of 18.73%, as reported by Dantas et al. [17]. This high thermal resistance of the IVS320 is consistent with some substances of the naphthoquinone class [34].

The TG curve of the HPMC exhibited a degradation process, in the temperature range of 250 to 395 °C (peak 355 °C), with about 61% mass loss (Δm). The SDs formed between the IVS320 and the HPMC exhibited, in the range 210 to 390 °C (peak 345 °C), a mass loss of 58.12% and 57.41% for the dispersions IVS320-HPMC/KN and IVS320-HPMC/RE, respectively. In the range of 390 to 700 °C, a gradual mass loss of around 10.9% (IVS320-HPMC/KN) and 9.81% (IVS320-HPMC/RE) was observed, while in the TG curve of HPMC, a mass loss of around 7.41% was observed (Figure 4b).

The PEG polymer showed major mass loss, with a mass loss of approximately 95.3% in the range of 290 to 450 °C (Figure 4c). The IVS320-PEG SDs showed variations in mass loss at temperatures from 220 to 315 °C, with mass losses of 9.27% (IVS320-PEG/KN) and 6.36% (IVS320-PEG/RE). On the other hand, in the range of 315 to 440 °C, higher mass losses were observed, with values of 52.3% for the SD of IVS320-PEG/KN and about 49.17% for the SD of IVS320-PEG/RE.

The TG curve of the PVP exhibited two mass losses: the first, in the range of 30 to 100 °C, with a mass variation of 8.77%, corresponding to the presence of water; the second, in the range of 357 to 480 °C, with a mass loss of 83.16%, attributed to the process of thermal polymer degradation, corroborating the literature [35,36]. For SDs with PVP, in the range of 137 to 245 °C, mass variations of

3.53% (IVS320-PVP/RE) and 1% (IVS320-PVP/KN) were found. In turn, in the range of 245 to 480 °C, mass losses of 49.26% and 39.75% were obtained, respectively, for the SDs of IVS320-PVP/RE and IVS320-PVP/KN (Figure 4d).

Analyzing the percentages of mass loss among all dispersions, those obtained with the PVP polymer presented the smallest variations in mass, mainly the IVS320-PVP/KN SD, suggesting that at higher temperatures, there was an increase in the thermal stability of the dispersion.

According to the TG and DSC curves of the IVS320-polymer systems, the materials obtained have a stable thermal profile, showing that the presence of IVS320 does not interfere with the thermal profile of the polymeric carriers, which is behavior that strongly suggests a good interface between compound and polymers [37].

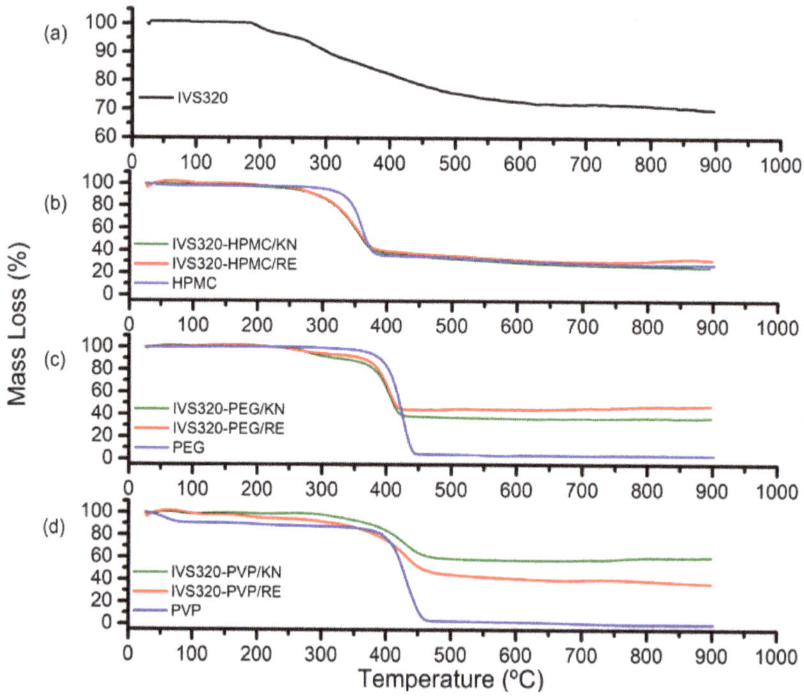

Figure 4. TG curves of (**a**) IVS320, (**b**) SDs of IVS320-HPMC, (**c**) SDs of IVS320-PEG, and (**d**) SDs of IVS320-PVP under N_2 atmosphere.

3.4. Powder X-ray Diffraction (PXRD)

The X-ray diffractograms of IVS320 presented high intensity peaks at 10.50°, 14.34°, 24.38°, and 28.22°, followed by a series of secondary reflections, indicating a crystalline profile, as already reported by Dantas et al. [17].

The PEG showed two intense crystalline reflections at 19.30° and 23.43° (Figure 5b), while HPMC and PVP showed an amorphous pattern, with no peak in the diffractogram [30,35].

For all IVS320-polymer systems, a reduction in the intensity of the standard peaks was observed, indicating a change in the amorphous state, both in the dispersions obtained by the kneading (KN) and rotary evaporation (RE) methods. Evidence of less intense peaks in polymeric systems may indicate the achievement of partially amorphous shapes due to the influence of polymers.

The physical mixtures showed peaks of the individual species, showing IVS320 crystallinity characteristics with a decrease in its intensity, which is attributed to the dilution and surface coverage of the IVS320 particles by the polymers [20]. On the other hand, the PXRD profile of the SDs showed a more marked reduction in the peaks, suggesting the formation of amorphous solid dispersions.

In SDs with HPMC, a more pronounced reduction in peaks was observed, particularly in the range of 20° to 30° (Figure 5a), indicating a change in the amorphous characteristic state of pure HPMC, thus suggesting that the polymer inhibits IVS320 crystallization, mainly in the SD of IVS320-HPMC/RE.

On the other hand, the SDs obtained with PEG were more crystalline than the IVS320-HPMC and IVS320-PVP (Figure 5c), possibly suggesting a recrystallization of the IVS320 during the process of preparing its dispersions, after evaporation of the solvent. This can be confirmed by the small displacements of the peaks corresponding to the PEG, bearing in mind that in the diffractogram, in relation to the polymer, crystalline reflections appeared at 19.30° and 23.43°, while in the SDs they were found at 18.95° and 23.07°.

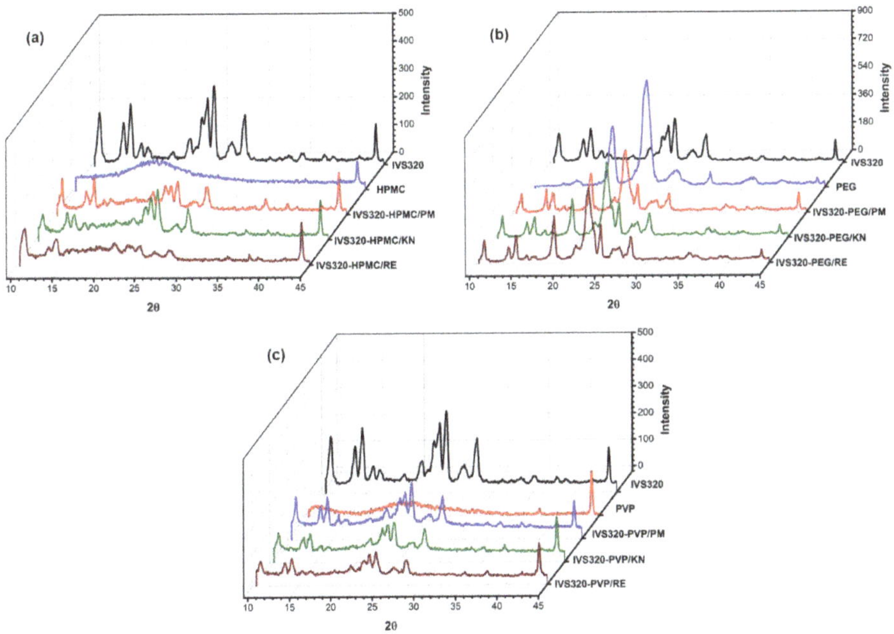

Figure 5. Powder X-ray diffraction (PXRD) of systems (**a**) IVS320-HPMC, (**b**) IVS320-PEG, and (**c**) IVS320-PVP.

3.5. Scanning Electron Microscopy (SEM)

The micrograph of the IVS320 displays very fine crystals with varying sizes and a predominance of the pyramidal shape and some crystals in the form of needles (Figure 6), corroborating the PXRD results.

HPMC exhibited an irregular cylindrical filament structure, while PEG exhibited a block-shaped structure, and the PVP particles showed irregular shapes with a smooth surface [20,30].

Images obtained for the IVS320-polymer systems by the physical mixture (PM) method showed the deposition of fragments of the IVS320 on the polymeric surface of the carriers, not verifying a uniform structure or surface that proves the interaction of the IVS320 in the polymeric network.

On the other hand, the micrographs for the IVS320-HPMC, IVS320-PEG and IVS320-PVP systems that were obtained through the kneading (KN) and rotary evaporation (RE) methods presented particles with irregular shapes and morphologies that were different from both components that originated it.

Images of the SDs of IVS320-HPMC/KN and IVS320-HPMC/RE showed rough surfaces, while the SDs of IVS320-PEG/KN and IVS320-PVP/KN exhibited images with amorphous structures of agglomerates. The IVS320-PEG/RE and IVS320-PVP/RE images suggest an association of the IVS320 with the polymer, composing a uniform matrix.

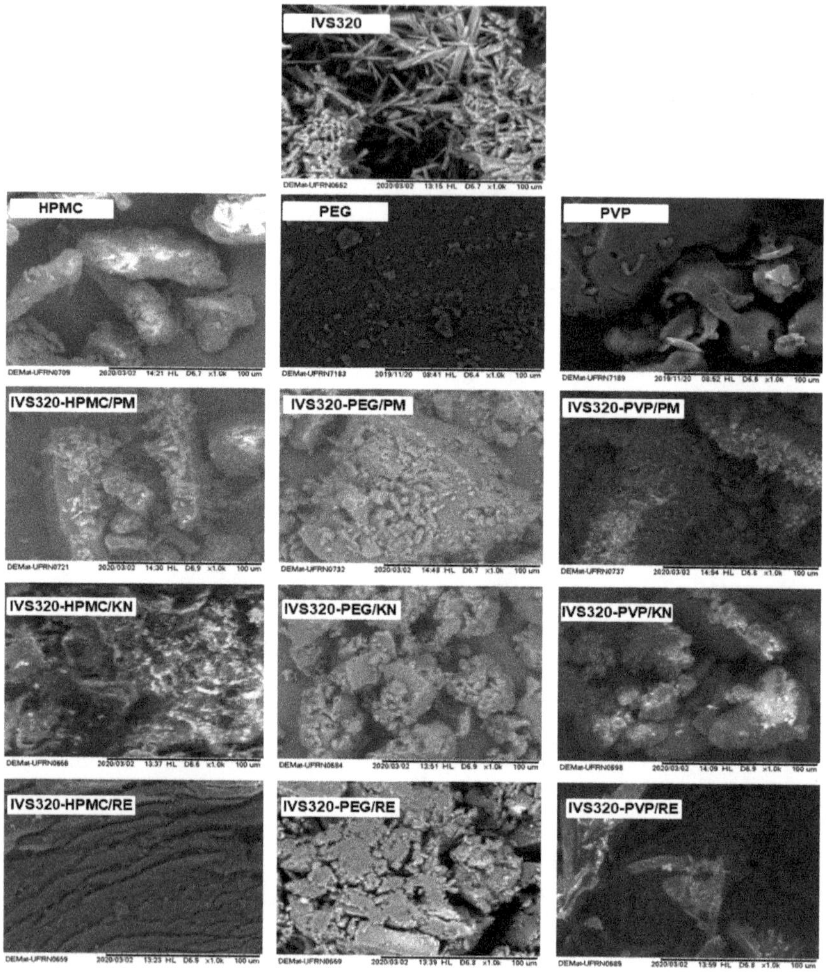

Figure 6. Scanning electron microscopy of IVS320, polymers (HPMC, PEG, and PVP), physical mixtures (PMs), and SDs obtained by kneading (KN) and rotary evaporation (RE).

3.6. *In Vitro Antichagasic Activity*

The antichagasic activity of the pure IVS320 is practically comparable to benznidazole (reference drug) at concentrations of 50 to 10 μg/mL, as shown in Figure 7. However, the IVS320-polymer SDs showed better activity, with a higher percentage of inhibitory growth (%) when compared to the pure IVS320, which presented a value of 37% for the highest concentration evaluated (50 μg/mL).

Figure 7 reports the inhibitory growth (%) as a function of the concentration of pure IVS320 and its equivalent concentrations in the dispersions.

The IVS320-HPMC/RE dispersion showed an activity higher to pure IVS320 for all concentrations assessed in the range of 50 to 2.5 μg/mL, presenting an inhibitory growth rate around 62% for the highest concentration.

Analyzing the IVS320-PEG system at concentrations of 50 and 25 μg/mL, the IVS320-PEG/PM and IVS320-PEG/RE dispersions exhibited a greater inhibitory effect, when compared to pure IVS320 at the same concentrations, showing inhibitory activity around approximately 55% for a concentration of 50 μg/mL.

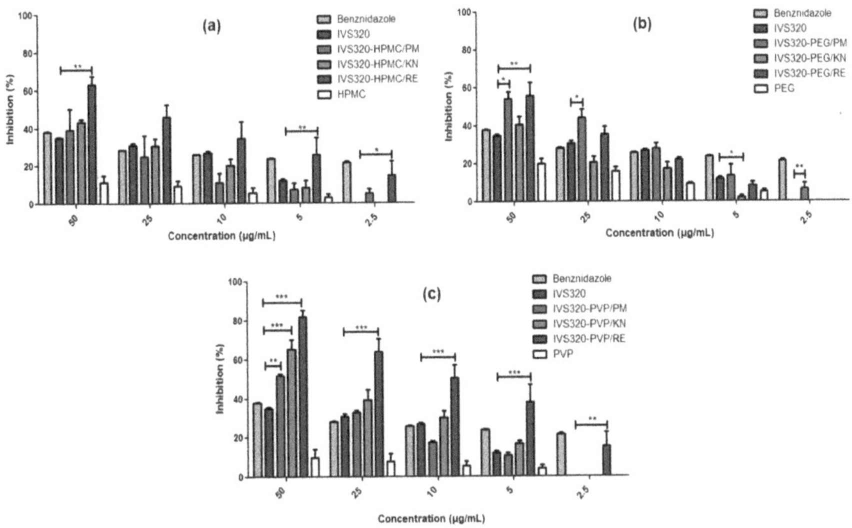

Figure 7. In vitro antichagasic activity of IVS320-polymer systems: (**a**) IVS320-HPMC, (**b**) IVS320-PEG, and (**c**) IVS320-PVP against epimastigotes of *T. cruzi* Y strain for 24 h. Values refer to % of control (untreated). *** $p < 0.0001$, ** $p < 0.001$ and * $p < 0.05$ compared with pure IVS320 at the same concentration. Statistical analysis was performed using ANOVA followed by Dunnett's *t*-test for multiple comparisons.

Similar to the IVS320-HPMC system, the results obtained for the IVS320-PVP system showed a better performance of the inhibitory activity for the SD of IVS320-PVP/RE, with an inhibitory rate around 82% and 64%, respectively, for the concentrations of 50 and 25 µg/mL. Additionally, it exhibited inhibitory activity higher than pure IVS320, even at the lowest concentrations, showing values of 15% inhibitory growth at the concentration of 2.5 µg/mL of IVS320 when incorporated in the SDs.

The SDs of the IVS320 with the HPMC and PVP polymers presented the best results for antichagasic activity when obtained by the rotary evaporation (RE) method. However, among the results, the IVS320-PVP/RE system stood out due to the 82% antichagasic activity rate and the improvement in inhibitory activity in relation to all the concentrations evaluated, when compared to pure IVS320. The SD of IVS320-PVP/KN also displays significant inhibitory activity, showing a rate of around 65% at a 50 µg/mL concentration.

The activity of the IVS320 incorporated in the dispersions was increased when compared to the pure IVS320, which can probably be explained as being due to the increment in the solubility of the compound in these polymeric matrices. Also, linked to this behavior, it can have the influence of the polymer. Despite the low inhibitory activity found in each of the polymers, even at higher concentrations, there may be a synergism of the inhibitory effect between the two constituents of the polymeric matrix (IVS320 and polymer), and it may also be another factor that explains the increase in the antichagasic activity of the SDs in relation to pure IVS320.

Studies carried out by Lima et al. [38] show that a reduction in particle size, an increase in surface area, and the interaction between polymeric carriers can result in the increase in the solubility of drugs.

4. Conclusions

Using different polymeric materials, it was possible to evaluate the interactions of carriers with the compound IVS320 to obtain SDs. The results showed changes in vibrational modes, as well as alterations in the thermal behavior and IVS320 crystallinity pattern, which in turn, led to an improvement of the IVS320 antichagasic activity, when present in polymeric matrices. The results indicated that the inhibitory growth (%) improved gradually with the increase in the concentration of the samples, and IVS320-polymer SDs showed potentiated activity in comparison with IVS320 in some concentrations. It was observed that the SD obtained by the rotary evaporation method with the PVP polymer resulted in the best antichagasic activity of the IVS320 incorporated in the SDs. This behavior shows that the addition of polymers alters the effectiveness of IVS320 in terms of its antichagasic activity, probably due to the improved solubility and bioavailability of IVS320.

Author Contributions: Conceptualization: Á.A.N.d.L., V.d.S.O., and E.D.D.; methodology: V.d.S.O., E.D.D., A.T.d.S.Q., J.W.d.F.O., M.d.S.d.S., P.G.F., F.d.C.d.S., and V.F.F.; formal analysis: Á.A.N.d.L. and V.d.S.O.; data curation: V.d.S.O. and E.D.D.; writing—original draft preparation: Á.A.N.d.L. and V.d.S.O.; writing—review and editing: Á.A.N.d.L., V.d.S.O., E.D.D., A.T.d.S.Q., J.W.d.F.O., M.d.S.d.S., P.G.F., F.d.C.d.S., and V.F.F. All authors have read and agreed to the published version of the manuscript.

Funding: This research was financed in part by the Coordenação de Aperfeiçoamento de Pessoal de Nível Superior-Brasil (CAPES)-Finance Code 001.

Acknowledgments: The authors wish to thank the Coordenação de Aperfeiçoamento de Pessoal de Nível Superior (CAPES) and Conselho Nacional de Desenvolvimento Científico e Tecnológico (CNPq) for grants and fellowships. V.d.S.O. and J.W.d.F.O. acknowledge the financial support (PhD and Postdoctoral fellowships) provided by CAPES, Brazil. M.d.S.d.S. acknowledges CNPq, Brazil, for the Research Grant. This work was partially funded by Global Health and Tropical Medicine and Fundação para a Ciência e Tecnologia, FCT, Portugal (Grant number IHMT-UID/multi/04413/2013 and Grant number PTDC/CVT/CVT/28908/2017).

Conflicts of Interest: The authors declare no conflict of interest.

References

1. Gómez, L.A.; Gutierrez, F.R.S.; Peñuela, O.A. Trypanosoma cruzi infection in transfusion medicine. *Hematol. Transfus. Cell Ther.* **2019**, *41*, 262–267. [CrossRef] [PubMed]
2. Morais, T.R.; Conserva, G.A.A.; Varela, M.T.; Costa-Silva, T.A.; Thevenard, F.; Ponci, V.; Fortuna, A.; Falcão, A.C.; Tempone, A.G.; Fernandes, J.P.S.; et al. Improving the drug-likeness of inspiring natural products—Evaluation of the antiparasitic activity against Trypanosoma cruzi through semi-synthetic and simplified analogues of licarin A. *Sci. Rep.* **2020**, *10*, 1–14. [CrossRef] [PubMed]
3. Gomes, C.; Almeida, A.B.; Rosa, A.C.; Araujo, P.F.; Teixeira, A.R.L. American trypanosomiasis and Chagas disease: Sexual transmission. *Int. J. Infect. Dis.* **2019**, *81*, 81–84. [CrossRef] [PubMed]
4. Ibis, C.; Tuyun, A.F.; Bahar, H.; Ayla, S.S.; Stasevych, M.V.; Ya Musyanovych, R.; Komarovska-Porokhnyavets, O.; Novikov, V. Nucleophilic substitution reactions of 1,4-naphthoquinone and biologic properties of novel S-, S,S-, N-, and N,S-substituted 1,4-naphthoquinone derivatives. *Med. Chem. Res.* **2014**, *23*, 2140–2149. [CrossRef]
5. Kapadia, G.; Rao, G.; Sridhar, R.; Ichiishi, E.; Takasaki, M.; Suzuki, N.; Konoshima, T.; Iida, A.; Tokuda, H. Chemoprevention of Skin Cancer: Effect of Lawsonia inermis L. (Henna) Leaf Powder and its Pigment Artifact, Lawsone in the Epstein- Barr Virus Early Antigen Activation Assay and in Two-Stage Mouse Skin Carcinogenesis Models. *Anticancer Agents Med. Chem.* **2013**, *13*, 1500–1507. [CrossRef] [PubMed]
6. Leyva, E.; Loredo Carrillo, S.E.; López, L.I.; Escobedo Avellaneda, E.G.; Navarro-Tovar, G. Importancia química y biológica de naftoquinonas. Revisión bibliográfica. *Afinidad Rev. Química Teórica Apl.* **2017**, *74*, 36–50.
7. López López, L.I.; Nery Flores, S.D.; Silva Belmares, S.Y.; Sáenz Galindo, A. Naphthoquinones: Biological properties and synthesis of lawsone and derivatives—A structured review. *Vitae* **2014**, *21*, 248–258.
8. Sánchez-Calvo, J.M.; Barbero, G.R.; Guerrero-Vásquez, G.; Durán, A.G.; Macías, M.; Rodríguez-Iglesias, M.A.; Molinillo, J.M.G.; Macías, F.A. Synthesis, antibacterial and antifungal activities of naphthoquinone derivatives: A structure–activity relationship study. *Med. Chem. Res.* **2016**, *25*, 1274–1285. [CrossRef]
9. Ramos-Peralta, L.; Lopez-Lopez, L.I.; Silva-Belmares, S.Y.; Zugasti-Cruz, A.; Rodriguez-Herrera, R.; Anguilar-Gonzalez, C.N. Naphthoquinone: Bioactivity and Green Synthesis. In *The Battle Against Microbial Pathogens: Basic Science, Technological Advances and Educational Programs*; Méndez-Vilas, A., Ed.; Formatex Research Center: Badajoz, Spain, 2015; pp. 542–550.
10. Rivera-Ávalos, E.; De Loera, D.; Araujo-Huitrado, J.G.; Escalante-García, I.L.; Muñoz-Sánchez, M.A.; Hernández, H.; López, J.A.; López, L. Synthesis of amino acid-naphthoquinones and in vitro studies on cervical and breast cell lines. *Molecules* **2019**, *24*, 4285. [CrossRef]

11. Sodero, A.C.R.; Abrahim-Vieira, B.; Torres, P.H.M.; Pascutti, P.G.; Garcia, C.R.S.; Ferreira, V.F.; da Rocha, D.R.; Ferreira, S.B.; Silva, F.P. Insights into cytochrome bc1 complex binding mode of antimalarial 2-hydroxy-1,4-naphthoquinones through molecular modelling. *Mem. Inst. Oswaldo Cruz* **2017**, *112*, 299–308. [CrossRef]
12. Sharma, U.; Katoch, D.; Sood, S.; Kumar, N.; Singh, B.; Thakur, A.; Gulati, A. Synthesis, antibacterial and antifungal activity of 2-amino-1,4- naphthoquinones using silica-supported perchloric acid ($HClO_4$-SiO_2) as a mild, recyclable and highly efficient heterogeneous catalyst. *Indian J. Chem. Sect. B Org. Med. Chem.* **2013**, *52*, 1431–1440. [CrossRef]
13. Naujorks, A.A.D.S.; Da Silva, A.O.; Lopes, R.D.S.; De Albuquerque, S.; Beatriz, A.; Marques, M.R.; De Lima, D.P. Novel naphthoquinone derivatives and evaluation of their trypanocidal and leishmanicidal activities. *Org. Biomol. Chem.* **2015**, *13*, 428–437. [CrossRef]
14. Ferreira, S.B.; Salomão, K.; De Carvalho Da Silva, F.; Pinto, A.V.; Kaiser, C.R.; Pinto, A.C.; Ferreira, V.F.; De Castro, S.L. Synthesis and anti-Trypanosoma cruzi activity of β-lapachone analogues. *Eur. J. Med. Chem.* **2011**, *46*, 3071–3077. [CrossRef]
15. Bourguignon, S.C.; Cavalcanti, D.F.B.; de Souza, A.M.T.; Castro, H.C.; Rodrigues, C.R.; Albuquerque, M.G.; Santos, D.O.; da Silva, G.G.; da Silva, F.C.; Ferreira, V.F.; et al. Trypanosoma cruzi: Insights into naphthoquinone effects on growth and proteinase activity. *Exp. Parasitol.* **2011**, *127*, 160–166. [CrossRef] [PubMed]
16. do Perpetuo Socorro Borges Carrio Ferreira, M.; do Carmo Cardoso, M.F.; de Carvalho da Silva, F.; Ferreira, V.F.; Lima, E.S.; Souza, J.V.B. Antifungal activity of synthetic naphthoquinones against dermatophytes and opportunistic fungi: Preliminary mechanism-of-action tests. *Ann. Clin. Microbiol. Antimicrob.* **2014**, *13*, 1–6. [CrossRef] [PubMed]
17. Dantas, E.D.; de Souza, F.J.J.; Nogueira, W.N.L.; Silva, C.C.; de Azevedo, P.H.A.; Aragão, C.F.S.; de Almeida, P.D.O.; do Carmo Cardoso, M.F.; de Carvalho da Silva, F.; de Azevedo, E.P.; et al. Characterization and trypanocidal activity of a novel pyranaphthoquinone. *Molecules* **2017**, *22*, 1631. [CrossRef] [PubMed]
18. Nicoletti, C.D.; Faria, A.F.M.; de Sá Haddad Queiroz, M.; dos Santos Galvão, R.M.; Souza, A.L.A.; Futuro, D.O.; Faria, R.X.; Ferreira, V.F. Synthesis and biological evaluation of β-lapachone and nor-β-lapachone complexes with 2-hydroxypropyl-β-cyclodextrin as trypanocidal agents. *J. Bioenerg. Biomembr.* **2020**. [CrossRef] [PubMed]
19. Sun, Z.; Zhang, H.; He, H.; Sun, L.; Zhang, X.; Wang, Q.; Li, K.; He, Z. Cooperative effect of polyvinylpyrrolidone and HPMC E5 on dissolution and bioavailability of nimodipine solid dispersions and tablets. *Asian J. Pharm. Sci.* **2019**, *14*, 668–676. [CrossRef]
20. Xiong, X.; Zhang, M.; Hou, Q.; Tang, P.; Suo, Z.; Zhu, Y.; Li, H. Solid dispersions of telaprevir with improved solubility prepared by co-milling: Formulation, physicochemical characterization, and cytotoxicity evaluation. *Mater. Sci. Eng. C* **2019**, *105*. [CrossRef]
21. Alves, L.D.S.; De La Roca Soares, M.F.; De Albuquerque, C.T.; Da Silva, É.R.; Vieira, A.C.C.; Fontes, D.A.F.; Figueirêdo, C.B.M.; Soares Sobrinho, J.L.; Rolim Neto, P.J. Solid dispersion of efavirenz in PVP K-30 by conventional solvent and kneading methods. *Carbohydr. Polym.* **2014**, *104*, 166–174. [CrossRef]
22. Lima, Á.A.N.; Sobrinho, J.L.S.; De Lyra, M.A.M.; Dos Santos, F.L.A.; Figueirêdo, C.B.M.; Neto, P.J.R. Evaluation of in vitro dissolution of benznidazole and binary mixtures: Solid dispersions with hydroxypropylmethylcellulose and β-cyclodextrin inclusion complexes. *Int. J. Pharm. Pharm. Sci.* **2015**, *7*, 371–375.
23. Chavan, R.B.; Rathi, S.; Jyothi, V.G.S.S.; Shastri, N.R. Cellulose based polymers in development of amorphous solid dispersions. *Asian J. Pharm. Sci.* **2019**, *14*, 248–264. [CrossRef] [PubMed]
24. Lima, A.A.N.; dos Santos, P.B.S.; de Lyra, M.A.M.; dos Santos, F.L.A.; Rolim-Neto, P.J. Solid dispersion systems for increase solubility: Cases with hydrophilic polymers in poorly water soluble drugs. *Braz. J. Pharm.* **2011**, *92*, 269–278.
25. Figueirêdo, C.B.M.; Nadvorny, D.; de Medeiros Vieira, A.C.Q.; de Medeiros Schver, G.C.R.; Soares Sobrinho, J.L.; Rolim Neto, P.J.; Lee, P.I.; de La Roca Soares, M.F. Enhanced delivery of fixed-dose combination of synergistic antichagasic agents posaconazole-benznidazole based on amorphous solid dispersions. *Eur. J. Pharm. Sci.* **2018**, *119*, 208–218. [CrossRef]
26. de Frana Almeida Moreira, C.D.L.; de Oliveira Pinheiro, J.G.; Silva-Júnior, W.F.; da Barbosa, E.G.; Lavra, Z.M.M.; Pereira, E.W.M.; Resende, M.M.; de Azevedo, E.P.; Quintans-Júnior, L.J.; de Souza Araújo, A.A.;

et al. Amorphous solid dispersions of hecogenin acetate using different polymers for enhancement of solubility and improvement of anti-hyperalgesic effect in neuropathic pain model in mice. *Biomed. Pharmacother.* **2018**, *97*, 870–879. [CrossRef]
27. Ribeiro, M.A.; Lanznaster, M.; Silva, M.M.P.; Resende, J.A.L.C.; Pinheiro, M.V.B.; Krambrock, K.; Stumpf, H.O.; Pinheiro, C.B. Cobalt lawsone complexes: Searching for new valence tautomers. *Dalton Trans.* **2013**, *42*, 5462–5470. [CrossRef]
28. Cardoso, S.H.; de Oliveira, C.R.; Guimarães, A.S.; Nascimento, J.; de Oliveira dos Santos Carmo, J.; de Souza Ferro, J.N.; de Carvalho Correia, A.C.; Barreto, E. Synthesis of newly functionalized 1,4-naphthoquinone derivatives and their effects on wound healing in alloxan-induced diabetic mice. *Chem. Biol. Interact.* **2018**, *291*, 55–64. [CrossRef]
29. Salunke-Gawali, S.; Pereira, E.; Dar, U.A.; Bhand, S. Metal complexes of hydroxynaphthoquinones: Lawsone, bis-lawsone, lapachol, plumbagin and juglone. *J. Mol. Struct.* **2017**, *1148*, 435–458. [CrossRef]
30. Jahangiri, A.; Barzegar-jalali, M.; Garjani, A.; Javadzadeh, Y.; Hamishehkar, H. European Journal of Pharmaceutical Sciences Evaluation of physicochemical properties and in vivo ef fi ciency of atorvastatin calcium / ezetimibe solid dispersions. *Eur. J. Pharm. Sci.* **2016**, *82*, 21–30. [CrossRef]
31. Zhai, X.; Li, C.; Lenon, G.B.; Xue, C.C.L.; Li, W. Preparation and characterisation of solid dispersions of tanshinone IIA, cryptotanshinone and total tanshinones. *Asian J. Pharm. Sci.* **2017**, *12*, 85–97. [CrossRef]
32. He, Y.; Liu, H.; Bian, W.; Liu, Y.; Liu, X.; Ma, S.; Zheng, X.; Du, Z.; Zhang, K.; Ouyang, D. Molecular interactions for the curcumin-polymer complex with enhanced anti-inflammatory effects. *Pharmaceutics* **2019**, *11*, 442. [CrossRef] [PubMed]
33. Chen, B.; Wang, X.; Zhang, Y.; Huang, K.; Liu, H.; Xu, D.; Li, S.; Liu, Q.; Huang, J.; Yao, H.; et al. Improved solubility, dissolution rate, and oral bioavailability of main biflavonoids from Selaginella doederleinii extract by amorphous solid dispersion. *Drug Deliv.* **2020**, *27*, 309–322. [CrossRef] [PubMed]
34. Masek, A.; Chrzescijanska, E.; Latos-brozio, M.; Zaborski, M. Voltammetry and Spectrophotometric Methods. *Food Chem.* **2019**, *301*, 125279. [CrossRef] [PubMed]
35. da Silva Júnior, W.F.; Pinheiro, J.G.D.O.; De Menezes, D.L.B.; De Sobral E Silva, N.E.; De Almeida, P.D.O.; Silva Lima, E.; Da Veiga Júnior, V.F.; De Azevedo, E.P.; De Lima, Á.A.N. Development, physicochemical characterization and in vitro anti-inflammatory activity of solid dispersions of α,β amyrin isolated from protium oilresin. *Molecules* **2017**, *22*, 1512. [CrossRef] [PubMed]
36. Mishra, R.; Varshney, R.; Das, N.; Sircar, D.; Roy, P. Synthesis and characterization of gelatin-PVP polymer composite scaffold for potential application in bone tissue engineering. *Eur. Polym. J.* **2019**, *119*, 155–168. [CrossRef]
37. Costa, A.R.D.M.; Marquiafável, F.S.; De Oliveira Lima Leite Vaz, M.M.; Rocha, B.A.; Pires Bueno, P.C.; Amaral, P.L.M.; Da Silva Barud, H.; Berreta-Silva, A.A. Quercetin-PVP K25 solid dispersions: Preparation, thermal characterization and antioxidant activity. *J. Therm. Anal. Calorim.* **2011**, *104*, 273–278. [CrossRef]
38. Lima, A.A.N.; Soares-Sobrinho, J.L.; Silva, J.L.; Corrêa-Júnior, R.A.C.; Lyra, M.A.M.; Santos, F.L.A.; Oliveira, B.G.; Hernandes, M.Z.; Rolim, L.A.; Rolim-Neto, P.J. The Use of Solid Dispersion Systems in Hydrophilic Carriers to Increase Benznidazole Solubility. *J. Pharm. Sci.* **2011**, *100*, 2443–2451. [CrossRef]

Publisher's Note: MDPI stays neutral with regard to jurisdictional claims in published maps and institutional affiliations.

pharmaceutics MDPI

Review

Pulmonary Drug Delivery of Antimicrobials and Anticancer Drugs Using Solid Dispersions

Hisham Al-Obaidi *, Amy Granger, Thomas Hibbard and Sefinat Opesanwo

The School of Pharmacy, University of Reading, Reading RG6 6AD, UK; a.granger@student.reading.ac.uk (A.G.); t.hibbard@pgr.reading.ac.uk (T.H.); s.t.opesanwo@student.reading.ac.uk (S.O.)
* Correspondence: h.al-obaidi@reading.ac.uk; Tel.: +44-1183-786-261

Abstract: It is well established that currently available inhaled drug formulations are associated with extremely low lung deposition. Currently available technologies alleviate this low deposition problem via mixing the drug with inert larger particles, such as lactose monohydrate. Those inert particles are retained in the inhalation device or impacted in the throat and swallowed, allowing the smaller drug particles to continue their journey towards the lungs. While this seems like a practical approach, in some formulations, the ratio between the carrier to drug particles can be as much as 30 to 1. This limitation becomes more critical when treating lung conditions that inherently require large doses of the drug, such as antibiotics and antivirals that treat lung infections and anticancer drugs. The focus of this review article is to review the recent advancements in carrier free technologies that are based on coamorphous solid dispersions and cocrystals that can improve flow properties, and help with delivering larger doses of the drug to the lungs.

Keywords: cocrystals; coamorphous; solubility; dry powder inhalers; pulmonary drug delivery

Citation: Al-Obaidi, H.; Granger, A.; Hibbard, T.; Opesanwo, S. Pulmonary Drug Delivery of Antimicrobials and Anticancer Drugs Using Solid Dispersions. *Pharmaceutics* **2021**, *13*, 1056. https://doi.org/10.3390/pharmaceutics13071056

Academic Editor: Carsten Ehrhardt

Received: 14 June 2021
Accepted: 6 July 2021
Published: 10 July 2021

1. Introduction

The respiratory tract is susceptible to a range of conditions, such as viral, bacterial and fungal infections [1], which in turn can result in an exacerbation of other existing conditions through inflammation [2]. Lower respiratory tract infections which include bronchitis, tuberculosis and pneumonia, are classed as two of the leading causes of death, while pneumonia is a leading cause of death in children globally [3]. While inflammation can affect the entire respiratory system, different pathogens will inhabit different parts of the respiratory tract. For example, tuberculosis causative microorganism *Mycobacterium tuberculosis* colonises the lungs, deep within, and also inside the lung's alveolar surfaces [4]. In cystic fibrosis (CF) patients, the bacterium, *Pseudomonas aeruginosa*, inhabits the conducting and respiratory zones of the lungs and is associated with recurrent infections. This is mainly caused by bacterial transformation to the biofilm producing mucoid strain, which exhibits increased resistance to both antibiotics and natural lung defence mechanisms, such as phagocytosis.

In addition to respiratory tract infections, lung cancer is a major cause of death. It is estimated that 1.6 million people die every year from lung cancer, making it one of the most fatal cancers [5]. Lung cancer remains difficult to cure using chemotherapy, as evident by the low long-term survival rate of patients [6]. Apart from the frequent low efficacy, cancer treatments have been associated with significant side effects. A study that included 449 cancer patients, revealed 86% of patients experienced at least one side effect from cancer chemotherapy, with 67% having experienced six or more side effects [7]. This calls for the exploration of new approaches, aimed at reducing the side effects, and in turn, improving patient tolerance [8], and further improve the quality of care provided to patients.

The pulmonary route has been used to treat different lung conditions, such as asthma and chronic obstructive pulmonary disease (COPD). Drugs delivered to the lungs are typically formulated in low doses; however, there is an increasing clinical need to deliver higher doses. For example, when antimicrobials are delivered directly to the lungs, higher doses ensure optimum lung concentration for tackling the infection [9]. Hence, a major challenge is to formulate these drugs with efficient deposition, while maximising the delivered dose. Current technologies are based on formulating the required doses of the active pharmaceutical ingredient (API) with the addition of excipients (known as carrier particles) [10]. These carrier particles serve to minimise undesired deposition into the oropharyngeal region, as well as reducing loss in the inhalation device itself. Among approaches to deliver larger doses of the drug to the lungs is solid dispersions. These are molecular mixtures of the drug and a miscible carrier by which properties can be tailored to achieve optimum outcomes, such as improved solubility [11]. The carrier can be a small molecule known as a coformer or a polymeric carrier leading to the formation of crystalline or amorphous dispersions.

Although oral formulations have a more enhanced therapeutic profile when it comes to treating systemic diseases, inhaled drug formulations are better at targeting lung conditions, such as infections and cancer. The aim of this review article is to explore the respiratory route for drug delivery, highlighting its advantages and challenges to deliver larger doses. While there have been different approaches to maximise the dose that can be delivered, the focus of this review is on recent advancements in the delivery of particulates prepared as solid dispersions. The use of solid dispersions represents a novel approach to deliver larger amounts of the drug while maintaining enhanced physicochemical properties. These physical molecular complexes (i.e., solid dispersions) can be engineered to modify properties, such as adhesion, aerodynamic diameter and morphology, allowing enhanced pulmonary drug delivery.

Physiology of the Lungs and Factors Affecting Particles Deposition

The pulmonary alveoli exhibit a large surface area of over 100 m^2 and thin walls of less than 1 μm, allowing fast absorption of drugs into the rich blood supply for systemic effect [12]. This would be beneficial for a number of drugs as the fast absorption would result in a more rapid onset of action compared to other administration routes, such as oral administration. As well as being advantageous for drugs acting both locally and systemically, it also has reduced systemic side effects that are common with other administration routes [13]. It is considered a non-invasive form of drug administration and often require lower doses compared with other systemic drug delivery routes [14]. The local administration also avoids first pass liver metabolism, which is detrimental for some drugs. Thus, the pulmonary route may be favoured over other parenteral routes, or when absorption via the gastrointestinal (GI) tract is inappropriate or ineffective [12,15].

The respiratory system comprise 23 generations (G0–G23) which have varying sizes, structures and functions [16]. Each generation splits into two smaller daughter branches to give the next generation. Altogether, these generations are divided into two zones: Conducting and respiratory. The conducting zone includes the structures from the trachea to the bronchioles from G0 to G16. The main role of the conducting zone is to carry the air into the lungs. The respiratory zone includes structures from the respiratory bronchioles to the alveolar ducts and alveoli from G17 to G23. This zone contains functional tissues where gaseous exchange occurs [17].

As a result of the varying size and structure of the lungs, inhaled particle size affects how deep into the lungs the drug and excipients can penetrate. For example, particles with a mass median aerodynamic diameter (MMAD) of 10 μm or larger tend to be deposited in the oropharyngeal region, whereas smaller particles with an MMAD of less than 3 μm can penetrate much deeper through the lungs into the alveoli [18]. MMAD is a measurement used to define the size of aerosol particles. The aerodynamic diameter of a particle relates to a sphere with the same density as water (1 g/cm^3) which settles at the same velocity as the particle of interest in still air [19]. Using the mass median value, the aerosol size distribution is divided in half [20].

Particles intended for pulmonary drug delivery are categorised according to their size. The ideal particle size for inhalation is said to be <5 µm [21], with <3 µm having an 80% chance of reaching the lower airways and 50–60% chance of reaching the alveolar regions [22]. Particles > 5 µm are identified as coarse particles; fine particles range from 0.1–5 µm, and ultrafine particles are <0.1 µm. A monodisperse aerosol is said to be highly desirable for maximum deposition and specific targeting in the lungs [23]. However, there have been reports showing minimum differences in deposition between polydisperse and monodisperse aerosols [24]. The deposition of inhaled particles within the regions of the lungs is dependent on a number of physiological and pharmaceutical factors, such as particle shape, particle size, surface morphology, the breathing rate of the patient, lung volume and health condition of the patient. The physicochemical factors that affect particles deposition in the respiratory system are shown in Table 1. Particle deposition can occur through the following mechanisms: sedimentation, impaction and diffusion. Diffusion is a fundamental mechanism of particle deposition for particles < 0.5 µm. The process is influenced by Brownian motion—in other words, motion increases with decreasing particle size and particles move from a higher concentration to a lower concentration leading to the deposition upon contact with the airway walls. This mechanism heavily influences deposition in the lower regions of the lungs and the alveoli. Gravitational sedimentation can happen at a later region, typically within the tracheobronchial region (approximately the last six generations), as a result of the relatively low air velocity within this region [25]. In fact, as the residence time is longer, a combination of both sedimentation and free diffusion can occur. It is vital for drug absorption to occur, that the particles are deposited before exhalation takes place. Typical size depends on the aerodynamic diameter, which is approximately above 0.5 µm for sedimentation and below that for diffusion [26,27]. The bigger the particle and the lower the airflow rate, the faster the sedimentation in which inertial impaction plays a significant role for particles > 5 µm [28].

Table 1. Physicochemical factors to be considered when designing inhaled formulations. Each parameter influences both aerosolisation and deposition. Adapted from the work by the authors of [29–31].

Property Type	Parameter
Aerosol	Air/Particle velocity Mass median aerodynamic diameter Fine particle fraction
Particle	Bulk density Tap density Shape Charge Surface energy * Surface texture * Surface composition *
Physicochemical	Solubility Hygroscopicity

Note—* Factors specifically affecting aerosolisation.

2. The Necessity to Deliver Larger Doses to Treat Lung Infections and Cancer

There are a variety of inhaled products for the treatment of conditions affecting the respiratory system (Table 2). As can be seen in Table 2, the maximum dose that has been successfully formulated for inhalation, thus far, is colistimethate sodium, with doses ranging from 80 to 125 mg, formulated as inhalation powder. Many of the higher inhalable doses are administered via the use of nebulisers, which may require guidance and can also be complicated for patients to use [32]. Desgrouas and Ehrmann reviewed the available evidence and called for developing inhaled antibiotics, especially for mechanically ventilated patients [33]. However, the majority of inhalable drug formulations are available in the dose range of micrograms and are indicated for conditions, such as asthma and COPD. In addition, there is a limited number of inhaled medications available as combinations, which are again mostly limited for use in asthma and COPD.

Table 2. Examples of currently available inhaled products in the market.

Drug	Quantity of API per Dose	Indication	Ref
Salbutamol	100–200 µg	Asthma	[34]
Fluticasone propionate	50–500 µg	Prophylaxis of asthma	[35]
Colistimethate sodium	80–125 mg	Treatment of pneumonia	[36]
Tiotropium	10–18 µg	Maintenance of COPD	[37]
Nedocromil sodium	2 mg	Prophylaxis of asthma	[38]
Zanamivir	5 mg	Treatment of influenza	[39]
Mannitol	5–40 mg	Treatment of cystic fibrosis as add-on therapy to standard care	[40]
Budesonide with formoterol	100–400 µg with 4.5–12 µg	Maintenance of asthma	[41]
Ciclesonide	80–160 µg	Prophylaxis of asthma	[42]

There are two major lung conditions that require high doses of medications, these being infections and malignancy. For lung infections, the causative bacteria and viruses can be found throughout the different structures of the lungs, which will alter the efficacy of treatment. The conducting zone of the respiratory system, which consists of the trachea, bronchi and bronchioles, plays a role in trapping microbes in mucus produced by goblet cells. The mucus is then transported by ciliated cells located on the epithelium to the oropharynx, where the mucus is either swallowed or removed via coughing [43]. However, in cases where the microbes go on to develop into infections, it is important to determine the specific site in the lungs where the infection has developed to effectively target and deliver the drug directly to the causative microbes. Table 3 lists common loci of prevalent microorganisms within the respiratory tract noting that several pathogens found in the respiratory tract are associated with the development of biofilms.

Upper respiratory tract infections typically occur in the conducting zone, examples of which include common cold, sinusitis and pharyngitis, whereas lower respiratory tract infections happen in the respiratory zone and include pneumonia and bronchitis. In addition, some microorganisms, such as *Haemophilus influenzae*, that are normally found in the conducting zone can go on to migrate to the respiratory zone, subsequently causing lower respiratory tract infections [44]. Some studies suggest that the conducting zone of the respiratory tract is the more common zone where bacteria and viruses are contained during infections. One study found that for *Pseudomonas aeruginosa* biofilms in cystic fibrosis patients, the majority of the bacteria were present in the conducting zone, which acted as a reservoir for the bacteria to multiply and form biofilms [45]. Consequently, it is important for inhaled formulations of antibacterials and antivirals to reach and deposit in the conducting zone of the respiratory tract to effectively treat the infections. With the ability to locally deliver antimicrobial drugs for treatment, there could also be a further advantage of reducing antimicrobial resistance. With oral and intravenous administration, there is a risk of an accumulation in infection-free sites, leading to the development of antimicrobial resistance. In addition, a number of antimicrobials, such as tobramycin, can be toxic when given in repeated high doses systemically, which can occur when antimicrobials are prescribed for recurrent infections [46]. Hence, utilising local drug delivery directly into the lungs could reduce the risk of toxicity.

Table 3. Location of infections causative microorganisms within the respiratory tract.

Organism	Type	Location/Generation	Reference
Bacteria	*Pseudomonas aeruginosa*	Non-mucoid strain located mainly in the conduction airways Mucoid strain present throughout the respiratory zone	[47]
	Staphylococcus aureus	Nasal cavity Generation 0	[48]
	Mycobacterium tuberculosis *Chlamydia pneumonia*	Alveolar surfaces Macrophages in lungs Generation 20–22 Generation 21–23 Alveolar type 2 cells	[4] [2]
Fungi	*Aspergillus* spp.	Terminal bronchioles, Terminal airways Generation 16–23	[49]
Viruses	Herpes Simplex Virus	Oropharynx–Generation 0	[48]

Similar to infectious conditions, lung cancer is another serious condition that requires large doses of drugs for treatment. Current survival rates for all stages of lung cancer include 40% of patients surviving for one year or more after diagnosis, 15% of patients surviving for five years or more after diagnosis and only 10% of patients surviving for 10 years or more after diagnosis [50]. The nature of lung cancer suggests that it is more effectively treated by direct delivery to the lungs. Sardeli et al., suggested the need to use inhaled immunotherapy as opposed to intravenous administration to avoid systemic side effects and achieve a localised effect [51]. Hence, pulmonary drug delivery provides a novel opportunity to avoid unwanted drug distribution and could achieve maximum deposition of the drug at the site of action. As such, there is the potential for higher concentrations of the drugs to reach the lungs compared with oral and intravenous administration. Lung cancer cells are prone to rapidly developing resistance to anticancer drugs, so higher doses are given to combat this challenge [52]. When administered systemically, chemotherapy drugs can cause toxic side effects affecting healthy organs, especially when administering large doses, which may be prevented by local administration of the drugs [53]. By targeting the lungs directly, this also means there is potential for an accumulation of the drug to build up in the tumour cells, rather than in the kidneys, liver and spleen, which is observed with systemic drug use [54]. This has the potential to beneficially impact patients' treatment and increase the likelihood of remission and survival.

3. Challenges Associated with Drug Delivery to the Lungs

Whilst pulmonary drug delivery has its advantages, to ensure advantageous deposition in the lungs, drug particles must be within the optimum size for lung deposition (Table 4 and Figure 1). If the particle size is relatively large, the deposition will occur in the larynx causing irritation, and if the particle size is relatively small, the particles will be immediately exhaled from the lungs and will not be deposited [55]. It is important to note that drug particles should be deposited in the lungs in a high enough proportion for the API to be absorbed and produce a therapeutic effect. Unsurprisingly, it has been shown that higher drug deposition in the lungs leads to enhanced clinical benefits [56]. As mentioned above, a common approach to improve the flow properties of powder for inhalation is to use carrier particles (such as lactose monohydrate). This is based on combining the drug with the larger lactose particles, which strip from the drug particles inside the device in response to the high velocity created by inhalation [57,58]. Any escaped lactose will deposit in the throat, leaving the drug particles to carry on with their journey towards the alveoli. Hence, the main challenge would be to deliver high doses of the API without significant loss in the oropharynx region. Another challenge is that inhalation devices do not have the capacity to accommodate high masses of powders [59].

As shown by the data collated and presented in Table 4, two different studies used a combination of beclomethasone dipropionate with formoterol labelled with technetium-99. The first dissolved the particles in hydrofluoroalkane to give an MMAD of 1.3 µm via pressurised metered-dose inhaler (pMDI) [60], the second delivered the solid particles with MMAD of 1.5 µm via a NEXThaler® dry powder inhaler (DPI) [61]. The lung deposition for the pMDI was 34.08% (standard deviation (SD) = 9.3%), and the DPI had a lung deposition of 55.2% (SD = 3.7%). This shows that even though the same APIs were used, and the particle sizes were similar, the way that the API is formulated plays a significant role in the lung deposition of the particles. It is apparent that the type of the API has an effect by comparing the lung deposition of albuterol and beclomethasone dipropionate with formoterol in the pMDI considered previously [62]. They both have an MMAD of 1.5 µm, are labelled with technium-99 and are administered using an pMDI; however, the lung deposition of albuterol was 56.3% (SD = 9.2%), which is considerably higher than the beclomethasone, which has a lung deposition of 34.08% (SD = 9.3%). This suggests that API type and formulation must also be carefully considered when formulating drugs for pulmonary drug delivery.

Furthermore, Table 4 shows that within the same administration device, the preparation method of the particle has an impact on the deposition within the lungs. One study compared two suspensions of beclomethasone dipropionate labelled with technetium-99 in a metered-dose inhaler, by which the first formulation was dissolved in hydrofluoroalkane, and the other was dissolved in chlorofluorocarbon [63], it demonstrated that the different preparation methods led to a difference in both MMAD and lung deposition. Therefore, this suggests that the type of excipients used must be carefully considered for optimum lung deposition.

Table 4. MMAD, lung deposition and preparation method of different inhaled APIs.

API	MMAD (µm)	Lung Deposition (%)	Preparation of API	Ref
Formoterol	0.8	31 ± 11	Labelled with technetium-99 and dissolved in hydrofluoroalkane (HFA) in an pMDI	[64]
Beclomethasone dipropionate	0.9	53 ± 7	Labelled with technetium-99 and dissolved in HFA in an pMDI	[63]
Fluticasone propionate	2	12 ± 7	Labelled with technetium-99 and dissolved in chlorofluorocarbon (CFC) in an pMDI	
Beclomethasone dipropionate	3.5	4 ± 11		
Albuterol (salbutamol)	1.5	56.3 ± 9.2	Labelled with technetium-99 in an pMDI	[62]
	3	51 ± 8.9		
	6	46 ± 13.6		
Beclomethasone dipropionate and formoterol	1.3	34.08 ± 9.3	Labelled with technetium-99 and dissolved in HFA in a pMDI	[60]
Ciclesonide	1	52 ± 11	Labelled with technetium-99 and dissolved in HFA in a pMDI	[65]
Beclomethasone dipropionate and formoterol fumarate	1.5	55.2 ± 3.7	Labelled with technetium-99 in a NEXThaler® DPI	[61]

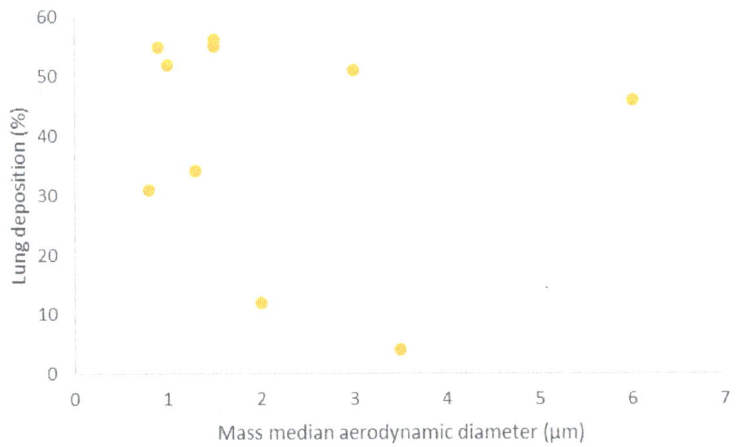

Figure 1. A scatter graph, comparing MMAD with corresponding lung deposition values in Table 4.

4. Impact of Drug Delivery Devices on the Extent of Pulmonary Drug Delivery

There are currently three major categories of delivery devices that are used for inhalation therapy: pressurised metered-dose inhalers (pMDIs), nebulisers and dry powder inhalers (DPIs). pMDIs are widely available, portable inhalers designed to releases a specific dose of the drug, following depression of an actuator, which the patient then inhales in a single inspiration or via a spacer device. Studies have found that many patients have difficulty in using these devices, due to the required coordination of pressing down on the actuator and inhaling the resultant mist simultaneously [66]. If patients do not have the required coordination to self-administer their medication using a pMDI, some of the drug is deposited in the mouth and at the back of the throat. As a result, less of the API will be inhaled than intended, so the correct dose will not be administered. This could potentially result in substandard treatment, leading to exacerbation or worsening of a patient's condition. pMDIs are formulated as a suspension or solution with one or more APIs along with a propellant. The propellant is used to generate the pressure required to form micron-scaled droplets for inhalation [67]. Furthermore, pMDIs deposit less of the drug into the lungs, due to the high particle exit velocity with the actuations [68]. This means that it is more difficult to formulate higher drug doses for use with this type of inhaler device.

DPIs are portable inhalers that operate by breath actuation, thus removing the need for coordination of inhalation with actuation required for pMDIs. Breath actuation involves the powder being released from the inhaler when the patient inhales with enough force, known as inspiratory flow. The inhalers are formulated to contain powder forms of drugs, and this solid-state gives the drug improved stability compared to pMDIs formulations [67]. The use of powders also allows for greater potential when formulating higher dose medications. Dry powder inhalers are available in three types: single-unit dose, multi-dose reservoir and multi-unit dose. The powder is stored in either capsules or sealed blisters that are broken during actuation. In general, patients tend to find that multi-dose DPIs are preferable to single-unit dose DPIs, due to the convenience of not having to replace the capsule or blisters with every use of the inhaler.

Nebulisers are less popular inhalation devices compared to the other two devices and are available in two forms. The most common nebuliser is the jet nebuliser, which works by passing a liquid formulation along with compressed air through a narrow tube into a wide chamber, which causes a reduction in pressure. This reduction in pressure then forms micron-sized droplets, which can then be inhaled. The alternative type of nebuliser is the ultrasonic nebuliser, which breaks down the solution of the drug into inhalable droplets by piezoelectric vibrations [68]. Both types of nebulisers require a face mask or mouthpiece, which are bulkier and less portable than the other two devices, and therefore, less convenient for frequent and repeated use [69]. The way nebulisers are formulated are unsuitable for some drugs, particularly for drugs that are unstable when in solution [70]. Nebulisers also take longer to administer the drug when compared to other drug delivery devices, due to the time taken for the drug to pass through the chamber into the mouthpiece or mask, and with inhalation over a prolonged period. The use of nebulisers can also result in the administration of less precise doses as there can be deposition of the API particles in the chamber and mouthpiece or face mask. In general, nebulisers tend to be used by patients who cannot use the other devices, due to difficulties with coordination associated with pMDIs and/or the inability to produce adequate inspiratory flow associated with DPIs. The main patient groups these affect are young children, elderly populations, patients with COPD and patients with CF [71].

There is a limited number of licensed inhaled antimicrobials that are only available for patients suffering from recurrent infections as in CF (Table 5). Other antibiotics, such as gentamicin and amikacin, can be nebulised for pulmonary drug delivery by using the solution for injection if required; however, these cases are for unlicensed use in CF patients only and not for use for respiratory infections [72]. As can be seen in Table 5, nebuliser solutions and inhalation powders can be used to formulate higher doses of drugs, and hence, these are the devices currently used on the market. Tobramycin, colistimethate sodium and zanamivir are all available as dry powders for inhalation. Tobramycin inhalation powder is produced via an oil-in-water emulsion-based spray-drying process [69], and each particle consists of an amorphous tobramycin sulphate, and a gel-phase phospholipid 1,2-distearoyl-sn-glycero-3-phosphocholine (DSPC). Tobramycin inhalation powder comes in a device known as a TOBI® Podhaler®, which consists of the engineered dry powder, the hard capsule, which is the primary packaging of the powder, and the device to administer the powder [70]. Other excipients used in the capsules include calcium chloride, which is used to stabilise the emulsion droplets during the spray drying and sulphuric acid for pH adjustment [71,73]. Each inhalation delivers 28 mg of tobramycin from the capsule to the lungs; however, four capsules are required for each dose, and therefore, eight inhalations per day. Colistimethate sodium is an inactive prodrug of colistin (polymyxin E), and is formulated as such to reduce the toxicity, typically nephrotoxicity, that is often experienced by patients when taking colistin. It is formulated to be administered using a Turbospin® inhaler, under the brand name Colobreathe® [74]. The powder is micronised and filled in polyethylene glycol (PEG)-gelatin hard capsules, which also contain purified water and sodium lauryl sulphate as excipients [75,76]. Zanamivir is an antiviral drug used for the treatment and prevention of influenza. It is formulated under the brand name Relenza® for its inhalation powder, and is produced via air-jet milling to contain 5 mg of zanamivir and 20 mg of lactose monohydrate particles as carrier particles in each double-foil blister [77].

Table 5. Licensed inhaled antibacterials currently available on the market.

Antibacterial Drug	Form	Available Strength	Ref
Tobramycin	Nebuliser liquid	300 mg/5 mL, 300 mg/4 mL, 170 mg/1.7 mL	[78]
	Inhalation powder	28 mg (1 dose = 4 $\times$ 28 mg inhalations)	
Colistimethate sodium	Inhalation powder	1,662,500 IU $\approx$ 125 mg	[79,80]
	Powder for nebuliser solution	1,000,000 IU $\approx$ 80 mg	
Aztreonam	Powder and solvent for nebuliser solution	75 mg	[81]

5. Pulmonary Drug Delivery Using Carrier Free Technology

Large carrier particles, such as alpha-lactose monohydrate, are added to the API to form an inhaled powder with enhanced formulation properties. These properties include the flowability and aerosol dispersion of the powder to maximise the number of API particles reaching the site of action. Lactose is available in a variety of grades for inhalation, with median diameters ranging from <5 μm to 250 μm [82]. During formulation, micronised API particles with an MMAD of between 1 μm and 5 μm are attached to the surface of these much larger carrier particles. Physical interactions hold the API in place, which prevents agglomeration of micronised API, due to cohesive forces. On inhalation, API is detached from the carrier and aerosolised, due to force exerted by the inhaler design in the form of friction, inertia and drag [83]. By retaining micronised API particle size, maximum dispersion of powder into the lungs is seen, due to increased aerosolisation [84]. This influences the drug administration in addition to preventing problems during the manufacturing process of the powder [85].

Carrier-based DPIs are the most prevalent formulation of DPIs and tend to be used to deliver relatively higher doses of API compared to pMDIs. However, certain mass ratios of API to carrier particles must be maintained to prevent agglomeration, as for example these can vary depending on the type of the drug, ranging from 1:2 to 1:4 in some cases [86]. For example, Relenza® requires a 25 mg powder to be inhaled so that to deliver 5 mg of zanamivir (i.e., a mass ratio of 1:5). These ratios create a maximum practical dose that can be delivered to the lungs through dry powder inhalation, making high dose delivery even more challenging. Even with carrier particles enhancing aerosolisation, deposition in the lungs is still relatively low, and as a result, the amount of API to reach the site of action is low. Fine particle fraction (FPF) refers to the fraction of particles with a size smaller than the respirable size divided by the total emitted dose of the inhaled API [87]. DPIs that do not contain carrier particles, therefore, have greater potential for delivering higher doses to the lungs.

Carrier free DPIs are a new-generation system that use special excipients and technologies, such as crystal engineering, to formulate micronised API particles whilst addressing the limitations explained above. These methods can increase the amount of lung deposition, with some formulations reporting FPF values of 63% [88]. One of the benefits of developing carrier free DPI formulations is that there can be an increased mass of API without having to retain large masses of the carrier particles. Therefore, this can maximise the dose available in each inhalation through maximising the amount of API that is deposited in the lungs. There are several dry powder formulations that have been or are currently being investigated to produce carrier free dry powder inhalers, as shown below in Table 6. As can be seen, different doses have been tested with some formulations delivering doses up to 50 mg, as displayed in Figure 2.

Many different methods have been used to develop dry powders for inhalation, also shown in Table 6, and each method produces variation to yield, cost of production and manufacture time. This data shows the difference in the size of the doses able to be formulated in DPIs for inhalation, compared to current doses available for inhalation (Figure 2). This also suggests that the use of carrier free inhalation technology can play a key factor in addressing the challenge of increasing doses of drugs for pulmonary drug delivery. Also noted is the wide range of drug classes that are being considered with regards to delivery via dry powder inhalation, including non-steroidal anti-inflammatory drugs (NSAIDs, i.e., ibuprofen and indomethacin), antimicrobials (i.e., tobramycin and netilmicin), antihistamines (i.e., ketotifen), and phosphodiesterase type 5 inhibitors (i.e., sildenafil). It was also noted that for many antimicrobials, there is the possibility of creating drug combinations within the same device—a useful tool for patients who require complex antimicrobial treatment.

Table 6. Carrier free DPIs and the technology used to produce the powders.

API	Dose in Capsule	Conditions Used	Ref
Indomethacin	3 mg	Spray drying an aqueous-based feed to form microparticles	[89]
Ethionamide + moxifloxacin	20 mg	Spray drying using a mini spray dryer	[90]
Ketotifen	20 mg	Spray drying with different solvents (water, ethanol and water-ethanol mix)	[91]
Sildenafil	5 mg	Spray drying using a mini spray dryer	[92]
Ibuprofen	50 mg	Air-jet milling to produce micronised samples	[93]
Sodium cromoglycate	20 mg	Pelletised	[94]
Tobramycin + clarithromycin	22.72 mg tobramycin, 2.27 mg clarithromycin	Spray drying	[95]
Salbutamol sulphate	5.1–7.1 mg	Gas-phase coating method to produce L-leucine coated powders	[96]
Tranexamic acid	38 mg	Spray drying	[97]
Roflumilast	20 mg	Spray drying with hydroxypropyl-β-cyclodextrin	[98]
Tobramycin	28 mg	Micronised using a Labomill jet milling system	[99]
Meloxicam potassium	1.3 mg	Cospray drying	[100]
Ciprofloxacin + colistin	10 mg	Cospray drying	[101]
Netilmicin	30 mg	Cospray drying	[102]

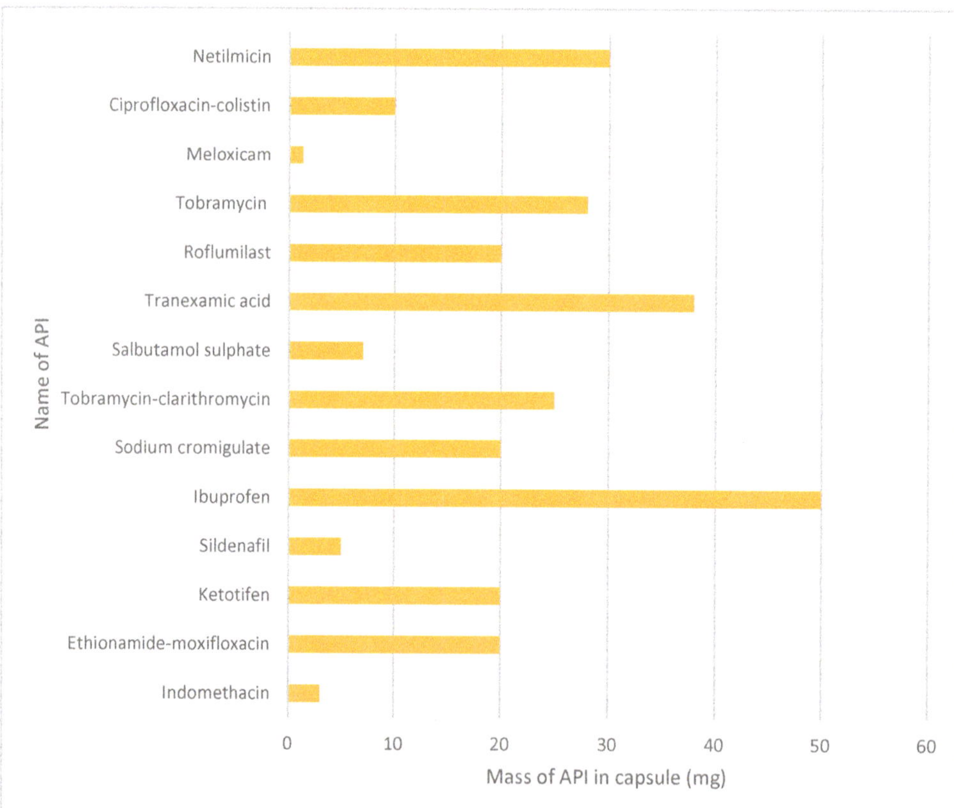

Figure 2. Bar graph for the weight of API in the capsule for each type of drug extracted from Table 6.

6. Pulmonary Drug Delivery on the Nanoscale

Nanoparticulate formulations maintain nanoscale particles of the drug through encapsulation within inhalable size particles to prevent aggregation and ensure deposition of nanoparticles in the lung. There are several particle engineering methods to achieve nanoparticle formulation, such as the use of liposomes, solid lipids and polymers, which have been reviewed elsewhere [103–105]. It is common for these formulations to be stored as suspensions then delivered to the lung through a nebuliser; however, carrier free, dry powder nanoparticle formulations are also being developed.

Zhu et al., described the formulation of poorly soluble ivacaftor (Iva) as bovine serum albumin nanoparticles, which were then spray freeze-dried with different ratios of soluble colistin (Co) matrix and L-leucine [106]. All subsequent formulations showed high emitted dose following inhalation for both APIs (>90%); however, a range of FPF values (%) were seen corresponding to initial solid content in the formulations. Dissolution rates of the best performing nanoformulation and a jet-milled physical mixture of Iva and Co were then compared. After 3 h, the nanoformulation showed three times greater dissolution of Iva compared to the physical mixture control, with the concentration of Iva dissolved equivalent to highly soluble Co. Enhanced dissolution of Iva was attributed to the albumin maintaining Iva nanoscale particle size and amorphous form. The use of Co as a water-soluble matrix both enhanced the dispersion of Iva nanoparticles, as well as broadening the formulation clinical scope for patients with cystic fibrosis.

Doxorubicin nanoparticles (DNPs) were formulated as a colloid through emulsion polymerisation then spray freeze-dried with lactose to form particles of inhalable size [107,108]. DNPs showed increased survival rates compared to IV doxorubicin controls (including IV DNPs) in a rat model and much lower cardiotoxicity compared to doxorubicin DPI control. DNPs were also formulated with sodium carbonate to enhance the release of doxorubicin from the formulation. When inhaled, the effervescent formulations showed increased survival rates compared to DNP controls showing the importance of an active release mechanism from the formulation. In a slightly different approach, often utilising nanoscale dimensions to enhance inhalation properties, the use of porous particles has been investigated and was shown to improve efficacy of inhalation [109–111].

7. The Design of Carrier Free Formulations Using Coamorphous Solid Dispersions (CACDs)

Most solid APIs exist in a crystalline state held together by strong intermolecular bonds, and therefore, display good stability profiles (Figure 3). However, the crystalline state often shows poor solubility, due to the high energy required to break the crystalline lattice which creates a major problem for developing new APIs [112,113]. There has been much interest in the process of 'drug amorphisation', to address poor solubility, which involves the conversion from a crystalline state to an amorphous solid state. The amorphous solid state offers an improved solubility and dissolution rates as a result of possessing higher entropy [114–116]. The advantages associated with higher energy forms are often negated, due to recrystallisation to a more thermodynamically stable form during processing and storage [117–119]. This can limit applications, and so the production and maintenance of amorphous drugs with adequate stability remain a challenge. Formulation strategies based on solid molecular dispersions are being explored, including polymeric amorphous solid dispersions (PASDs), and more recently, coamorphous solid dispersions (CASDs).

PASDs incorporate low API loading within a compatible polymer to maintain the solubility advantage of amorphous systems through the formation of strong intermolecular attractions [120]. Examples of PASDs applications are seen in the combination of paclitaxel and polyvinylpyrrolidone (PVP), ritonavir and PVP, ciprofloxacin and polyvinyl alcohol [121–123]. However, the main limitation of this approach is the low level of API loading within the formulation, requiring increased dosing. In addition, many polymeric carriers are hygroscopic, which can lead to API recrystallisation while in some cases polymers can be unsuccessful in maintaining a good stability profile [123]. The applications and limitations of PASDs have been highlighted in a number of reviews [124–126].

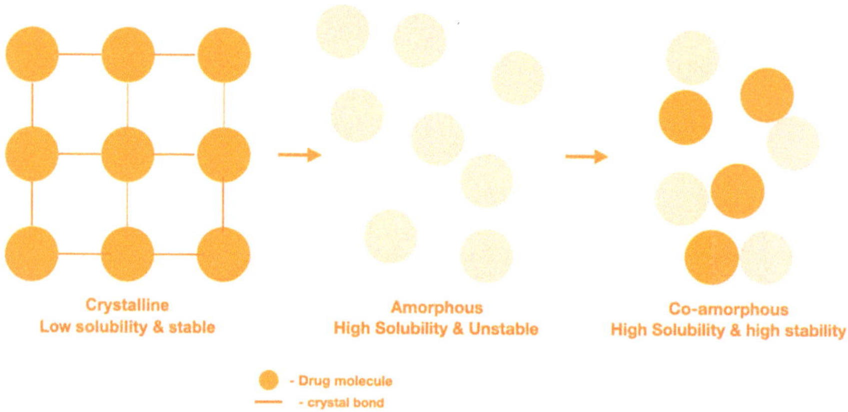

Figure 3. Illustration of the crystalline state, amorphous state and coamorphous state and their characteristics.

CASDs are a relatively new formulation approach through which the crystalline drug is amorphised thermally or mechanically and stabilised with a small molecular weight coformer. These systems are often stabilised through strong intermolecular hydrogen bonding and exhibit apparent stability, due to an increased glass transition temperature [124]. Glass transition temperature refers to the temperature at which below, the system exists in an unstable glassy state and above, the system exists in a rubbery state [125]. CASDs can be obtained through different preparation methods, such as freeze drying, quench cooling, milling and solvent evaporation methods [126,127]. Milling is an example of mechanical amorphisation where a direct mechanical impact on the drug causes a disturbance in the crystalline structure, forming an amorphous state. Quench cooling and solvent evaporation are examples of thermal amorphisation where a crystalline solid state is prevented from reforming, due to molecular level interactions. Here, the drug is either in a liquid state following melting or dissolved to form a solution then rapidly cooled, or solvent removed rapidly. CASDs are believed to be able to provide an increase in drug kinetic solubility thanks to the high energy of the amorphous state driving faster dissolution [128]. In addition, improved physical stability is offered through an increase in the glass transition temperature (T_g) and the presence of intermolecular interactions between the API and coformer, as reported by several research papers [124,129]. CASDs are said to reduce hygroscopicity, as API hydrogen bonding regions are strongly bound to the coformer. Therefore, hydrogen bonding with water is less favourable, which helps to maintain stability and particle integrity.

CASDs provide the opportunity for many types of formulations, such as drug-excipient and drug-drug combinations, and many have been reported as being successful. An example of a drug-drug combination is the spray drying of colistin with azithromycin to treat infections caused by multi-drug resistant bacteria [130]. A study conducted by Wang et al., showed an enhanced solubility profile between two APIs with low aqueous solubility: lacidipine and spironolactone [131]. Many other drug-excipient combinations are listed in Table 7, by which commonly used coformers in CASDs for pulmonary drug delivery include mannitol, sugars and amino acids, such as leucine [132,133].

The spray drying technique is based on solvent evaporation which uses the processes of liquid atomisation, gas/droplet mixing and drying to create microparticles [134]. First, a liquid feed consisting of mixed components, typically pure API, excipient and a common solvent, is converted into smaller droplets via atomisation. Next, the droplets are sprayed downwards into a vertical drying chamber exposed to air or nitrogen at a temperature higher than the solvent boiling point [135]. Within this drying chamber, the droplets rapidly shrink in size to produce dry microparticles.

Through spray drying, aspects of particle morphology, such as shape, size and surface properties, can be highly controlled, which deems the technique appropriate for formulating inhalation powders. Many papers have controlled particle morphology through variations to the inlet temperature, the spray nozzle diameter and properties of the feed solution, such as concentration and viscosity [136]. The inlet temperature (also known as the drying temperature) and the feed rate are parameters of high importance as they control the rate of evaporation and drying load, which determine particle properties and product yield [137]. It is common to see outlet temperatures ranging from 70 °C to 105 °C with inlet temperatures ranging from 100 °C to 210 °C. A faster drying rate could lead to a higher product yield, as there are fewer particles adhering to the chamber walls, which may also implicate that solvent evaporation is not complete. The spray drying process is highly applicable for large scale manufacture, where the integrity of formulation is retained despite 'upscaling' [138]. Altogether, spray drying offers a great opportunity to incorporate a number of excipients into one formulation whilst improving the physical stability of the particles.

Table 7. Application of coamorphous solid dispersions prepared using different preparation methods. The inclusion of PASD of ciprofloxacin was shown to identify the apparent benefits of polymeric solid dispersions compared to coamorphous solid dispersions.

API	Prime Excipient	Preparation Method	Fine Particle Fraction (FPF) (%)	Ref
Cyclosporin A (CsA)	Lactose, methylcellulose and erythritol	Jet-milling and freeze drying	54	[139]
Ciprofloxacin	Polyvinyl alcohol PVA	Spray drying	25 ± 2.1 after 6 months	[121]
Ciprofloxacin	No excipient	Spray drying	67.35 ± 1.1 after 6 months	[121]
Ciprofloxacin	Leucine	Spray drying	79.78 ± 1.2 after 6 months	[121]
Ciprofloxacin	Hydroxypropyl-beta-cyclodextrin	Spray drying	36.32 ± 1.3 after 6 months	[121]
Colistin	No excipient	Spray drying	43.8 ± 4.6%	[140]
Colistin	No excipient	Jet milling	28.4 ± 6.7 %	[140]
Colistin	L-leucine	Spray drying	43.8 ± 4.6% (no difference from Spray dried alone)	[140]
Ciprofloxacin	No excipient	Spray drying	28.0 ± 3.2%	[141]
Ciprofloxacin	Lactose, sucrose, trehalose, L-leucine	Spray drying	Lactose (43.5 ± 3.3%), sucrose (44.0 ± 4.3%), trehalose (44.0 ± 1.9%), L-leucine (73.5 ± 7.1%)	[141]
Thymopentin	Lactose/mannitol, Leucine, poloxamer 188	Spray drying	44.8%, 45.6%, 44.9%, 43.8%	[142]
CsA	Inulin	Spray freeze drying	>50	[143]
Tacrolimus	Mannitol	Thin-film freeze-drying	83.3	[144]
Tacrolimus	Raffinose	Thin-film freeze-drying	69.2	[145]
Tacrolimus	Lactose	Thin-film freeze-drying	68.7	[145]

8. Examples of Coformers as Components of CAMs

L-leucine has been widely used in many studies to improve the aerosolisation properties of particles, due to its antihygroscopic effect and its ability to generate coarse particles, following surface enrichment. Momin et al., formulated spray dried kanamycin with various concentrations of L-leucine, showing the 5% L-leucine formulation to have the best improvement in aerosolisation properties [146]. Likewise, significant increases in aerosolisation properties were recorded for the other ratios of L-leucine. Other papers also reported L-leucine as a well-rounded excipient for improving the aerosolisation properties of powder particles. Interestingly Mangal et al., formulated spray dried azithromycin with various concentrations of L-leucine yet reported no significant change to powder morphology or FPF ($p > 0.05$) [132]. In addition, there was no significant increase seen in the percentage emitted dose ($p > 0.05$); however, an increased in vitro dissolution rate was reported. A lack of change to FPF and emitted dose can be explained when considering particle surface morphology and composition. It is believed that the more corrugated a particle surface is, the greater the resulting aerosolisation properties. This process is attributed to the success of L-leucine in improving aerosolisation. Chen et al., reported how changes to the composition of PASDs result in differences to the surface composition of particles, and therefore, flowability and aggregation [136]. Perhaps the surface enrichment of azithromycin with L-leucine for this formulation was poor, resulting in little change to particle surface morphology when compared to the control. When using L-leucine as a coformer, it is important that drying causes the distribution to the outer regions of droplets, causing recrystallisation of L-leucine on the particle surface, creating a corrugated effect [147]. This concept has also been documented by McShane et al. [138]. The impact of spray drying conditions on particles morphologies was recently highlighted in a different study by which spray dried coamorphous ciprofloxacin tartrate salt was shown to exhibit improved properties compared to ciprofloxacin alone [148].

The distribution of the components within the droplet during spray drying is said to be dependent on physicochemical properties of the components, such as solubility, the diffusivity of components and hydrophobicity [149]. These properties ultimately result in the final composition of the surface of spray dried particles. In the example mentioned above of azithromycin with L-leucine, the solubility of azithromycin is very low, and it likely had an impact on the distribution of L-leucine within the CASD. Li et al., conducted in vitro study to assess deposition, by which, spray drying azithromycin with mannitol resulted in low FPF values ranging between 38–42% [150]. Signifying the impact of preparation method and locality of excipients, Padhi et al., suggested that by increasing the ratio of L-leucine may improve surface enrichment, which would result in a substantial increase in the FPF [151]. In a different study, it was proposed that the presence of hydrophobic azithromycin helps to control humidity to prevent reduction in FPF [130].

Hassan et al., suggested that surface enrichment of spray dried materials is dependent on solvent evaporation rate and diffusion coefficients of the solutes [152]. Based on computational fluid dynamics, it was shown that the maximum rate of solvent evaporation proportionally affected MMAD [153]. It is believed the flow of the solute equates to that of the solvent; therefore, the movement of solute within a droplet is highly linked to solvent flow. Lower evaporation rates have a more significant impact on redistribution to droplet surface for solutes with high diffusion coefficients [149,152]. Shetty et al., when using water as a solvent, concluded that a low inlet temperature (<120 °C) was not considered, due to the possibility of it not being sufficient for drying [141]. Chen et al., also documented lower surface enrichment occurring, due to a reduction in the inlet temperature [136]. It is believed that the inlet temperature influences the particle shape, as well as surface enrichment. The inlet temperature indirectly affects the outlet temperature, and a lower outlet temperature is believed to result in a more spherical particle shape, which is detrimental to FPF [154,155].

Benke et al., reported spray dried meloxicam potassium (MXLspd) that was compared to the carrier-based meloxicam and lactose monohydrate InhaLac® (μMXL + IH70) [21]. As expected MXLspd showed an improvement in FPF < 5 μm (59.47%) compared to μMXL + IH70 with an FPF < 5 μm (24.99%). In addition, μMXL + IH70 had an MMAD of 7.18 μm, therefore unfavourable for pulmonary drug delivery. When assessing the impact of humidity on FPF, authors observed a significant decline ($p < 0.0001$) observed in the FPF (<4.9 μm) between cospray dried ciprofloxacin with lactose when stored at 20% RH and 50% RH after 10 days. This may be due to the caking observed, which led to larger powder particles formed. Similar observations were reported by Guenette et al., which showed that larger lactose particles led to reduced flow properties [156]. The larger particle size may have occurred, due to particle agglomeration when stored at a higher RH. These examples highlight that for highly aerosolised powder, it is necessary to maintain the stability of spray dried CASDs when hygroscopic coformers are used. This is because of the presence of the excipient such as L-leucine, which has been reported to combat moisture-related stability issues, due to its hydrophobic nature [157]. Part of the reason why CASDs are successful is that the coformer changes from being a passive carrier to an essential constituent in the formulation [138].

In a separate study, Lababidi et al., further examined the success CASDs had in pulmonary drug delivery by which azithromycin and ciprofloxacin were spray dried with L-leucine and n-acetyl cysteine (NAC), an amino acid derivative and known mucolytic [129]. Mucolytics are used in combination with antibiotics to dissolve thick mucus found in the CF lung [157,158]. When mucolytics, such as NAC, are used as CASDs coformers, the formulation benefit becomes two-fold, creating a more effective treatment with better formulation characteristics. As mentioned before, the addition of L-leucine is aimed at reducing particle cohesion and increasing aerosolisation properties. However, in the study conducted by Lababidi et al., the main aim of adding NAC was to the formulation was to dissolve mucus, enhancing antibiotic effect against the bacteria. The authors showed that azithromycin/n-acetyl cysteine combination reduced *P. aeruginosa* biofilm by 25% at a concentration of 0.3 μm/mL compared to azithromycin and n-acetyl cysteine alone.

Forming amorphous dispersions composed of antimicrobials through complex formation with metal cations has been explored as a method for enhancing formulation characteristics of inhaled antimicrobials. Lamy et al., produced inhalable dry powder microparticles of ciprofloxacin, complexed with calcium or copper ions (Cip-Ca; Cip-Cu), through spray drying [159,160]. The resulting powders were amorphous, and both calcium and copper formulations showed similar shell-like morphology, characteristic of spray dried materials, and similar in vitro solution properties. When comparing pharmacokinetics in vivo, both formulations showed enhanced lung retention compared to the ciprofloxacin control, suggesting that forming metal counterion complexes can reduce the rate of transcellular diffusion from the lung fluid. It was also seen that the ratio of ciprofloxacin in epithelial lining fluid to plasma was five times greater for the Cip-Cu than Cip-Ca formulations. The authors linked this with the strength of complex which had formed, corresponding to: The enhanced stability of the Cip-Cu, a slower rate of absorption into plasma and longer elimination half-life. Thus, antibiotic lung retention can be controlled depending on the metal ion selected, leading to an optimised concentration in lung fluid.

The influence of metal salts on the inhalation properties of levofloxacin has also been analysed. Barazesh et al., cospray dried levofloxacin with four metal chloride salts with and without leucine [161]. Sodium (Na) and potassium (K) monovalent salts were compared with magnesium (Mg) and calcium (Ca) divalent salts. When measuring aerosol properties, most effects was seen for formulations containing the highest measured percentage of metal ion (i.e., 20% *w*/*w*). It was also observed that the addition of divalent salts reduced the FPF of formulations and caused higher water retention post spray drying. The authors concluded that the highly hygroscopic divalent metal salts are likely to uptake more water, leading to local crystallisation and particle aggregation. Monovalent salts at 20% *w*/*w* showed a higher FPF compared to lower concentrations. With respect to leucine formulations, FPF was significantly higher, corresponding to the characteristic corrugated effect seen when spray drying with leucine.

9. The Design of Carrier Free Formulations Using Cocrystals

The formulation of carrier free dry powders can be achieved by a variety of methods. These include the formation of salts, such as the generation of a sildenafil-citrate salt [92], forming complexes with cationic groups, such as the generation of indomethacin and polylysine microparticles [89], and forming CASDs, such as with ciprofloxacin and colistin [101]. Another route being investigated is the generation of cocrystals and seems to be a promising method of improving physicochemical properties if other methods are not suitable. For example, with the formation of salts, if the API is neutral, and therefore, does not contain an acidic or basic group that can be ionised, salts cannot be practically formed [162]. Cocrystallisation may also be seen as a more desirable method of physicochemical enhancement than others, for example, CASDs, due to enhanced physical and chemical stability [163].

Pharmaceutical cocrystals are multicomponent crystals formed between an API and pharmaceutically acceptable coformer. API and coformer exist in a fixed stoichiometric ratio held together through [164] non-covalent and non-ionic interactions. These interactions are intermolecular in nature and include hydrogen bonding, halogen bonding, π-π stacking and van der Waal forces [165]. Hydrogen bonds are considered the 'key interaction' for cocrystallisation, due to their strength, directionality and being commonly found in organic molecules [166]. Therefore, cocrystals are distinct from polymorphs, due to multiple components, and salts, due to lack of ionisation, and CASDs, due to their crystalline nature. The formulation of APIs as cocrystals is regarded as a physical modification to improve physicochemical properties, which maintains the API pharmacodynamic profile [167]. Since the cocrystal is a unique crystalline entity, it has unique physical properties, including solubility, dissolution, flowability and stability. In general, coformers are chosen based on molecular compatibility, as well as possession of desired physical property change. For example, to improve an API's solubility, one would choose a coformer with compatible functional groups, as well as high solubility [168].

Cocrystallisation is currently attracting considerable interest, due to the improved physicochemical properties that can be introduced to the API [169,170]. For example, cocrystals of meloxicam–succinic acid were combined with PEG 4000 to enhance aqueous solubility [171]. Cocrystals have largely been studied with a focus on improving oral drug delivery, and it is only recently that studies have begun to explore how cocrystals can be used for other administration routes, including pulmonary drug delivery. It is important to focus on all changes to API physical properties rather than flowability and dispersion alone. For example, altering solubility and dissolution will influence retention and epithelium wall permeation which will determine either a local or systemic effect. Selecting the correct coformer can increase aerosolisation properties, as well as maintaining the balance between API retention in the lung and API clearance through various mechanisms [172].

M Karashima et al., produced micronised powder formulations of itraconazole for inhalation using cocrystallisation and jet-milling [163]. These cocrystal formulations showed superior aerodynamics compared with itraconazole control put through the same processing and amorphous itraconazole spray dried with mannitol corresponding with reduced particle size. Also tested were the intrinsic dissolution rates of cocrystal formulations compared to itraconazole and amorphous itraconazole in mock lung surfactant. After 1 h, the dissolved concentration of itraconazole from cocrystal formulations were between 5 and 10 times higher than itraconazole and the amorphous form. The authors reported higher plasma concentrations (C_{max}) compared to itraconazole and the amorphous form of itraconazole. Comparing this finding with intrinsic dissolution data, supports the conclusion that the extent of dissolution in the lungs influences the extent of systemic partitioning.

Alhalaweh et al., produced several inhalable powders of theophylline through cocrystallisation with nicotinamide, urea and saccharin [173]. These were compared against theophylline control, but also against cocrystals formulated with lactose carrier. The authors comment that cocrystal formulations showed more favourable aerosolisation with less API loss, due to impaction. However, not all cocrystal formulations were better aerodynamically compared to the control. This highlights the unpredictable nature of coformer selection in cocrystal formation and the importance of considering all aspects of physical property changes. Tanaka et al., produced a theophylline:oxalic acid cocrystal through a combination of spray and freeze drying [174]. When compared to the control, the resulting formulation showed resistance to hygroscopicity and polymorphic transition whilst maintaining good pulmonary delivery. The authors attribute this stability to the low recorded energy state and close intermolecular interactions within the cocrystal.

API solution properties are a key component to determining API destination following inhalation. One study investigated how the properties of 5-fluorouracil differ from three different cocrystals created with gentisic acid (FUGA), 4,5 dihydroxybenzoic acid (FUBA) and 4-aminopyridine (FUPN). This study determined that the intrinsic dissolution of all three cocrystals were increased compared to pure 5-fluorouracil in both pHs of 1.2 and 6.8, with the largest difference being between 5-fluoruracil and FUPN. 5-fluorouracil had an intrinsic dissolution rate (IDR) of 0.12 mg cm^{-2}/s in pH 1.2 and 0.18 cm^{-2}/s in pH 6.8, whereas FUPN had an IDR of 0.7 mg cm^{-2}/s in pH 1.2 and 0.92 mg cm^{-2}/s in pH 6.8, showing more rapid dissolution from the cocrystal forms of 5-fluoruracil than the pure API [175]. Increased dissolution has also been observed in three diacerein cocrystals with isonicotinamide, nicotinamide and theophylline (DA-ISO), (DA-NIC) and (DA-THE). Diacerein had an IDR of 0.065 mg cm^{-2}/min compared to 0.216 mg cm^{-2}/min, 0.284 mg cm^{-2}/min and 0.137 mg cm^{-2}/min for DA-ISO, DA-NIC and DA-THE, respectively. This again shows that each cocrystal had a higher IDR value than the pure API molecule, suggesting a faster rate of dissolution [176].

Another study demonstrated the effects that cocrystals have on the pharmacokinetic profile of a drug. Daidzein and three cocrystals, with isonicotinamide (DIS), cytosine (DCYT) and theobromine (DTB) as the coformers, were compared. It was found that all three cocrystals had higher Cmax values, than daidzein (870.5 ng/mL), with DIS possessing the highest Cmax (1848.7 ng/mL). In addition, the time taken to reach the Cmax (Tmax) values for each cocrystal was 3 h compared to 4 h for pure daidzein [177]. Table 8 summarises a variety of cocrystal formulations, the method of manufacture used and the difference between API and coformer molecular weight. Shown is that cocrystallisation applied to a great variety of different APIs produced through a wide range of different production methods. Of note are the studies which use methods applicable for producing inhalable dry powders, such as spray drying, spray freeze drying and jet milling. It also gives an example of the wide variety of successful coformers available for cocrystallisation, suggesting a great variety in physical property variations even for the same API. Examples of drug:drug cocrystals are listed to show the possibility of creating synergistic therapies through a single phase crystalline powder, i.e., cocrystals. As shown in Table 8 and Figure 4, all coformers have molecular weights lower than alpha-lactose monohydrate (342.3 g/mol) and either much smaller or similar molecular weight to the API. This demonstrates how formulations containing coformers take up less volume in comparison to carrier-based delivery systems. As discussed earlier, the volume which lactose particles take up in formulations restricts the quantity of API that can be practically delivered. By avoiding the use of carrier particles, through coformer based methods, greater amounts of API particles can be formulated within a DPI device, possibly increasing the dose of API per actuation.

Table 8. Cocrystalline APIs, their coformers, technology used and molecular weights.

API	Coformer	Technology Used	Molecular Weight (g/mol)		Ref
			API	Coformer	
Itraconazole	Succinic acid	Jet-milling	705.63	118.09	[163]
	L-tartaric acid			150.09	
Levofloxacin	Metacetamol	Grinding and heating	361.37	151.16	[178]
Pyrazinamide	3-Hydroxy benzoic acid	Slow evaporation and neat grinding	123.11	138.12	[179]
Dapsone	Caffeine	Slow evaporation, liquid-assisted grinding, spray drying	248.30	194.19	[180]
Nitrofurantoin	Melamine	Slow evaporation	238.16	126.12	[181]
Telaprevir	4-aminosalicylic acid	Ball milling	679.85	153.14	[182]
Trimethoprim	Glutarimide	Slow evaporation	290.32	113.11	[183]
Telmisartan	Gentisic acid	Slurry approach	514.62	154.12	[184]
	Maleic acid			116.07	
Sulfadimidine	4-aminosalicylic acid	Liquid-assisted comilling	278.33	153.14	[185]
Isoniazid	Ferulic acid	Liquid-assisted grinding	137.14	194.18	[186]
	Caffeic acid			180.16	
	Vanillic acid			168.15	
Adefovir	Stearic acid	Antisolvent precipitation	273.19	284.50	[187]
Acyclovir	Fumaric acid	Liquid-assisted grinding	225.20	116.07	[188]
Meloxicam	Aspirin	Solution, slurry and solvent drop methods	351.40	180.16	[189]
Theophylline	Oxalic acid	Spray freeze drying	180.16	90.03	[174]
Niclosamide	Nicotinamide	Spray drying	327.12	122.12	[190]
Lomefloxacin	Barbituric acid	Slow evaporation	351.35	128.09	[191]
	Isophthalic acid			166.13	
Enoxacin	Oxalic acid	Slow evaporation	320.32	90.03	[192]
	Malonic acid			104.06	
	Fumaric acid			116.07	
Sulfaguanidine	Thiobarbutaric acid	Slow evaporation	214.24	144.15	[193]
	1,10-phenanthroline			180.20	

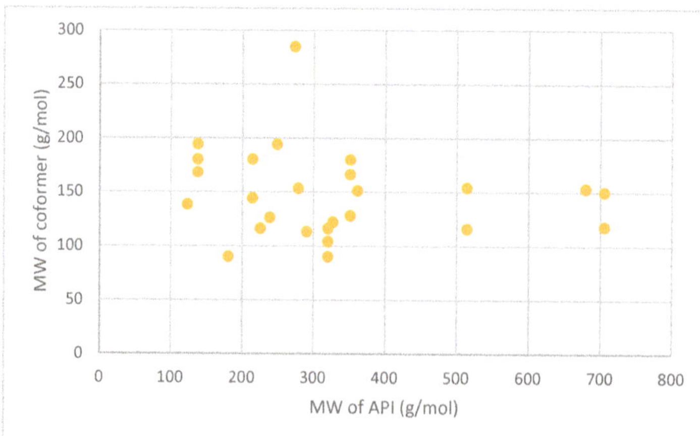

Figure 4. A scatter graph of the data from Table 8 comparing the molecular weights of the APIs and their coformers.

10. Conclusions

Coamorphous solid dispersions and cocrystals are attractive physical structures that have been used to improve the physicochemical properties of inhaled APIs to target infections and lung cancer. When developing pharmaceutical cocrystals/coamorphous for inhalation, careful selection of the coformer should be followed to improve the efficacy to deliver drugs to the lungs. It is likely that by creating cocrystals/coamorphous dispersions, the need for carrier particles in DPIs can be eliminated, and therefore, a greater amount of the API can be delivered. This would result in an increased dose in each inhalation, allowing lower potency drugs that require higher doses to be formulated for inhalation. It is important to understand the complexities of pulmonary drug delivery when considering the formulation of drugs for inhalation to ensure the drug particles have adequate aerosolization properties. This is necessary to ensure the particles can reach deep into the lungs at the site of action, without being aggregated or immediately exhaled from the lungs. It is important to consider all steps throughout drug development process, including the excipients that are used and the devices to administer the drug, because some of the devices have shown clear advantages when it comes to overcoming the challenges with formulating inhaled APIs. The complexity of diseases imposes greater challenges demanding more studies to optimise particles properties for better deposition. Ultimately, it is crucial to match the properties of the particles to the desired site of action, such as the case when targeting bacterial biofilms and cancer.

Author Contributions: Conceptualization, H.A.-O.; investigation, H.A.-O., A.G., T.H., S.O.; data curation, H.A.-O., A.G., T.H., S.O.; writing—H.A.-O., T.H.; writing—review and editing, H.A.-O.; supervision, H.A.-O.; project administration, H.A.-O. All authors have read and agreed to the published version of the manuscript.

Funding: This research received no external funding.

Institutional Review Board Statement: Not applicable.

Informed Consent Statement: Not applicable.

Data Availability Statement: Not applicable.

Conflicts of Interest: The authors declare no conflict of interest.

References

1. Moss, N.H.; Thompson, S.D.; Bauer, H.C.; Clark, J.J.; Clark, W.R. Diseases of the Respiratory System. *New Complet. Med. Health Encycl.* **1997**, *2*, 474–476.
2. Rupp, J.; Droemann, D.; Goldmann, T.; Zabel, P.; Solbach, W.; Vollmer, E.; Branscheid, D.; Dalhoff, K.; Maass, M. Alveolar epithelial cells type II are major target cells for C. pneumoniae in chronic but not in acute respiratory infection. *FEMS Immunol. Med. Microbiol.* **2004**, *41*, 197–203. [CrossRef]
3. Goyal, K.A.; Garg, T.; Bhandari, S.; Rath, G. Advancement in pulmonary drug delivery systems for treatment of tuberculosis. In *Nanostructures for Drug Delivery*; Andronescu, E., Grumezescu, M.A., Eds.; Elsevier: Amsterdam, The Netherlands, 2017; pp. 669–695.
4. Smith, I. Mycobacterium tuberculosis pathogenesis and molecular determinants of virulence. *Clin. Microbiol. Rev.* **2003**, *16*, 463–496. [CrossRef] [PubMed]
5. Forum of International Respiratory Societies. *Global Impact of Respiratory Disease*; Forum of International Respiratory Societies: Sheffield, UK, 2017; pp. 12–26.
6. Goel, A.; Baboota, S.; Sahni, J.K.; Ali, J. Exploring targeted pulmonary delivery for treatment of lung cancer. *Int. J. Pharm. Investig.* **2013**, *3*, 8–14. [CrossRef]
7. Pearce, A.; Haas, M.; Viney, R.; Pearson, S.A.; Haywood, P.; Brown, C.; Ward, R. Incidence and severity of self-reported chemotherapy side effects in routine care: A prospective cohort study. *PLoS ONE* **2017**, *12*, e0184360. [CrossRef]
8. Nurgali, K.; Jagoe, R.T.; Abalo, R. Editorial: Adverse Effects of Cancer Chemotherapy: Anything New to Improve Tolerance and Reduce Sequelae? *Front. Pharmacol.* **2018**, *9*, 245. [CrossRef]
9. Rangaraj, N.; Pailla, S.; Sampathi, S. Insight into pulmonary drug delivery: Mechanism of drug deposition to device characterization and regulatory requirements. *Pulm. Pharmacol. Ther.* **2019**, *54*, 1–21. [CrossRef]
10. Chaurasiya, B.; Zhao, Y.-Y. Dry powder for pulmonary delivery: A comprehensive review. *Pharmaceutics* **2021**, *13*, 31. [CrossRef]
11. Al-Obaidi, H.; Buckton, G. The characterization and dissolution properties of griseofulvin solid dispersions with HPMCAS. *J. Pharm. Pharmacol.* **2006**, *1*, A18–A19.

12. Newman, S. Drug delivery to the lungs: Challenges and opportunities. *Ther. Deliv.* **2017**, *8*. [CrossRef]
13. Islam, N.; Ferro, V. Recent Advances in Chitosan-Based Nanoparticulate Pulmonary Drug Delivery. *Nanoscale* **2016**, *8*, 14341–14358. [CrossRef]
14. Luczak-Wozniak, K.; Dabrowska, M.; Domagala, I.; Miszczuk, M.; Lubanski, W.; Leszczynski, A.; Krenke, R. Mishandling of pMDI and DPI inhalers in asthma and COPD- repetitive and non-repetitive errors. *Pulm. Pharmacol. Ther.* **2018**, *51*, 65–72. [CrossRef] [PubMed]
15. Parumasivam, T.; Chang, R.; Abdelghany, S.; Ye, T.; Britton, W.; Chan, H. Dry powder inhalable formulations for anti-tubercular therapy. *Adv. Drug Deliv. Rev.* **2016**, *102*, 83–101. [CrossRef] [PubMed]
16. Patwa, A.; Shah, A. Anatomy and physiology of respiratory system relevant to anaesthesia. *Indian J. Anaesth* **2015**, *59*, 533–541. [CrossRef] [PubMed]
17. Hakim, A.; Usmani, O.S. Structure of the Lower Respiratory Tract. In *Reference Module in Biomedical Sciences*; Elsevier: Amsterdam, The Netherlands, 2014.
18. Carvalho, T.; Peters, J.; Williams, R.r. Influence of particle size on regional lung deposition- What evidence is there? *Int. J. Pharm.* **2011**, *406*, 1–10. [CrossRef]
19. DeCarlo, P.; Slowik, J.; Worsnop, D.; Davidovits, P.; Jimenez, J. Particle Morphology and Density Characterization by Combined Mobility and Aerodynamic Diameter Measurements. Part 1: Theory. *Aerosol Sci. Technol.* **2004**, *38*, 1185–1205. [CrossRef]
20. Laube, B.; Janssens, H.; de Jongh, F.; Devadason, S.; Dhand, R.; Diot, P.; Everard, M.; Horvath, I.; Navalesi, P.; Voshaar, T.; et al. What the pulmonary specialist should know about the new inhalation therapies. *Eur. Respir. J.* **2011**, *37*, 1308–1417. [CrossRef]
21. Benke, E.; Farkas, Á.; Szabó-Révész, P.; Ambrus, R. Development of an Innovative, Carrier-Based Dry Powder Inhalation Formulation Containing Spray-Dried Meloxicam Potassium to Improve the In Vitro and In Silico Aerodynamic Properties. *Pharmaceutics* **2020**, *12*, 535. [CrossRef]
22. Labiris, N.R.; Dolovich, M.B. Pulmonary drug delivery. Part I: Physiological factors affecting therapeutic effectiveness of aerosolized medications. *Br. J. Clin. Pharmacol.* **2003**, *56*, 588–599. [CrossRef]
23. Verma, K.R.; Ibrahim, M.; Garcia-Contreras, L. *Lung Anatomy and Physiology and Their Implications for Pulmonary Drug Delivery*; Wiley: Chichester, UK, 2015; pp. 2–14.
24. Kim, C.S.; Rosati, J.A. Comparison of Monodisperse and Polydisperse Aerosol Deposition in a Packed Bed. Available online: https://cfpub.epa.gov/si/si_public_record_Report.cfm?Lab=NHEERL&dirEntryID=62538 (accessed on 18 April 2021).
25. Patton, J.S. Mechanisms of macromolecule absorption by the lungs. *Adv. Drug Deliv. Rev.* **1996**, *19*, 3–36. [CrossRef]
26. Heyder, J. Deposition of inhaled particles in the human respiratory tract and consequences for regional targeting in respiratory drug delivery. *Proc. Am. Thorac. Soc.* **2004**, *1*, 315–320. [CrossRef]
27. Cheng, Y.S. Mechanisms of pharmaceutical aerosol deposition in the respiratory tract. *AAPS PharmSciTech* **2014**, *15*, 630–640. [CrossRef] [PubMed]
28. Weers, J.; Clark, A. The Impact of Inspiratory Flow Rate on Drug Delivery to the Lungs with Dry Powder Inhalers. *Pharm. Res.* **2017**, *34*, 507–528. [CrossRef]
29. Ibrahim, M.; Verma, R.; Garcia-Contreras, L. Inhalation drug delivery devices: Technology update. *Med. Devices* **2015**, *8*, 131–139. [CrossRef]
30. Momin, M.A.M.; Tucker, I.G.; Das, S.C. High dose dry powder inhalers to overcome the challenges of tuberculosis treatment. *Int. J. Pharm.* **2018**, *550*, 398–417. [CrossRef] [PubMed]
31. Sallam, A.S.; Nazzal, S.; AlKhatib, S.H.; Darwazeh, N. *Quality by Design: Concept for Product Development of Dry-Powder Inhalers*, 1st ed.; Wiley: Chichester, UK, 2015; pp. 322–324.
32. Alhaddad, B.; Smith, F.; Robertson, T.; Watman, G.; Taylor, K. Patients' practices and experiences of using nebuliser therapy in the management of COPD at home. *BMJ Open Respir. Res.* **2015**, *2*, e000076. [CrossRef]
33. Desgrouas, M.; Ehrmann, S. Inhaled antibiotics during mechanical ventilation—Why it will work. *Ann. Transl. Med.* **2020**, *9*, 598. [CrossRef]
34. NICE. Salbutamol. Available online: https://bnf.nice.org.uk/medicinal-forms/salbutamol.html (accessed on 5 November 2020).
35. NICE. Fluticasone. Available online: https://bnf.nice.org.uk/drug/fluticasone.html (accessed on 4 November 2020).
36. Kalil, A.; Metersky, M.; Klompas, M.; Muscedere, J.; Sweeney, D.; Palmer, L.; Napolitano, L.; O'Grady, N.; Bartlett, J.; Carratalà, J.; et al. Management of Adults with Hospital-Acquired and Ventilator-Associated Pneumonia: 2016 Clinical Practice Guidelines by the Infectious Diseases Society of America and the American Thoracic Society. *Clin. Infect. Dis.* **2016**, *63*, e61–e111. [CrossRef]
37. NICE. Tiotropium. Available online: https://bnf.nice.org.uk/medicinal-forms/tiotropium.html (accessed on 5 November 2020).
38. NICE. Nedocromil Sodium. Available online: https://bnf.nice.org.uk/drug/nedocromil-sodium.html#medicinalForms (accessed on 5 November 2020).
39. NICE. Zanamivir. Available online: https://bnf.nice.org.uk/medicinal-forms/zanamivir.html (accessed on 5 November 2020).
40. NICE. Mannitol. Available online: https://bnf.nice.org.uk/drug/mannitol.html#medicinalForms (accessed on 5 November 2020).
41. NICE. Budesonide with Formoterol. Available online: https://bnf.nice.org.uk/drug/budesonide-with-formoterol.html (accessed on 28 March 2021).
42. NICE. Ciclesonide. Available online: https://bnf.nice.org.uk/drug/ciclesonide.html (accessed on 28 March 2021).
43. Rogers, D.F. Physiology of Airway Mucus Secretion and Pathophysiology of Hypersecretion. *Respir. Care* **2007**, *52*, 1134–1149.
44. Paul, K. Haemophilus influenzae and the lung (Haemophilus and the lung). *Clin. Transl. Med.* **2012**, *1*, 10. [CrossRef]
45. Bjarnsholt, T.; Jensen, P.; Fiandaca, M.; Pedersen, J.; Hansen, C.; Andersen, C.; Pressler, T.; Givskov, M.; Høiby, N. Pseudomonas aeruginosa biofilms in the respiratory tract of cystic fibrosis patients. *Paediatr. Pulmonol.* **2009**, *44*, 547–558. [CrossRef] [PubMed]

46. Ho, D.; Nichols, B.; Edgar, K.; Murgia, X.; Loretz, B.; Lehr, C. Challenges and strategies in drug delivery systems for treatment of pulmonary infections. *Eur. J. Pharm. Biopharm.* **2019**, *144*, 110–124. [CrossRef] [PubMed]
47. Cramer, N.; Wiehlmann, L.; Tümmler, B. Clonal epidemiology of Pseudomonas aeruginosa in cystic fibrosis. *Int. J. Med. Microbiol.* **2010**, *300*, 526–533. [CrossRef] [PubMed]
48. Robinson, J. Colonization and infection of the respiratory tract: What do we know? *Paediatr Child. Health* **2004**, *9*, 21–24. [CrossRef] [PubMed]
49. Carpagnano, G.E.; Lacedonia, D.; Palladino, G.P.; Logrieco, G.; Crisetti, E.; Susca, A.; Logrieco, A.; Foschino-Barbaro, M.P. Aspergillus spp. colonization in exhaled breath condensate of lung cancer patients from Puglia Region of Italy. *BMC Pulm Med.* **2014**, *14*, 22. [CrossRef] [PubMed]
50. Cancer Research UK. Survival. Available online: https://www.cancerresearchuk.org/about-cancer/lung-cancer/survival (accessed on 8 April 2021).
51. Sardeli, C.; Zarogoulidis, P.; Kosmidis, C.; Amaniti, A.; Katsaounis, A.; Giannakidis, D.; Koulouris, C.; Hohenforst-Schmidt, W.; Huang, H.; Bai, C.; et al. Inhaled chemotherapy adverse effects: Mechanisms and protection methods. *Lung Cancer Manag.* **2020**, *8*, LMT19. [CrossRef]
52. Garbuzenko, O.; Saad, M.; Pozharov, V.; Reuhl, K.; Mainelis, G.; Minko, T. Inhibition of lung tumor growth by complex pulmonary delivery of drugs with oligonucleotides as suppressors of cellular resistance. *Proc. Natl. Acad. Sci. USA* **2010**, *107*, 10737–10742. [CrossRef]
53. Garbuzenko, O.; Mainelis, G.; Taratula, O.; Minko, T. Inhalation treatment of lung cancer: The influence of composition, size and shape of nanocarriers on their lung accumulation and retention. *Cancer Biol. Med.* **2014**, *11*, 44–55. [CrossRef]
54. Lee, W.; Loo, C.; Ghadiri, M.; Leong, C.; Young, P.; Traini, D. The potential to treat lung cancer via inhalation of repurposed drugs. *Adv. Drug Deliv. Rev.* **2018**, *133*, 107–130. [CrossRef]
55. Gandhimathi, C.; Venugopal, J.; Sundarrajan, S.; Sridhar, R.; Tay, S.; Ramakrishna, S.; Kumar, S. Breathable Medicine: Pulmonary Mode of Drug Delivery. *J. Nanosci. Nanotechnol.* **2015**, *15*, 2591–2604. [CrossRef]
56. Reychler, G.; San Miguel-Pagola, M.; Aubriot, A.; Herrero-Cortina, B.; Lecocq, V.; Hesse, M.; Liistro, G.; Jamar, F. Targeted Lung Deposition from Nebulization Is Not Improved in the Lateral Decubitus Position in Healthy Volunteers. *Respir. Care* **2019**, *64*, 1537–1544. [CrossRef]
57. Healy, A.M.; Amaro, M.I.; Paluch, K.J.; Tajber, L. Dry powders for oral inhalation free of lactose carrier particles. *Adv. Drug Deliv. Rev.* **2014**, *75*, 32–52. [CrossRef]
58. Kaialy, W.; Martin, G.P.; Larhrib, H.; Ticehurst, M.D.; Kolosionek, E.; Nokhodchi, A. The influence of physical properties and morphology of crystallised lactose on delivery of salbutamol sulphate from dry powder inhalers. *Colloids Surf. B Biointerfaces* **2012**, *89*, 29–39. [CrossRef]
59. Sibum, I.; Hagedoorn, P.; de Boer, A.; Frijlink, H.; Grasmeijer, F. Challenges for pulmonary delivery of high powder doses. *Int. J. Pharm.* **2018**, *548*. [CrossRef] [PubMed]
60. De Backer, W.; Devolder, A.; Poli, G.; Acerbi, D.; Monno, R.; Herpich, C.; Sommerer, K.; Meyer, T.; Mariotti, F. Lung Deposition of BDP/Formoterol HFA pMDI in Healthy Volunteers, Asthmatic, and COPD Patients. *J. Aerosol Med. Pulm. Drug Deliv.* **2010**, *23*, 137–148. [CrossRef]
61. Virchow, J.; Poli, G.; Herpich, C.; Kietzig, C.; Ehlich, H.; Braeutigam, D.; Sommerer, K.; Häussermann, S.; Mariotti, F. Lung Deposition of the Dry Powder Fixed Combination Beclometasone Dipropionate Plus Formoterol Fumarate Using NEXThaler Device in Healthy Subjects, Asthmatic Patients and, COPD Patients. *J. Aerosol Med. Pulm. Drug Deliv.* **2018**, *31*, 269–280. [CrossRef] [PubMed]
62. Usmani, O.; Biddiscombe, M.; Barnes, P. Regional Lung Deposition and Bronchodilator Response as a Function of β_2-Agonist Particle Size. *Am. J. Respir. Crit. Care Med.* **2005**, *172*, 1497–1504. [CrossRef]
63. Leach, C.; Davidson, P.; Hasselquist, B.; Boudreau, R. Lung Deposition of Hydrofluoroalkane-134a Beclomethasone is Greater Than That of Chlorofluorocarbon Fluticasone and Chlorofluorocarbon Beclomethasone: A Cross-Over Study in Healthy Volunteers. *Chest* **2002**, *122*, 510–516. [CrossRef] [PubMed]
64. Häussermann, S.; Acerbi, D.; Brand, P.; Herpich, C.; Poli, G.; Sommerer, K.; Meyer, T. Lung deposition of formerol HFA (Atimos/Forair) in healthy volunteers, asthmatic and COPD patients. *J. Aerosol Med.* **2007**, *20*, 331–341. [CrossRef] [PubMed]
65. Leach, C.; Bethke, T.; Boudreau, R.; Hasselquist, B.; Drollmann, A.; Davidson, P.; Wurst, W. Two-dimensional and three-dimensional imaging show ciclesonide has high lung deposition and peripheral distribution: A nonrandomized study in healthy volunteers. *J. Aerosol Med.* **2006**, *19*, 117–126. [CrossRef] [PubMed]
66. Javadzadeh, Y.; Yaqoubi, S. Therapeutic nanostructures for pulmonary drug delivery. In *Nanostructures for Drug Delivery*; Andronescu, E., Grumezescu, A.M., Eds.; Elsevier: Amsterdam, The Netherlands, 2017; pp. 619–638.
67. Peng, T.; Lin, S.; Niu, B.; Wang, X.; Huang, Y.; Zhang, X.; Li, G.; Pan, X.; Wu, C. Influence of physical properties of carrier on the performance of dry powder inhalers. *Acta Pharm. Silica B* **2016**, *4*, 308–318. [CrossRef] [PubMed]
68. Martin, A.; Finlay, W. Nebulizers for Drug Delivery to the Lungs. *Expert Opin. Drug Deliv.* **2015**, *12*, 889–900. [CrossRef] [PubMed]
69. Hamed, K.; Debonnett, L. Tobramycin inhalation powder for the treatment of pulmonary *Pseudomonas aeruginosa* in patients with cystic fibrosis: A review based on clinical evidence. *Ther. Adv. Respir. Dis.* **2017**, *11*, 193–209. [CrossRef] [PubMed]
70. Miller, D.; Tan, T.; Nakamura, J.; Malcolmson, R.; Tarara, T.; Weers, J. Physical Characterization of Tobramycin Inhalation Powder. II. State Diagram of an Amorphous Engineered Particle Formulation. *Mol. Pharm.* **2017**, *14*, 1950–1960. [CrossRef]
71. Emc. Tobi Podhaler 28 mg Inhalation Powder, Hard Capsules. Available online: https://www.medicines.org.uk/emc/product/4757 (accessed on 19 November 2020).

72. Quon, B.; Goss, C.; Ramsey, B. Inhaled Antibiotics for Lower Airway Infections. *Annu. Am. Thorac. Soc.* **2014**, *11*, 425–434. [CrossRef]
73. Miller, D.; Tan, T.; Tarara, T.; Nakamura, J.; Malcolmson, R.; Weers, J. Physical Characterization of Tobramycin Inhalation Powder. I. Rational Design of a Stable-Engineered Particle Formulation for Delivery to the Lungs. *Mol. Pharm.* **2015**, *12*. [CrossRef]
74. Conole, D.; Keating, G. Colistimethate Sodium Dry Powder for Inhalation: A Review of Its Use in the Treatment of Chronic *Pseudomonas aeruginosa* Infection in Patients with Cystic Fibrosis. *Drugs* **2014**, *74*, 377–387. [CrossRef]
75. Emc. Colobreathe. Available online: https://www.medicines.org.uk/emc/product/3063 (accessed on 19 November 2020).
76. Schwarz, C. Colobreathe for the Treatment of Cystic Fibrosis-Associated Pulmonary Infections. *Pulm. Ther.* **2015**, *1*, 19–30. [CrossRef]
77. Cai, X.; Yang, Y.; Xie, X.; Yu, F.; Yang, Y.; Yang, Z.; Zhang, T.; Mei, X. Preparation, characterization and pulmonary pharmacokinetics of a new inhalable zanamivir dry powder. *Drug Deliv.* **2016**, *23*, 1962–1971. [CrossRef]
78. NICE. Tobramycin. Available online: https://bnf.nice.org.uk/medicinal-forms/tobramycin.html (accessed on 5 November 2020).
79. Emc. Colomycin 1 Million International Units (IU) Powder for Solution, Injection, Infusion or Inhalation. Available online: https://www.medicines.org.uk/emc/product/1094/smpc (accessed on 5 November 2020).
80. NICE. Colistimethate Sodium. Available online: https://bnf.nice.org.uk/medicinal-forms/colistimethate-sodium.html (accessed on 5 November 2020).
81. NICE. Aztreonam. Available online: https://bnf.nice.org.uk/medicinal-forms/aztreonam.html (accessed on 5 November 2020).
82. Pharma, D. Dry Powder Inhalation. Available online: https://dfepharma.com/Excipients/Expertise/Dry-Powder-Inhalation (accessed on 31 March 2021).
83. Demoly, P.; Hagedoorn, P.; De Boer, A.H.; Frijlink, H.W. The clinical relevance of dry powder inhaler performance for drug delivery. *Respir. Med.* **2014**, *108*, 1195–1203. [CrossRef]
84. Tan, B.M.J.; Chan, L.W.; Heng, P.W.S. Improving Dry Powder Inhaler Performance by Surface Roughening of Lactose Carrier Particles. *Pharm. Res.* **2016**, *33*. [CrossRef]
85. Lee, H.; Lee, H.; Kwon, Y.; Kim, J.; Rhee, Y.; Chon, J.; Park, E.; Kim, D.; Park, C. The role of lactose carrier on the powder behaviour and aerodynamic performance of bosentan microparticles for dry powder inhalation. *Eur. J. Pharm. Sci.* **2018**, *117*, 279–289. [CrossRef]
86. Yeung, S.; Traini, D.; Tweedie, A.; Lewis, D.; Church, T.; Young, P. Limitations of high dose carrier based formulations. *Int. J. Pharm.* **2018**, *544*, 141–152. [CrossRef] [PubMed]
87. Pacławski, A.; Szlęk, J.; Lau, R.; Jachowicz, R.; Mendyk, A. Empirical modeling of the fine particle fraction for carrier-based pulmonary delivery formulations. *Int J. Nanomed.* **2015**, *10*, 801–810. [CrossRef]
88. Ambrus, R.; Benke, E.; Farkas, Á.; Balásházy, I.; Szabó-Révész, P. Novel dry powder inhaler formulation containing antibiotic using combined technology to improve aerodynamic properties. *Eur. J. Pharm. Sci.* **2018**, *123*, 20–27. [CrossRef] [PubMed]
89. Ceschan, N.; Bucalá, V.; Mateos, M.; Smyth, H.; Ramírez-Rigo, M. Carrier free indomethacin microparticles for dry powder inhalation. *Int. J. Pharm.* **2018**, *549*, 169–178. [CrossRef] [PubMed]
90. Momin, M.; Sinha, S.; Tucker, I.; Das, S. Carrier-free combination dry powder inhaler formulation of ethionamide and moxifloxacin for treating drug resistant tuberculosis. *Drug Dev. Ind. Pharm.* **2019**, *45*, 1321–1331. [CrossRef] [PubMed]
91. Azari, F.; Ghanbarzadeh, S.; Safdari, R.; Yaqoubi, S.; Adibkia, K.; Hamishehkar, H. Development of a Carrier Free Dry Powder Inhalation Formulation of Ketotifen for Pulmonary Drug Delivery. *Drug Res.* **2020**, *70*, 26–32. [CrossRef]
92. Nguyen, T.; Yi, E.; Hwang, K.; Cho, C.; Park, C.; Kim, J.; Rhee, Y.; Park, E. Formulation and evaluation of carrier-free dry powder inhaler containing sildenafil. *Drug Deliv. Transl. Res.* **2019**, *9*, 319–333. [CrossRef]
93. Yazdi, A.; Smyth, H. Carrier-free high-dose dry powder inhaler formulation of ibuprofen: Physicochemical characterization and in vitro aerodynamic performance. *Int. J. Pharm.* **2016**, *511*, 403–414. [CrossRef]
94. Edwards, A.; Chambers, A. Comparison of a lactose-free formulation of sodium cromoglycate and sodium cromoglycate plus lactose in the treatment of asthma. *Curr. Med. Res. Opin.* **1989**, *11*, 283–292. [CrossRef] [PubMed]
95. Pilcer, G.; De Bueger, V.; Traina, K.; Traore, H.; Sebti, T.; Vanderbist, F.; Amighi, K. Carrier-free combination for dry powder inhalation of antibiotics in the treatment of lung infections in cystic fibrosis. *Int. J. Pharm.* **2013**, *451*, 112–120. [CrossRef]
96. Raula, J.; Lahde, A.; Kauppinen, E. Aerosolization behaviour of carrier-free L-leucine coated salbutamol sulphate powders. *Int. J. Pharm.* **2009**, *365*, 18–25. [CrossRef] [PubMed]
97. Haghi, M.; van den Oetelaar, W.; Moir, L.; Zhu, B.; Phillips, G.; Crapper, J.; Young, P.; Traini, D. Inhalable tranexamic acid for haemoptysis treatment. *Eur. J. Pharm. Biopharm.* **2015**, *93*, 311–319. [CrossRef]
98. Suzuki, É.; Amaro, M.; de Almeida, G.; Cabral, L.; Healy, A.; de Sousa, V. Development of a new formulation of roflumilast for pulmonary drug delivery to treat inflammatory lung conditions. *Int. J. Pharm.* **2018**, *550*, 89–99. [CrossRef]
99. Zhu, B.; Padroni, M.; Colombo, G.; Phillips, G.; Crapper, J.; Young, P.; Traini, D. The development of a single-use, capsule-free multi-breath tobramycin dry powder inhaler for the treatment of cystic fibrosis. *Int. J. Pharm.* **2016**, *514*, 392–398. [CrossRef] [PubMed]
100. Chvatal, A.; Farkas, Á.; Balásházy, I.; Szabó-Révész, P.; Ambrus, R. Aerodynamic properties and in silico deposition of meloxicam potassium incorporated in a carrier-free DPI pulmonary system. *Int. J. Pharm.* **2017**, *520*, 70–78. [CrossRef]
101. Shetty, N.; Ahn, P.; Park, H.; Bhujbal, S.; Zemlyanov, D.; Cavallaro, A.; Mangal, S.; Li, J.; Zhou, Q. Improved Physical Stability and Aerosolization of Inhalable Amorphous Ciprofloxacin Powder Formulations by Incorporating Synergistic Colistin. *Mol. Pharm.* **2018**, *15*, 4004–4020. [CrossRef]
102. Cui, Y.; Zhang, X.; Wang, W.; Huang, Z.; Zhao, Z.; Wang, G.; Cai, S.; Jing, H.; Huang, Y.; Pan, X.; et al. Moisture-Resistant Co-Spray-Dried Netilmicin with l-Leucine as Dry Powder Inhalation for the Treatment of Respiratory Infections. *Pharmaceutics* **2018**, *10*, 252. [CrossRef] [PubMed]

103. Kumar, R.; Dalvi, S.V.; Siril, P.F. Nanoparticle-Based Drugs and Formulations: Current Status and Emerging Applications. *ACS Appl. Nano Mater.* **2020**, *3*, 4944–4961. [CrossRef]
104. Mangal, S.; Gao, W.; Li, T.; Zhou, Q. Pulmonary delivery of nanoparticle chemotherapy for the treatment of lung cancers: Challenges and opportunities. *Acta Pharmacol. Sin.* **2017**, *38*, 782–797. [CrossRef]
105. Huang, Z.; Kłodzińska, S.N.; Wan, F.; Nielsen, H.M. Nanoparticle-mediated pulmonary drug delivery: State of the art towards efficient treatment of recalcitrant respiratory tract bacterial infections. *Drug Deliv. Transl. Res.* **2021**, *11*, 1634–1654. [CrossRef]
106. Zhu, C.; Chen, J.; Yu, S.; Que, C.; Taylor, L.S.; Tan, W.; Wu, C.; Zhou, Q.T. Inhalable Nanocomposite Microparticles with Enhanced Dissolution and Superior Aerosol Performance. *Mol. Pharm.* **2020**, *17*, 3270–3280. [CrossRef]
107. Roa, W.H.; Azarmi, S.; Al-Hallak, M.H.D.K.; Finlay, W.H.; Magliocco, A.M.; Löbenberg, R. Inhalable nanoparticles, a non-invasive approach to treat lung cancer in a mouse model. *J. Control. Release* **2011**, *150*, 49–55. [CrossRef]
108. Azarmi, S.; Tao, X.; Chen, H.; Wang, Z.; Finlay, W.H.; Löbenberg, R.; Roa, W.H. Formulation and cytotoxicity of doxorubicin nanoparticles carried by dry powder aerosol particles. *Int. J. Pharm.* **2006**, *319*, 155–161. [CrossRef] [PubMed]
109. Indermun, S.; Govender, M.; Kumar, P.; Choonara, Y.E.; Pillay, V. Porous particulate platforms for enhanced pulmonary delivery of bioactives. In *Targeting Chronic Inflammatory Lung Diseases Using Advanced Drug Delivery Systems*; Elsevier: Amsterdam, The Netherlands, 2020; pp. 359–373.
110. Chvatal, A.; Ambrus, R.; Party, P.; Katona, G.; Jójárt-Laczkovich, O.; Szabó-Révész, P.; Fattal, E.; Tsapis, N. Formulation and comparison of spray dried non-porous and large porous particles containing meloxicam for pulmonary drug delivery. *Int. J. Pharm.* **2019**, *559*, 68–75. [CrossRef] [PubMed]
111. Ogienko, A.G.; Bogdanova, E.G.; Trofimov, N.A.; Myz, S.A.; Ogienko, A.A.; Kolesov, B.A.; Yunoshev, A.S.; Zubikov, N.V.; Manakov, A.Y.; Boldyrev, V.V.; et al. Large porous particles for respiratory drug delivery. Glycine-based formulations. *Eur. J. Pharm. Sci.* **2017**, *110*, 148–156. [CrossRef]
112. Al-Obaidi, H.; Majumder, M.; Bari, F. Amorphous and crystalline particulates: Challenges and perspectives in drug delivery. *Curr. Pharm. Des.* **2016**, *23*, 350–361. [CrossRef]
113. Al-Obaidi, H.; Lawrence, M.J.; Buckton, G. Atypical effects of incorporated surfactants on stability and dissolution properties of amorphous polymeric dispersions. *J. Pharm. Pharmacol.* **2016**, *68*, 1373–1383. [CrossRef]
114. Hancock, B.C.; Parks, M. What is the true solubility advantage for amorphous pharmaceuticals? *Pharm. Res.* **2000**, *17*, 397–404. [CrossRef] [PubMed]
115. Bhanderi, A.; Bari, F.; Al-Obaidi, H. Evaluation of the impact of surfactants on miscibility of griseofulvin in spray dried amorphous solid dispersions. *J. Drug Deliv. Sci. Technol.* **2021**, *64*, 102606. [CrossRef]
116. Baghel, S.; Cathcart, H.; O'Reilly, N.J. Polymeric Amorphous Solid Dispersions: A Review of Amorphization, Crystallization, Stabilization, Solid-State Characterization, and Aqueous Solubilization of Biopharmaceutical Classification System Class II Drugs. *J. Pharm. Sci.* **2016**, *105*, 2527–2544. [CrossRef]
117. Grzybowska, K.; Paluch, M.; Wlodarczyk, P.; Grzybowski, A.; Kaminski, K.; Hawelek, L.; Zakowiecki, D.; Kasprzycka, A.; Jankowska-Sumara, I. Enhancement of amorphous celecoxib stability by mixing it with octaacetylmaltose: The molecular dynamics study. *Mol. Pharmacol.* **2012**, *9*, 894–904. [CrossRef]
118. Al-Obaidi, H.; Brocchini, S.; Buckton, G. Anomalous properties of spray dried solid dispersions. *J. Pharm. Sci.* **2009**, *98*, 4724–4737. [CrossRef]
119. Al-Obaidi, H.; Buckton, G. Evaluation of griseofulvin binary and ternary solid dispersions with HPMCAS. *AAPS PharmSciTech* **2009**, *10*, 1172–1177. [CrossRef]
120. AboulFotouh, K.; Zhang, Y.; Maniruzzaman, M.; Williams, R.O.; Cui, Z. Amorphous solid dispersion dry powder for pulmonary drug delivery: Advantages and challenges. *Int. J. Pharm.* **2020**, *587*, 119711. [CrossRef]
121. Karimi, K.; Katona, G.; Csóka, I.; Ambrus, R. Physicochemical stability and aerosolization performance of dry powder inhalation system containing ciprofloxacin hydrochloride. *J. Pharm. Biomed. Anal.* **2018**, *148*, 73–79. [CrossRef] [PubMed]
122. Moes, J.; Koolen, S.; Huitema, A.; Schellens, J.; Beijnen, J.; Nuijen, B. Development of an oral solid dispersion formulation for use in low-dose metronomic chemotherapy of paclitaxel. *Eur. J. Pharm. Biopharm.* **2013**, *83*, 87–94. [CrossRef] [PubMed]
123. Purohit, H.S.; Taylor, L.S. Phase Behavior of Ritonavir Amorphous Solid Dispersions during Hydration and Dissolution. *Pharm. Res.* **2017**, *34*, 2842–2861. [CrossRef] [PubMed]
124. Mizoguchi, R.; Waraya, H.; Hirakura, Y. Application of Co-Amorphous Technology for Improving the Physicochemical Properties of Amorphous Formulations. *Mol. Pharm.* **2019**, *16*, 2142–2152. [CrossRef] [PubMed]
125. Newman, A.; Zografi, G. Commentary: Considerations in the Measurement of Glass Transition Temperatures of Pharmaceutical Amorphous Solids. *Am. Assoc. Pharm. Sci.* **2019**, *21*, 26. [CrossRef]
126. Wostry, M.; Plappert, H.; Grohganz, H. Preparation of Co-Amorphous Systems by Freeze-Drying. *Pharmaceutics* **2020**, *12*, 941. [CrossRef]
127. Lim, A.W.; Löbmann, K.; Grohganz, H.; Rades, T.; Chieng, N. Investigation of physical properties and stability of indomethacin-cimetidine and naproxen-cimetidine co-amorphous systems prepared by quench cooling, coprecipitation and ball milling. *J. Pharm. Pharmacol.* **2016**, *68*, 36–45. [CrossRef]
128. Karagianni, A.; Kachrimanis, K.; Nikolakakis, I. Co-Amorphous Solid Dispersions for Solubility and Absorption Improvement of Drugs: Composition, Preparation, Characterization and Formulations for Oral Delivery. *Pharmaceutics* **2018**, *10*, 98. [CrossRef]

129. Lababidi, N.; Ofosu Kissi, E.; Elgaher, W.A.M.; Sigal, V.; Haupenthal, J.; Schwarz, B.C.; Hirsch, A.K.H.; Rades, T.; Schneider, M. Spray-drying of inhalable, multifunctional formulations for the treatment of biofilms formed in cystic fibrosis. *J. Control. Release* **2019**, *314*, 62–71. [CrossRef]
130. Zhou, Q.T.; Loh, Z.H.; Yu, J.; Sun, S.P.; Gengenbach, T.; Denman, J.A.; Li, J.; Chan, H.K. How Much Surface Coating of Hydrophobic Azithromycin Is Sufficient to Prevent Moisture-Induced Decrease in Aerosolisation of Hygroscopic Amorphous Colistin Powder? *Am. Assoc. Pharm. Sci.* **2016**, *18*, 1213–1224. [CrossRef]
131. Wang, Z.; Sun, M.; Liu, T.; Gao, Z.; Ye, Q.; Tan, X.; Hou, Y.; Sun, J.; Wang, D.; He, Z. Co-amorphous solid dispersion systems of lacidipine-spironolactone with improved dissolution rate and enhanced physical stability. *Asian J. Pharm. Sci.* **2019**, *14*, 95–103. [CrossRef] [PubMed]
132. Mangal, S.; Nie, H.; Xu, R.; Guo, R.; Cavallaro, A.; Zemlyanov, D.; Zhou, Q.T. Physico-Chemical Properties, Aerosolization and Dissolution of Co-Spray Dried Azithromycin Particles with L-Leucine for Inhalation. *Pharm. Res.* **2018**, *35*. [CrossRef] [PubMed]
133. Kasten, G.; Löbmann, K.; Grohganz, H.; Rades, T. Co-former selection for co-amorphous drug-amino acid formulations. *Int. J. Pharm.* **2019**, *557*, 366–373. [CrossRef]
134. Su, M.; Xia, Y.; Shen, Y.; Heng, W.; Wei, Y.; Zhang, L.; Gao, Y.; Zhang, J.; Qian, S. A novel drug–drug coamorphous system without molecular interactions: Improve the physicochemical properties of tadalafil and repaglinide. *RSC Adv.* **2020**, *10*, 565–583. [CrossRef]
135. Carvahlo, R.S.; Watts, B.A.; Peters, I.J.; Williams, O.R. *Dry Powder Inhalation for Pulmonary Delivery: Recent Advances and Continuing Challenges*; Wiley: Chichester, UK, 2015; pp. 36–48.
136. Chen, Z.; Yang, K.; Huang, C.; Zhu, A.; Yu, L.; Qian, F. Surface Enrichment and Depletion of the Active Ingredient in Spray Dried Amorphous Solid Dispersions. *Pharm. Res.* **2018**, *35*, 38. [CrossRef]
137. Al-Obaidi, H.; Ke, P.; Brocchini, S.; Buckton, G. Characterization and stability of ternary solid dispersions with PVP and PHPMA. *Int. J. Pharm.* **2011**, *419*, 20–27. [CrossRef]
138. Vehring, R. Pharmaceutical particle engineering via spray drying. *Pharm. Res.* **2008**, *25*, 999–1022. [CrossRef]
139. Onoue, S.; Sato, H.; Kawabata, Y.; Mizumoto, T.; Hashimoto, N.; Yamada, S. In vitro and in vivo characterization on amorphous solid dispersion of cyclosporine A for inhalation therapy. *J. Control. Release* **2009**, *138*, 16–23. [CrossRef]
140. Jong, T.; Li, J.; Morton, D.A.; Zhou, Q.T.; Larson, I. Investigation of the Changes in Aerosolization Behavior Between the Jet-Milled and Spray-Dried Colistin Powders Through Surface Energy Characterization. *J. Pharm. Sci.* **2016**, *105*, 1156–1163. [CrossRef]
141. Shetty, N.; Park, H.; Zemlyanov, D.; Mangal, S.; Bhujbal, S.; Zhou, Q.T. Influence of excipients on physical and aerosolization stability of spray dried high-dose powder formulations for inhalation. *Int. J. Pharm.* **2018**, *544*, 222–234. [CrossRef] [PubMed]
142. Wang, L.; Zhang, Y.; Tang, X. Characterization of a new inhalable thymopentin formulation. *Int. J. Pharm.* **2009**, *375*, 1–7. [CrossRef] [PubMed]
143. Zijlstra, G.S.; Rijkeboer, M.; Jan van Drooge, D.; Sutter, M.; Jiskoot, W.; van de Weert, M.; Hinrichs, W.L.; Frijlink, H.W. Characterization of a cyclosporine solid dispersion for inhalation. *AAPS J.* **2007**, *9*, E190–E199. [CrossRef] [PubMed]
144. Wang, Y.-B.; Watts, A.B.; Peters, J.I.; Liu, S.; Batra, A.; Williams, R.O., 3rd. In vitro and in vivo performance of dry powder inhalation formulations: Comparison of particles prepared by thin film freezing and micronization. *AAPS PharmSciTech* **2014**, *15*, 981–993. [CrossRef]
145. Watts, A.B.; Wang, Y.-B.; Johnston, K.P.; Williams, R.O. Respirable Low-Density Microparticles Formed In Situ from Aerosolized Brittle Matrices. *Pharm. Res.* **2013**, *30*, 813–825. [CrossRef]
146. Momin, M.A.M.; Sinha, S.; Tucker, G.I.; Doyle, C.; Das, C.S. Dry powder formulation of kanamycin with enhanced aerosolization efficiency for drug-resistant tuberculosis. *Int. J. Pharm.* **2017**, *528*, 107–117. [CrossRef]
147. McShane, P.J.; Weers, J.G.; Tarara, T.E.; Haynes, A.; Durbha, P.; Miller, D.P.; Mundry, T.; Operschall, E.; Elborn, J.S. Ciprofloxacin Dry Powder for Inhalation (ciprofloxacin DPI): Technical design and features of an efficient drug-device combination. *Pulm. Pharmacol. Ther.* **2018**, *50*, 72–79. [CrossRef]
148. Mohammed, A.; Zurek, J.; Madueke, S.; Al-Kassimy, H.; Yaqoob, M.; Houacine, C.; Ferraz, A.; Kalgudi, R.; Zariwala, M.G.; Hawkins, N.; et al. Generation of High Dose Inhalable Effervescent Dispersions against Pseudomonas aeruginosa Biofilms. *Pharm. Res.* **2020**, *37*, 150. [CrossRef]
149. Porowska, A.; Dosta, M.; Fries, L.; Gianfrancesco, A.; Heinrich, S.; Palzer, S. Predicting the surface composition of spray-dried particle by modelling component reorganisation in a drying droplet. *Chem. Eng. Res. Des.* **2016**, *110*, 131–140. [CrossRef]
150. Li, X.; Vogt, F.G.; Hayes, D.; Mansour, H.M. Design, characterization, and aerosol dispersion performance modeling of advanced co-spray dried antibiotics with mannitol as respirable microparticles/nanoparticles for targeted pulmonary delivery as dry powder inhalers. *Pharm. Drug Deliv. Pharm. Technol.* **2014**, *103*, 2937–2949. [CrossRef] [PubMed]
151. Padhi, B.K.; Chougule, M.B.; Misra, A. Optimization of formulation components and characterization of large respirable powders containing high therapeutic payload. *Pharm. Dev. Technol.* **2006**, *11*, 465–475. [CrossRef] [PubMed]
152. Hassan, A.; Farkas, D.; Longest, W.; Hindle, M. Characterization of excipient enhanced growth (EEG) tobramycin dry powder aerosol formulations. *Int. J. Pharm.* **2020**, *591*, 120027. [CrossRef]
153. Longest, P.W.; Farkas, D.; Hassan, A.; Hindle, M. Computational Fluid Dynamics (CFD) Simulations of Spray Drying: Linking Drying Parameters with Experimental Aerosolization Performance. *Pharm. Res.* **2020**, *37*, 101. [CrossRef] [PubMed]
154. Yu, J.; Chan, H.K.; Gengenbach, T.; Denman, J.A. Protection of hydrophobic amino acids against moisture-induced deterioration in the aerosolization performance of highly hygroscopic spray-dried powders. *Eur. J. Pharm. Biopharm.* **2017**, *119*, 224–234. [CrossRef] [PubMed]

155. Maa, Y.F.; Costantino, H.R.; Nguyen, P.A.; Hsu, C.C. The effect of operating and formulation variables on the morphology of spray-dried protein particles. *Pharm. Dev. Technol.* **1997**, *2*, 213–223. [CrossRef]
156. Guenette, E.; Barrett, A.; Kraus, D.; Brody, R.; Harding, L.; Magee, G. Understanding the effect of lactose particle size on the properties of DPI formulations using experimental design. *Int. J. Pharm.* **2009**, *380*, 80–88. [CrossRef]
157. Tulane University. Mucolytics. Available online: https://tmedweb.tulane.edu/pharmwiki/doku.php/mucolytics (accessed on 14 April 2021).
158. British National Formulary (Online). Cystic Fibrosis. Available online: https://bnf.nice.org.uk/treatment-summary/cystic-fibrosis.html (accessed on 18 April 2021).
159. Lamy, B.; Tewes, F.; Serrano, D.R.; Lamarche, I.; Gobin, P.; Couet, W.; Healy, A.M.; Marchand, S. New aerosol formulation to control ciprofloxacin pulmonary concentration. *J. Control. Release* **2018**, *271*, 118–126. [CrossRef]
160. Tewes, F.; Brillault, J.; Lamy, B.; O'Connell, P.; Olivier, J.-C.; Couet, W.; Healy, A.M. Ciprofloxacin-Loaded Inorganic–Organic Composite Microparticles To Treat Bacterial Lung Infection. *Mol. Pharm.* **2016**, *13*, 100–112. [CrossRef] [PubMed]
161. Barazesh, A.; Gilani, K.; Rouini, M.; Barghi, M.A. The effect of metal salts on aerosol performance of spray dried carrier-free formulations of levofloxacin. *Daru* **2020**, *28*, 75–85. [CrossRef] [PubMed]
162. Walsh, D.; Serrano, D.; Worku, Z.; Norris, B.; Healy, A. Production of cocrystals in an excipient matrix by spray drying. *Int. J. Pharm.* **2018**, *536*, 467–477. [CrossRef]
163. Karashima, M.; Sano, N.; Yamamoto, S.; Arai, Y.; Yamamoto, K.; Amano, N.; Ikeda, Y. Enhanced pulmonary absorption of poorly soluble itraconazole by micronized cocrystal dry powder formulations. *Eur. J. Pharm. Biopharm. Off. J. Arb. Pharm. Verfahr. EV* **2017**, *115*, 65–72. [CrossRef]
164. Emami, S.; Adibkia, K.; Barzegar-Jalali, M.; Siahi-Shadbad, M. Piroxicam cocrystals with phenolic coformers: Preparation, characterisation and dissolution properties. *Pharm. Dev. Technol.* **2018**, *24*, 199–210. [CrossRef] [PubMed]
165. Wicker, J.; Crowley, L.; Robshaw, O.; Little, E.; Stokes, S.; Cooper, R.; Lawrence, S. Will they cocrystallize? *CrysEngComm* **2017**, *19*, 5336–5340. [CrossRef]
166. Fukte, S.; Wagh, M.; Rawat, S. Coformer selection: An important tool in cocrystal formation. *Int. J. Pharm. Pharm. Sci.* **2014**, *6*, 9–14.
167. Cvetkovski, A.; Ferretti, V.; Bertolasi, V. New Pharmaceutical Salts Containing Pyridoxine. *Acta Crystallogr. Sect. C Struct. Chem.* **2017**, *73*, 1064–1070. [CrossRef]
168. Chadha, R.; Bhalla, Y.; Nandan, A.; Chadha, K.; Karan, M. Chrysin Cocrystals: Characterization and Evaluation. *J. Pharm. Biomed. Anal.* **2017**, *134*, 361–371. [CrossRef] [PubMed]
169. Perlovich, G.L. Thermodynamic characteristics of cocrystal formation and melting points for rational design of pharmaceutical two-component systems. *CrysEngComm* **2015**, *17*, 7019–7028. [CrossRef]
170. Panzade, P.; Shendarkar, G. Superior Solubility and Dissolution of Zaltoprofen via Pharmaceutical Cocrystals. *Turk. J. Pharm. Sci.* **2019**, *16*, 310–316. [CrossRef]
171. Ogienko, A.G.; Myz, S.A.; Ogienko, A.A.; Nefedov, A.A.; Stoporev, A.S.; Mel'gunov, M.S.; Yunoshev, A.S.; Shakhtshneider, T.P.; Boldyrev, V.V.; Boldyreva, E.V. Cryosynthesis of Co-Crystals of Poorly Water-Soluble Pharmaceutical Compounds and Their Solid Dispersions with Polymers. The "Meloxicam–Succinic Acid" System as a Case Study. *Cryst Growth Des.* **2018**, *18*, 7401–7409. [CrossRef]
172. Zellnitz, S.; Roblegg, E.; Pinto, J.; Fröhlich, E. Delivery of Dry Powders to the Lungs: Influence of Particle Attributes from a Biological and Technological Point of View. *Curr. Drug Deliv.* **2019**, *16*, 180–194. [CrossRef]
173. Alhalaweh, A.; Kaialy, W.; Buckton, G.; Gill, H.; Nokhodchi, A.; Velaga, S.P. Theophylline cocrystals prepared by spray drying: Physicochemical properties and aerosolization performance. *AAPS PharmSciTech* **2013**, *14*, 265–276. [CrossRef]
174. Tanaka, R.; Hattori, Y.; Otsuka, M.; Ashizawa, K. Application of spray freeze drying to theophylline-oxalic acid cocrystal engineering for inhaled dry powder technology. *Drug Dev. Ind. Pharm.* **2020**, *46*, 179–187. [CrossRef]
175. Gautam, M.; Besan, M.; Pandit, D.; Mandal, S.; Chadha, R. Cocrystal of 5-Fluorouracil: Characterization and Evaluation of Biopharmaceutical Parameters. *AAPS PharmSciTech* **2019**, *20*, 149. [CrossRef]
176. Tomar, S.; Chakraborti, S.; Jindal, A.; Grewal, M.; Chadha, R. Cocrystals of diacerein: Towards the development of improved biopharmaceutical parameters. *Int. J. Pharm.* **2020**, *574*, 118942. [CrossRef]
177. Bhalla, Y.; Chadha, K.; Chadha, R.; Karan, M. Daidzein cocrystals: An opportunity to improve its biopharmaceutical parameters. *Heliyon* **2019**, *5*, e02669. [CrossRef]
178. Shinozaki, T.; Ono, M.; Higashi, K.; Moribe, K. A Novel Drug-Drug Cocrystal of Levofloxacin and Metacetamol: Reduced Hygroscopicity and Improved Photostability of Levofloxacin. *J. Pharm. Sci.* **2019**, *108*, 2383–2390. [CrossRef]
179. Wang, Q.; Xue, J.; Hong, Z.; Du, Y. Pharmaceutical Cocrystal Formation of Pyrazinamide with 3-Hydroxybenzoic Acid: A Terahertz and Raman Vibrational Spectroscopies Study. *Molecules* **2019**, *24*, 488. [CrossRef]
180. do Amaral, L.; do Carmo, F.; Amaro, M.; de Sousa, V.; da Silva, L.; de Almeida, G.; Rodrigues, C.; Healy, A.; Cabral, L. Development and Characterization of Dapsone Cocrystal Prepared by Scalable Production Methods. *AAPS PharmSciTech* **2018**, *19*, 2687–2699. [CrossRef]
181. Khan, E.; Shukla, A.; Jhariya, A.; Tandon, P.; Vangala, V. Nitrofurantoin-melamine monohydrate (cocrystal hydrate): Probing the role of H-bonding on the structure and properties using quantum chemical calculations and vibrational spectroscopy. *Spectrochim. Acta Part. A Mol. Biomol. Spectrosc.* **2019**, *221*, 117170. [CrossRef]

182. Stavropoulos, K.; Johnston, S.; Zhang, Y.; Rao, B.; Hurrey, M.; Hurter, P.; Topp, E.; Kadiyala, I. Cocrystalline Solids of Telaprevir with Enhanced Oral Absorption. *J. Pharm. Sci.* **2015**, *104*, 3343–3350. [CrossRef]
183. Ton, Q.; Egert, E. Cocrystals of the antibiotic trimethoprim with glutarimide and 3,3-dimethylglutarimide held together by three hydrogen bonds. *Acta Crystallogr. Sect. C Struct. Chem.* **2015**, *71*, 75–79. [CrossRef]
184. Haneef, J.; Arora, P.; Chadha, R. Implication of Coformer Structural Diversity on Cocrystallization Outcomes of Telmisartan with Improved Biopharmaceutical Performance. *AAPS PharmSciTech* **2019**, *21*, 10. [CrossRef] [PubMed]
185. Serrano, D.; Persoons, T.; D'Arcy, D.; Galiana, C.; Dea-Ayuela, M.; Healy, A. Modelling and shadowgraph imaging of cocrystal dissolution and assessment of in vitro antimicrobial activity for sulfadimidine/4-aminosalicylic acid cocrystals. *Eur. J. Pharm. Sci.* **2016**, *89*, 125–136. [CrossRef] [PubMed]
186. Swapna, B.; Maddileti, D.; Nangia, A. Cocrystals of the Tuberculosis Drug Isoniazid: Polymorphism, Isostructurality, and Stability. *Cryst. Growth Des.* **2014**, *14*, 5991–6005. [CrossRef]
187. Seo, J.; Hwang, K.; Lee, S.; Kim, D.; Park, E. Preparation and characterization of adefovir dipivoxil-stearic acid cocrystal with enhanced physicochemical properties. *Pharm. Dev. Technol.* **2017**, *23*, 890–899. [CrossRef]
188. Cai, Q.; Xue, J.; Wang, Q.; Du, Y. Solid-state cocrystal formation between acyclovir and fumaric acid: Terahertz and Raman vibrational spectroscopic studies. *Spectrochim. Acta Part. A Mol. Biomol. Spectrosc.* **2017**, *186*, 29–36. [CrossRef] [PubMed]
189. Cheney, M.; Weyna, D.; Shan, N.; Hanna, M.; Wojtas, L.; Zaworotko, M. Coformer selection in pharmaceutical cocrystal development: A case study of a meloxicam aspirin cocrystal that exhibits enhanced solubility and pharmacokinetics. *J. Pharm. Sci.* **2011**, *100*, 2172–2181. [CrossRef] [PubMed]
190. Ray, E.; Vaghasiya, K.; Sharma, A.; Shukla, R.; Khan, R.; Kumar, A.; Verma, R. Autophagy-Inducing Inhalable Co-crystal Formulation of Niclosamide-Nicotinamide for Lung Cancer Therapy. *AAPS PharmSciTech* **2020**, *21*, 260. [CrossRef] [PubMed]
191. Zhang, Z.; Zhang, Q.; Zhang, Q.; Chen, C.; He, M.; Chen, Q.; Song, G.; Xuan, X.; Huang, X. From a binary salt to salt co-crystals of antibacterial agent lomefloxacin with improved solubility and bioavailability. *Acta Crystallogr. Sect. B Struct. Sci. Cryst. Eng. Mater.* **2015**, *71*, 437–446. [CrossRef]
192. Liu, L.; Zou, D.; Zhang, Y.; Zhang, Q.; Feng, Y.; Guo, Y.; Liu, Y.; Zhang, X.; Cheng, G.; Wang, C.; et al. Pharmaceutical salts/cocrystals of enoxacin with dicarboxylic acids: Enhancing in vitro antibacterial activity of enoxacin by improving the solubility and permeability. *Eur. J. Pharm. Biopharm.* **2020**, *154*, 62–73. [CrossRef] [PubMed]
193. Abidi, S.; Azim, Y.; Khan, S.; Khan, A. Sulfaguanidine cocrystals: Synthesis, structural characterization and their antibacterial and hemolytic analysis. *J. Pharm. Biomed. Anal.* **2018**, *149*, 351–357. [CrossRef]

pharmaceutics

MDPI

Review

Drug-Rich Phases Induced by Amorphous Solid Dispersion: Arbitrary or Intentional Goal in Oral Drug Delivery?

Kaijie Qian [1], Lorenzo Stella [2,3], David S. Jones [1], Gavin P. Andrews [1,4], Huachuan Du [5,6,*] and Yiwei Tian [1,*]

1 Pharmaceutical Engineering Group, School of Pharmacy, Queen's University Belfast, 97 Lisburn Road, Belfast BT9 7BL, UK; kqian02@qub.ac.uk (K.Q.); D.Jones@qub.ac.uk (D.S.J.); g.andrews@qub.ac.uk (G.P.A.)
2 Atomistic Simulation Centre, School of Mathematics and Physics, Queen's University Belfast, 7–9 College Park E, Belfast BT7 1PS, UK; l.stella@qub.ac.uk
3 David Keir Building, School of Chemistry and Chemical Engineering, Queen's University Belfast, Stranmillis Road, Belfast BT9 5AG, UK
4 School of Pharmacy, China Medical University, No.77 Puhe Road, Shenyang North New Area, Shenyang 110122, China
5 Laboratory of Applied Mechanobiology, Department of Health Sciences and Technology, ETH Zurich, Vladimir-Prelog-Weg 4, 8093 Zurich, Switzerland
6 Simpson Querrey Institute, Northwestern University, 303 East Superior Street, 11th Floor, Chicago, IL 60611, USA
* Correspondence: huachuan.du@hest.ethz.ch (H.D.); y.tian@qub.ac.uk (Y.T.); Tel.: +41-446339049 (H.D.); +44-2890972689 (Y.T.)

Citation: Qian, K.; Stella, L.; Jones, D.S.; Andrews, G.P.; Du, H.; Tian, Y. Drug-Rich Phases Induced by Amorphous Solid Dispersion: Arbitrary or Intentional Goal in Oral Drug Delivery? *Pharmaceutics* **2021**, *13*, 889. https://doi.org/10.3390/pharmaceutics13060889

Academic Editors: Vitaliy Khutoryanskiy and Hisham Al-Obaidi

Received: 20 May 2021
Accepted: 10 June 2021
Published: 15 June 2021

Abstract: Among many methods to mitigate the solubility limitations of drug compounds, amorphous solid dispersion (ASD) is considered to be one of the most promising strategies to enhance the dissolution and bioavailability of poorly water-soluble drugs. The enhancement of ASD in the oral absorption of drugs has been mainly attributed to the high apparent drug solubility during the dissolution. In the last decade, with the implementations of new knowledge and advanced analytical techniques, a drug-rich transient metastable phase was frequently highlighted within the supersaturation stage of the ASD dissolution. The extended drug absorption and bioavailability enhancement may be attributed to the metastability of such drug-rich phases. In this paper, we have reviewed (i) the possible theory behind the formation and stabilization of such metastable drug-rich phases, with a focus on non-classical nucleation; (ii) the additional benefits of the ASD-induced drug-rich phases for bioavailability enhancements. It is envisaged that a greater understanding of the non-classical nucleation theory and its application on the ASD design might accelerate the drug product development process in the future.

Keywords: amorphous solid dispersion; drug-rich phase; liquid-liquid phase separation; permeability enhancement; bioavailability enhancement

1. Introduction

Modern drug discovery has excelled through powerful computational chemistry and high-throughput screening technologies. However, it is widely accepted that efficacious delivery of these new chemical entities (NCEs) can be extremely difficult [1–3]. The oral absorption of a drug is a complex process that can be affected by various factors such as physicochemical factors of the drug and formulation, and physiological factors of the patients [4]. From the formulation and drug delivery perspectives, the Biopharmaceutics Classification System (BCS) is established based on two key parameters, the drug solubility in the media in relation to its maximum dose and the drug permeability through the gastrointestinal membrane. Statistically, approximately 40% of the marketed drugs and up to 90% of the new drug candidates have been revealed to be poorly soluble in aqueous media (BCS II and IV) [5–7]. More importantly, around 25% to 40% of all approved drugs are also suffering from low permeability (BCS IV) [6–8]. To combat these challenges, various formulation strategies such as amorphous solids [9,10], nanocrystals [11,12], liposomes [13,14], micro/nano-emulsifying systems [15] and co-crystals [16] have been developed.

Amorphous solids and amorphous solid dispersions (ASDs) have drawn increased attention due to their continuous commercial successes in the past decade [17]. The high free energy state and the disordered structure of the amorphous solids can lead to a remarkable enhancement of the drug solubility as well as the dissolution rate. For example, amorphous glibenclamide displayed 14 times higher solubility than its crystal form in an aqueous buffer [18]. The apparent solubility of amorphous indomethacin is higher than the γ-crystal API over the temperature range from 5 to 45 °C [18]. Amorphous pranlukast enhanced apparent solubility approximately 5.8 times in water and 19.4 times in phosphate buffered saline, compared with the crystal substance [19]. However, the inherent instability of amorphous solids remains the major concern in terms of its wider adoption in modern medicine.

Amorphous solid dispersions (ASDs) are, in general, homogeneous dispersions of amorphous drug molecules within a solid excipient [20]. To overcome the instability of amorphous drugs caused by the high free energy, certain polymers are utilized as the anti-plasticizer and/or the stabilizer to maintain the amorphous structure of the drug molecules during storage [21–23]. Polymers can increase formulation stability through various mechanisms such as the physical barrier, configurational entropy, molecular mobility, chemical potential, glass transition temperature and drug-polymer interaction [24]. Nontoxic polymers including polyvinylpyrrolidone (PVP), hypromellose (HPMC), polyvinylpyrrolidone/vinyl acetate (PVP/VA), hypromellose acetate succinate (HPMCAS) have been approved to use as excipients by the United States Food and Drug Administration (FDA) for oral dosage forms [20,24]. The ASD products approved by FDA in the last 5 years are briefly summarized in Table 1, indicating an ascending phase of this technology for wide adoptions in the pharmaceutical industry [20,24,25].

Table 1. Summary of ASD products granted FDA approval from 2015 to 2020.

Brand Name	Generic Name	Company	Manufacture Technique	FDA Approval	Dosage Form
Kalydeco®	ivacaftor	Vertex	SD	2015	granule
Orkambi®	lumacaftor; ivacaftor	Vertex	SD	2015	tablet
Epclusa®	sofosbuvir; velpatasvir	Gilead Sciences	SD	2016	tablet
Venclexta®	venetoclax	AbbVie	HME	2016	tablet
Viekira XR™	dasabuvir sodium; ombitasvir; paritaprevir; ritonavir	AbbVie	HME	2016	tablet
Zepatier®	elbasvir; grazoprevir	Merck	SD	2016	tablet
Lynparza®	olaparib	Astrazeneza	HME	2017	tablet
Norvir®	ritonavir	AbbVie	HME	2017	powder
Mavyret™	glecaprevir; pibrentasvir	AbbVie	HME	2017	tablet
Prograf®	tacrolimus	AbbVie	SD	2018	granule; capsule
Tibsovo®	ivosedinib	AbbVie	/	2018	tablet
symdeko®	tezacaftor; ivacaftor	Vertex	/	2018	tablet
Trikafta®	elexacaftor; tezacaftor; ivacaftor	Vertex	/	2019	tablet
Harvoni®	ledipasvir; sofosbuvir	Gilead Sciences	SD	2019	pellet

This summary is adapted from several references [26–29]. Details of the approval year, active ingredients, companies and dosage forms are derived from the FDA drug database and annual approval reports [17]. SD represents spray drying, HME represents hot melt extrusion.

It should also be highlighted here that the choices of excipients for the ASD system are not narrowly limited to the polymers. Indeed, certain small molecular additives, drugs [30–33] or ionic liquids (ILs) [34,35] have also been explored as the excipients in ASD design [36]. For example, the co-amorphous drug system of atorvastatin calcium-carvedilol and atorvastatin calcium-glibenclamide exhibited greater solubility than that of single amorphous component [33]. Amino acids, such as arginine and phenylalanine, were also implemented to stabilize indomethacin and enhance its solubility, reaching a level approximately 200 times greater than its pure amorphous counterpart [37]. Ionic liquid, defined as salts with melting/glass transition temperatures below 100 °C, can also be used to solubilize the drug [38,39]. The targeted biological properties of the drug-IL system have been categorized as the third evolution of ILs in the history of their development [40]. In the context of ILs' pharmaceutical application, the ideal drug-ILs are expected to be liquid at the body temperature for improved dissolution properties [41,42]. Solubilized drug-ILs were revealed to increase the drug's apparent solubility, dissolution rate and membrane transport properties for oral administration [34,35,43–45]. Several excellent reviews have already been published for the solubilized drug-IL amorphous systems, demonstrating the promising future of this molecular complex approach in the pharmaceutical field [41,42,46].

Size reduction is another important technique to increase the dissolution and oral absorption of poorly water-soluble drugs. The synergetic effects of size reduction and amorphous structure are expected to further improve the performance of the formulation. A range of existing scalable manufacturing methods can be used to produce amorphous nano-sized drugs, such as microfluidics, ultrasonication, antisolvent precipitation, electrospraying, and the supercritical fluid process [47–51]. Apart from amorphous drug nanoparticles prepared during formulation processes, various drug-rich amorphous nanoparticles or nanodrops were also reported during the dissolution of the ASDs. High apparent drug solubility and bioavailability enhancement were commonly associated with the presence of such drug-rich phases [52–57]. However, the formation of these phases and their corresponding stability in the solution are still poorly understood. Hence, we suggest that a better understanding of such drug-rich phases will be highly beneficial to predict the performance of different formulation designs and thereby rationalize them for the final implementation of ASD formulation strategy in industry.

In this review, we highlight: (i) the thermodynamics and kinetics associated with the formation of drug-rich phases from the dissolution (Section 2); (ii) in vitro permeability enhancement and in vivo bioavailability enhancement in the context of ASD formulation (Section 3). Other aspects associated with the drug-rich phases, such as the dissolution model [58], the liquid-liquid phase separation (LLPS) identification and screening technique [59], and the impact of surfactants and polymers in the nanodroplet uptake [60,61], can be found in other excellent studies in the literature.

2. Dissolution Pathways of ASD in Water

ASD has been widely revealed as a promising strategy in improving drug bioavailability and therapeutic windows for poorly water-soluble drugs. However, a mechanistic understanding of its phase separation process during dissolution and storage is still difficult to obtain. In this section, we elucidate the solubility, dissolution and several possible phase separation pathways of pure drug and drug-excipient ASD in water from a simple thermodynamic viewpoint. We would like to imply that the non-classical nucleation pathways discovered with recent advances in biomineralization and protein precipitation, including the formation of different metastable transient phases and their subsequent transformation into crystals, might be a suitable model to describe the phase separation process during excipient-assisted amorphous drug dissolution [62–64].

2.1. Thermodynamics of the Pure Drug in the Dissolution

A typical drug-water binary phase diagram is depicted in Figure 1A. The solubility of a crystalline drug in an aqueous solution is determined by the thermodynamic equilibrium between the dissolved free drug molecules and the non-dissolved crystals in the solution. The equilibrium boundary can be represented by the crystalline solubility line in the phase diagram [65]. The drug crystalline solubility ($C_{crystalline}$) at 300 K could be obtained using the intersection of the solubility line at this temperature (Figure 1, point a). If the number of free drug molecules dissolved in water is below the corresponding value of $C_{crystalline}$ (Figure 1A, region 1), the solution is homogeneous, and no drug molecules will precipitate out from the solution. By contrast, if the amount of dissolved drug is above $C_{crystalline}$ (Figure 1A, region 2 and region 3)—specifically, if it is supersaturated with respect to the solubility of the crystalline drug—the solution tends to phase separate due to its thermodynamic instability and, therefore, the drug precipitates out. When the supersaturation is moderate (region 2), the system undertakes a classical nucleation pathway to reduce its free energy, where solid crystalline nuclei exceeding the critical size form through the thermal fluctuation of the solution and subsequently grow via the molecule-by-molecule attachment process (Pathway (i) in Figure 1C). As a result, the free drug concentration in the solution will eventually decrease to the value of $C_{crystalline}$.

Remarkably, if the supersaturation of drug solution is very high, drug-rich transient liquid or amorphous solid phases are often observed to form in the solution prior to the formation of crystalline drugs [66]. These drug-rich phases were suggested to form through the liquid-liquid phase separation (LLPS) of the drug-water binary system only if the concentration of free drug molecules in solution exceeds the critical value at the binodal line. Indeed, similar solute-rich liquid phases generated by the LLPS process have been extensively reported in the protein, inorganic ion and organic molecule solutions [67]. As the free energy barrier to form metastable liquid phases is smaller than that to form crystalline solid nuclei, a two-step nucleation pathway, via the formation of metastable drug-rich liquid phases (pathway (ii) in Figure 1C), is thermodynamically favored over the classical one. More recently, a wide variety of metastable transient precursors including complexes, prenucleation clusters, liquid phases, amorphous solid particles and nanocrystals have been demonstrated in the inorganic system [64,68]. The presence of one or multiple metastable precursors in the crystallization process is expected to further alter its free energy landscape and, therefore, the crystallization pathways. For example, when the supersaturation of solution is sufficiently high, the formation of metastable solute clusters or complexes and their subsequent aggregation could become thermodynamically favored, thereby rendering the free energy landscape with multiple local minima (pathways (iii) in Figure 1C) [64]. Interestingly, a recent study demonstrates that the liquid phases are likely the dynamic aggregates of clusters in the $CaCO_3$ system, suggesting a possible correlation between different non-classical nucleation pathways [69]. However, more investigations must be conducted to identify the possible transient precursor species generated during the dissolution process of ASD and their corresponding nucleation pathways.

Using the above phase diagram, the general behaviors of the dissolution of the drug in water as well as the subsequent phase separation of the drug-water system can be qualitatively estimated. For example, if crystalline drug particles with an amount above the value of corresponding $C_{crystalline}$ are suspended into the water, the concentration of free drug in aqueous solution gradually increases and eventually reaches the crystalline equilibrium solubility value (Figure 1, point a) with an extended time, as illustrated in a black solid curve in Figure 1B. If amorphous drug particles are suspended in solution, the drug concentration can temporarily reach beyond the value of $C_{crystalline}$ because of their higher Gibbs free energy and, therefore, can achieve higher solubility relative to their crystalline counterparts [70]. The resulting high supersaturation of drug solution will lead to phase separation of the system. This phase separation process is typically expected to undergo via a classical nucleation and growth process (Figure 1, region 2), in which the drug concentration eventually approaches to the value of $C_{crystalline}$. Hence, the pure amorphous drug is often revealed to display a "spring" dissolution profile (Figure 1B, brown curve). Interestingly, some recent studies suggested that the critical concentration of drug at the binodal line (Figure 1A, point b) is similar to the intrinsic solubility of amorphous drugs [71]. In line with this suggestion, if the dissolution of pure amorphous drug particles reaches the thermodynamic equilibrium state in the solution, we should expect LLPS to occur in the system. However, as the supersaturated drug solution readily forms crystalline nuclei due to its metastability and the dissolution rate of amorphous drug particles is not sufficiently high, this critical concentration might be difficult to achieve in a pure amorphous drug-water system. Indeed, the experimental observations of LLPS are typically reported in the systems where they rapidly enter the binodal regime with minimum interference of the nucleation-growth pathway using alternative strategies. For example, $C_{crystalline}$ in the solution is rapidly decreased by switching the pH [72], temperature [73] and solvent [66] of a solution, resulting in an abrupt elevation of supersaturation extent of the system. Alternatively, certain polymeric or small molecular additives can be added into the system to delay the nucleation-growth pathway during the dissolution of the amorphous drug. In addition, as demonstrated in the biomineralization field, the amorphous phases of even the same mineral could possibly possess a broad range of local structures and water contents, which in return reflects on their solubility, dissolution and kinetic stability [68,74]. By analogy, the presence of possible structural or compositional variations in the drug-rich metastable phases might also influence the dissolution, solubility and therefore the phase separation phenomenon of the amorphous drug in water, yet the experimental evidence is still missing.

2.2. Thermodynamics of ASD in Dissolution

In the ASD system, different amorphous polymeric or small molecular excipients are utilized to stabilize the amorphous drug during the storage as well as to enhance the solubility and dissolution rate of the drug in solution [21–24]. While amorphous drug molecules are considered to be homogeneously dispersed within the excipients in an ideal ASD system, an amorphous-amorphous phase separation (AAPS) phenomenon is often observed during the manufacturing or storage due to the imperfect miscibility of drug and excipients [29,75–77]. The AAPS phenomenon can be normally interpreted by the schematic phase diagram of ASD illustrated in Figure 2A, where the solid blue line indicates the binodal line. At this binodal line, the single homogeneous ASD phase splits into drug-rich and drug-lean amorphous phases. Theoretically, the drug volume fractions of these two phases are expected to be Φ1 and Φ2 at the specific temperature T1 under the equilibrium state (Figure 2A). However, in most ASD systems, particularly those with polymers as the excipients, the restricted mobility of polymer chains hinders the system to reach the equilibrium state. Indeed, the AAPS phenomenon is more pronounced at lower amounts of excipient or at a temperature range above the glass transition temperature of the system, where the mobility of drug and excipient molecules is higher [75].

Recently, more investigations have been carried out to correlate AAPS to the stability of ASD systems during their preparation and storage. In particular, the impacts of water/solvent on the AAPS process have been extensively studied in cases in which ASD is stored in a humid condition [76,78–83]. The Gibbs free energy surface of the drug-excipient-water ternary system at a constant temperature and the corresponding binodal lines at various temperatures in the phase diagram may be schematically illustrated in Figure 2B,C. The composition of ASD, temperature and water amount are expected to affect the Gibbs free energy landscape of the ternary system, thereby influencing the AAPS process and the stability of ASD. More recently, a more quantitative thermodynamic model was reported for the drug-polymer-water ternary system based on the Flory-Huggins (F-H) theory [65]. This ternary phase model described the drug-polymer-water interaction and the phase separation situation at various temperatures and composition concentrations. Similarly, important tool such as Perturbed-Chain Statistical Associating Fluid Theory (PC-SAFT) modelling has also been used for the quantitative analysis of these systems [84]. Other hybrid models for drug-polymer binary systems and the subsequent extension of ternary or quaternary systems have been reported for the design and understanding of ASD formulations [85–88]. The combination of these insightful thermodynamic and kinetic models will certainly provide a more informative guide for the AAPS process in presence of moisture/solvent and therefore a better understanding of the ASD stability during its preparation and storage.

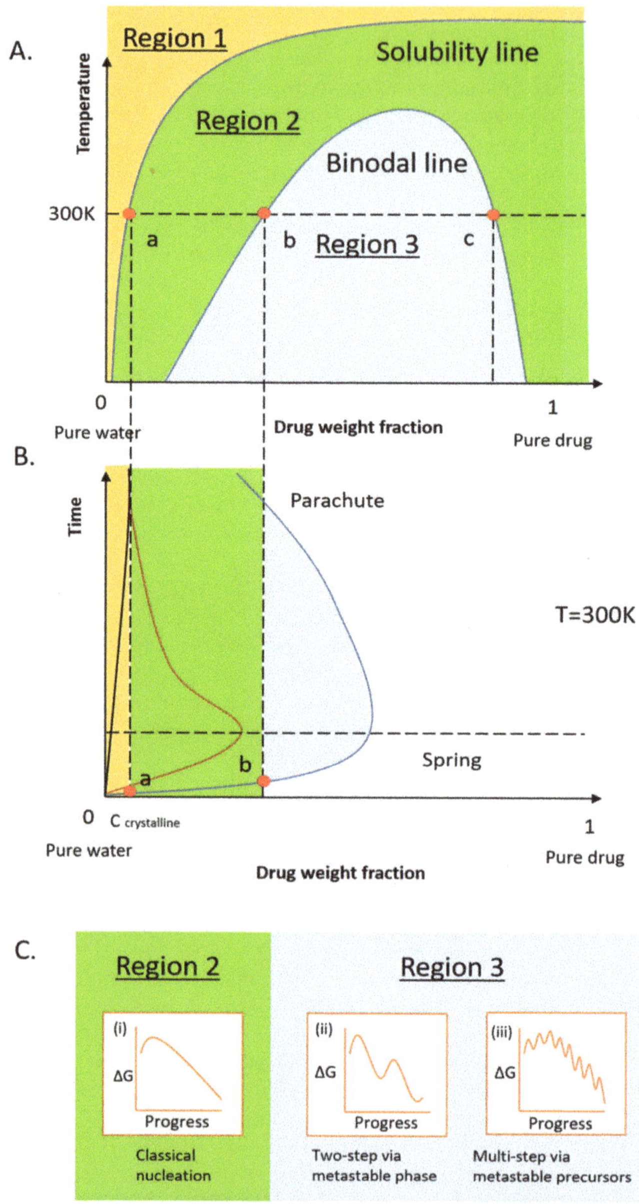

Figure 1. (**A**) Schematic temperature-composition phase diagram of the drug-water binary system, (**B**) the schematic dissolution diagram of a crystalline drug (the black solid curve), pure amorphous drug "spring" (the brown solid curve) and ASD formulation "spring and parachute" (the blue solid curve). (**C**) Gibbs free energy landscapes of dissolved drug molecules forming a stable bulk crystal through (i) classical nucleation and growth pathway, (ii) the two-step pathway via the metastable drug-rich liquid phase, and (iii) the aggregation of thermodynamically metastable particles and possible pathways [64]. Point a is the intersection of the solubility line and the horizontal temperature line, which reflects drug crystalline solubility at this temperature. Points b and c are intersections of the binodal line, indicating the drug and water weight fraction in drug-rich phases and drug-lean phases.

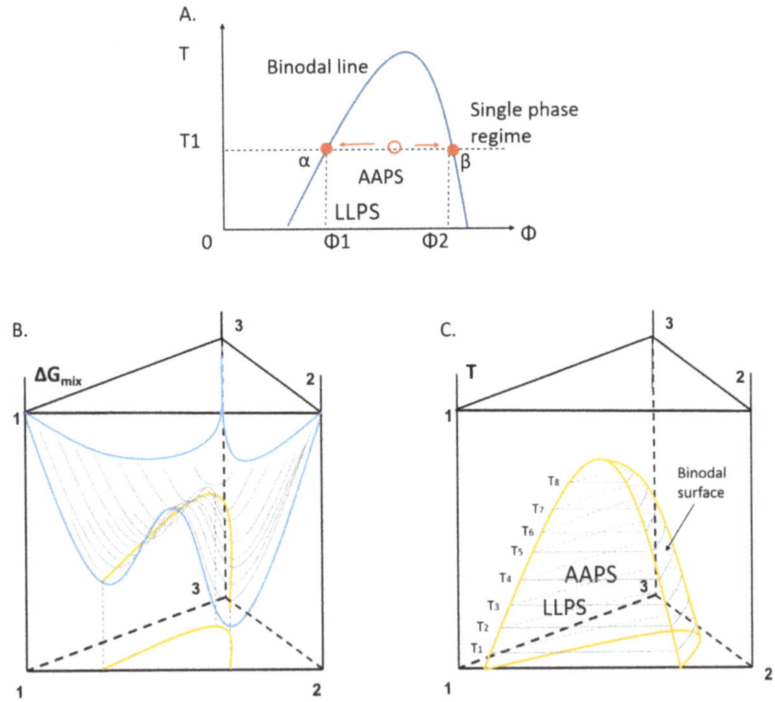

Figure 2. (**A**) Schematic diagram of the binodal line reflecting the boundary of a single homogeneous phase and AAPS/LLPS for an amorphous-amorphous binary system consisting of an upper critical solution temperature (UCST). (**B**) Gibbs free energy landscape of drug-polymer-water ternary system at a constant temperature, where the solid blue line describes the free energy surface, and the solid yellow line represents the binodal line at this temperature. (**C**) Composition-temperature phase diagram of the ternary system. The binodal surfaces are labeled with the solid yellow lines (components 1, 2 and 3 represent drug, polymer, and water, respectively).

With the same concept, a similar drug-excipient-water ternary system can be obtained if ASD is dissolved into water, where the amount of water in this system is nevertheless significantly higher than that absorbed from the moisture. Interestingly, the formation of metastable drug-rich amorphous phases, in the form of either nanoparticles or nano-sized liquid phases, are more frequently observed in this system compared to the pure drug-water binary system, as summarized in Table 2. These results indicate that the presence of excipients can play an important role in the phase separation of the drug in solution. For example, one can postulate that the introduction of the AAPS process of drug-excipient system to the LLPS process of drug-water system would result in a non-inferior [89] or lower supersaturation extent [88]. However, many hydrophilic polymer excipients can significantly enhance the rates of hydration and intrinsic dissolution performances of the drug from ASD [90]. It is evidenced that these approaches can facilitate the system to cross the binodal boundary and form different metastable drug-rich amorphous precursors following the non-classical nucleation pathways. Indeed, it remains unknown how the polymeric excipients assist the LLPS of the drug after ASD is exposed to an excess amount of water. Additionally, certain excipients can effectively prevent the aggregation of these metastable phases or reduce the nucleation and growth rate of drug crystals within the drug-rich metastable phases [23,91]. As a result, the kinetic stability (parachute stage), namely the extended time associated with the high apparent drug concentration in the solution, of the drug-rich metastable phase could be remarkably increased. This kinetic stabilization effect enables researchers to access the metastable drug-rich phases using the current characterization tools and therefore more frequently observe their presence. Moreover, the thermodynamically favored formation of drug-rich metastable phases and the kinetic stabilization of these phases during the dissolution process of ASD can eventually lead to a "spring and parachute" concentration profile in dissolution assays (Figure 1B, the blue solid curve). In this case, the drug remains at a much higher concentration for a more extended time in solution, as compared to that observed in the dissolution of pure crystalline or amorphous drug. Thus, such forced steps have been frequently used to screen the potential pharmaceutical excipients for such purposes.

2.3. Kinetic Stability of Drug-Rich Phases

If the drug-rich metastable phases are kinetically stabilized by certain excipients for a considerable amount of time, the high concentration of a drug in a solution can significantly promote the drug's oral absorption. Consequently, the "reservoir effect" of drug-rich phases followed by the drug replenishing may release drug molecules to the medium without the precipitation of crystals. These application potentials of drug-rich phases have driven extensive research on the kinetics of their transformation and the corresponding kinetic stability against the transformation. Different characterization tools, such as ultraviolet-visible (UV) spectroscopy [92,93], fluorescence spectroscopy [71,94], dynamic light scattering (DLS) [72,95], atomic force microscopy (AFM) [96], scanning electron microscope (SEM) [95], transmission electron microscope (TEM) [96,97], nuclear magnetic resonance (NMR) [98], tunable resistive pulse sensing (TRPS), analytical ultracentrifugation (AUC) and liquid cell TEM have been utilized to access the information at different stages of drug-rich phases [99–103].

As the transient drug-rich phases are thermodynamically metastable, they would eventually transform into thermodynamically stable crystals. Several possible transformation pathways that have been revealed are schematically summarized in Figure 3A, even though a full picture of drug-rich phase transformation is still missing. The drug-rich phases might continue to grow, coalesce, or aggregate to further reduce the free energy of the system [66,104]. For example, the diameter of drug-rich phases in the supersaturated danazol aqueous solution was revealed to increase over time when monitored using dynamic light scattering (DLS) techniques (Figure 3B) [23]. Similarly, Ralm et al. showed that the metastable amorphous particles formed in the supersaturated phenytoin-HPMCAS aqueous solution aggregated into irregularly shaped nanoparticles prior to the formation of crystals, as measured with Cryo-TEM and SAXS [105]. Although the growth, coalescence or aggregation of metastable phases is favored over the formation of crystalline nuclei due to the smaller free energy barrier, the nucleation events will still occur, most likely within the drug-rich phases where a higher supersaturation with respect to the crystalline solubility is present. These nucleation events will trigger the transformation of drug-rich metastable phases into thermodynamically stable crystalline phases. For example, the transformation of metastable amorphous probucol nanoparticles into crystalline ones was indicated by the gradual increase in the particle stiffness measured by atomic force microscopy (AFM) (Figure 3C) [96,106]. Similarly, the crystallization within transient drug-rich nanodroplets was revealed in the nifedipine-HPMCAS supersaturated solution using NMR (Figure 3D) [107]. Upon the onset of crystallization, the concentration of the drug decreased in the solution, whereas the concentration of polymer increased. Interestingly, these results indicate the HPMCAS polymer was distributed in the drug-rich phase, which might assist the formation of such phase as well as inhibit the crystallization within it.

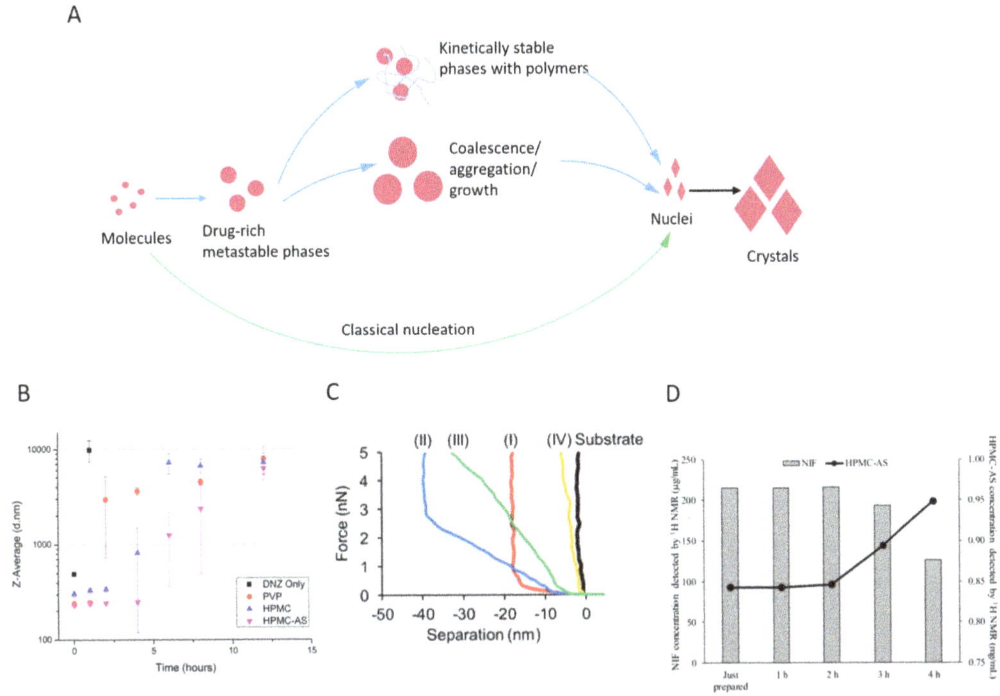

Figure 3. (**A**) Schematic illustration of the possible nucleation pathways (classical or non-classical) of drug in a solution. (**B**) Danazol Z—average diameters over time for systems with or without polymers. Reproduced with permission from [23], American Chemical Society, 2014. (**C**) Four schematic AFM force-distance curves for probucol (PBC)-HPMC-dodecyl sulfate systems over the storage time, (I) 1 d, (II) 2 d, (III) 4d, and (IV) 7d. Reproduced with permission from [106], American Chemical Society, 2015. (**D**) Solution 1H NMR detectable nifedipine and HPMCAS concentration over time. Reproduced with permission from [107], American Chemical Society, 2017.

The transformation of metastable drug-rich phases into crystals is a thermodynamically inevitable process, but the kinetics of this process can be significantly altered by the presence of certain excipients, such that the precipitation of crystalline drugs in ASD solution may not be observed within a prolonged experimental timescale. For example, Keisuke and Lynne suggested that the coalescence of the drug-rich nanodrops could be suppressed by the steric repulsion and electrostatic repulsion after certain polymers are adsorbed on the surfaces [108]. The suppressing effects can be varied by the choice of excipient materials, as demonstrated with the PVP, HPMC and its derivative HPMCAS in the danazol-rich nanodrop system (Figure 3B) [23]. Moreover, many polymers were observed to inhibit the formation of crystalline nuclei or their subsequent growth within the drug-rich phases. Interestingly, Prateek and Ronald investigated the drug and excipient performance in aqueous solutions using all-atom molecular dynamics simulations, where the effects of excipient on the mobility of drugs can be screened [109]. They found that the aggregation and diffusivity of the phenytoin drug can be reduced by as much as five orders of magnitude by the presence of HPMC and HPMCAS, indicating a significant role of excipient in reducing the transformation kinetics of metastable drug-rich phases during the dissolution of ASD.

Table 2. Reported drug-rich phases during ASD dissolution over the last decade.

Drugs	Drug Weight Fraction (%)	Excipients	Drug-rich Phase Formation Concentration (μg/mL)	Size (nm)	Metastable Phase [a]	Apparent Solubility/Binodal Point Concentration	References
Ritonavir	10	PVP	27–28.5	/	1	0.69	[93,95,110]
		PVPVA	26.8–27.5	/	1	2.96	
		HPMCAS	27.5	/	1	1.27	
	50	PVP	27–28.5	/	0	0.69	
		PVPVA	26.8–27.5	/	0	0.37	
		HPMCAS	27.5	/	1	0.76	
	10	pure drug	18.2	188–830	/	~1	
		PVP	18.3	188–631	1	~1	
		PAA [b]	18.3	213–922	1	~1	
		HPMC	18.7	203–714	1	~1	
		HPMCAS	18.2	235–352	1	~1	
		CAAdP 0.85 [c]	21.3	203–248	1	~1	
	10–30	PVPVA	~30	218	1	8.67–3.33	
	35–50			/	0	0.3–0.5	
clotrimazole	/	pure drug	490.2 (pH 4)	100–400	1	0.98	[72]
			7.7 (pH 8)		1	0.87	
nicardipine	/	pure drug	105.9 (pH 6)	100–400	1	1.1	[72]
			5.5 (pH 9)		1	0.69	
atazanavir	/	pure drug	668.4 (pH 3.5)	100–400	1	0.95	[72,111]
			65.8 (pH 9)		1	0.91	
	/	pure drug	/	204–226	/	/	
	10	HPMCAS	78	294	1	3.21	
	30			/	0	1.28	
	50			/	0	1	
	10	HPMCS	78	326	1	2.95	
	30			/	0	1	
	50			/	0	1	
	10	PVPVA	94	/	0	1.17	
	30			/	0	0.96	
	50			/	0	0.74	

Table 2. *Cont.*

Drugs	Drug Weight Fraction (%)	Excipients	Drug-rich Phase Formation Concentration (μg/mL)	Size (nm)	Metastable Phase [a]	Apparent Solubility/Binodal Point Concentration	References
danazol	/	pure drug	8.0 (estimated)	267	/	/	[112]
	10	PVP	mean 6.0	256	1	1.58	
	50		mean 8.5	/	1	1.12	
	10	HPMC	mean 8.5	284	1	1.18	
	50		mean 8.5	/	1	1.18	
	10	HPMCAS	mean 6.5	246	1	2.54	
	50		8.0 (estimated)	/	0	0.75	
nilvadipine	5–10	PVPVA	30–31.9	237–246	1	~3	[113]
	15–20			/	0	~0.5	
cilnidipine	5–15	PVPVA	0.5–0.6	255–366	1	~158	[113]
	20–25			/	0	~1–2	
glibenclamide	33.3	HPMC	~150	/	0	1	[114]
		HPMCAS-LF	~150	/	1	2.4	
		HPMCAS-HF	~150	/	0	1	
enzalutamide	/	pure drug	42–43	/	/	/	[53]
	10	PVPVA	42	/	1	1	
	50			/	0	0.95	
	10	HPMCAS	43	/	/	0.24	
	50			/	/	1	
Lopinavir	50	HPMC	17.4	/	/	1	[70]
itraconazole	25	HPMCAS-HF	~0.1	/	1	1600	[54,115]
		HPMCAS-LF	~0.1	/	1	4500	
		HPMCAS 716HP	0.1 (0% SIF[l])	170	1	4530	
			6 (0.5% SIF[l])	150	1	81.16	
			20 (2% SIF[l])	200	1	31.1	
		HPMCAS 126HP	0.1 (0% SIF[l])	140	1	1540	
			6 (0.5% SIF[l])	160	1	31	
			20 (2% SIF[l])	170	1	8.5	
		HPMCAS 716HP; HPMCAS HF	0.1 (0% SIF[l])	210	1	4030	
			6 (0.5% SIF[l])	200	1	93.33	
			20 (2% SIF[l])	190	1	29.05	

Table 2. *Cont.*

Drugs	Drug Weight Fraction (%)	Excipients	Drug-rich Phase Formation Concentration (μg/mL)	Size (nm)	Metastable Phase [a]	Apparent Solubility/Binodal Point Concentration	References
Telaprevir	10	PVPVA	100	156	1	1.22	[116]
	30				0	0.8	
	50				0	0.75	
	10	HPMC	96	147	1	1.88	
	30				1	1.1	
	10	HPMCAS	102	99	1	1.57	
	30				1	1.17	
	50				0	0.88	
	50	HPMCAS + 5% SDS [d]			1	1.08	
	10	CA Sub [e]	111	76	1	1.44	
phenytoin	10	HPMCAS	/	15	1	/	[105,117]
	25			15	1	/	
	50			/	0	/	
	10	C2-PNIPA-m-7 [f]	/	1.8–2.0 (pure polymer)	1	<1	
		C12-PNIPA-m-7 [f]	/	7.6–7.9(pure polymer)	1	<1	
		C12-PNIPA-m-30 [f]	/	12.0–12.7 (pure polymer)	1	<1	
		C12-PNIPA-m-98 [f]	/	24.5–32.1;7.9–8.6 (pure polymer)	1	<1	
probucol	10	HPMCAS	<1	16–20	1	/	[96,105,106,118–120]
	25			70	1	/	
	50			180	1	/	
	25	HPMC; SDS (weight ratio of 1.75:1.25)	<1	25–93.9 (0–7days)	1	>500	
	14.3	HPMC; SDS (weight ratio of 2:1)	<1	25.3–138.3 (0–12days)	1	>500	
	0	PDMA	/	14.5	1	<1	
	10			54	1	<1	
	25			6.3	1	<1	
	50			13.1	1	<1	

Table 2. *Cont.*

Drugs	Drug Weight Fraction (%)	Excipients	Drug-rich Phase Formation Concentration (μg/mL)	Size (nm)	Metastable Phase [a]	Apparent Solubility/Binodal Point Concentration	References
	0	P(DMA-grad-MAG)	/	35	1	<1	
	10			48	1	<1	
	25			61	1	<1	
	50			8	1	<1	
	0	PEP-PDMA	/	25.8	1	<1	
	10			25.1	1	<1	
	25			65	1	<1	
	50			77	1	<1	
	10;25	PND34-C2	~2000 (without excipient)	5.7	1	<1	
		PND34-C12		15.9	1	<1	
		PND34-b-PS2-C12		19.7	1	<1	
		PND34-b-PS9-C12		22.7	1	<1	
		PND34-b-PS14-C12		27.0	1	<1	
	10	PNIPAm; 5.0 mol % BIS [h, i]	/	42–46	1	/	
		PNIPAm; 2.5 mol % BIS [h, i]		43–44	1	/	
		PNIPAm; 0.5 mol % BIS [h, i]		41–43	1	/	
nilutamide	10;25	PND34-C2	~1400 (without excipient)	5.7	1	<1	[119]
		PND34-C12		15.9	1	<1	
		PND34-b-PS2-C12 [g]		19.7	1	<1	
		PND34-b-PS9-C12 [g]		22.7	1	<1	
		PND34-b-PS14-C12 [g]		27.0	1	<1	

Table 2. *Cont.*

Drugs	Drug Weight Fraction (%)	Excipients	Drug-rich Phase Formation Concentration (μg/mL)	Size (nm)	Metastable Phase [a]	Apparent Solubility/Binodal Point Concentration	References
nifedipine	10	HPMC-E5 LV [j]	110–156 (5–25 °C, without excipient)	~200	1	1.05	[121]
	20				1	1.15	
	10	PVPVA			1	1.1	
	20				1	1	
anacetrapib	20	copovidone; TPGS	<1	50–200 (10–2% TPGS) [k]	1	~90 /	[122,123]

[a] 1 represents the presence of the metastable phase, 0 represents the absence of the metastable phase; [b] PAA represents the poly(acrylic acid); [c] CAAdP 0.85 represents the cellulose acetate propionate 504–0.2 adipate 0.85; [d] SDS represents the Sodium dodecyl sulfate; [e] CA Sub represents the cellulose derivative cellulose acetate suberate; [f] C12-PNIPA-m represents the dodecyl (C12)-tailed poly(Nisopropylacrylamide); [g] PND-b-PS-C12 represents the poly(N-isopropylacrylamide-co-N,N-dimethyl- acrylamide)-b-polystyrene; [h] BIS represents the N,N′- methylenebis(acrylamide); [i] PNIPAm represents the poly(N-isopropylacrylamide); [j] HPMC-E5 LV represents the Hydroxypropyl methylcellulose E5 Premium LV; [k] TPGS represents the D-α-tocopheryl polyethylene glycol 1000 succinate; [l] SIF represents the bile salt.

3. In Vitro Permeability Enhancement Achieved by the ASD Solution

3.1. Drug Solubility-Permeability Interplay

In the past decades, various enabling formulations were developed to enhance the delivery of poorly water-soluble drugs; a reduction instead of an enhancement of the oral absorption was often observed for certain formulation strategies [124–127]. The lack of correlation between solubility and permeability is known as the "solubility-permeability interplay", resulting in the increased apparent solubility alone not being sufficient to predict and evaluate the drug's oral absorption [128–132]. The true driving force of mass transport through the membrane has been widely investigated as an efficient tool for the development of poorly water-soluble drugs. Importantly, the drug's thermodynamic activity was suggested to be the driving force of the increased membrane permeability [133]. The relationship between the drug influx across the membrane and its thermodynamic activity has been described by Raina et al. [133]:

$$J = \frac{dM}{dt} = \frac{DAa_{aq}}{\gamma_m h} \quad (1)$$

where J represents the drug flux across the membrane, dM/dt is the change of solute mass in unit time. The drug flux depends on the drug diffusion coefficient (D), membrane area (A), drug thermodynamic activity in the aqueous solution (a_{aq}), activity coefficient of the drug in the membrane (γ_m), and the thickness of the membrane (h). If we define:

$$B = \frac{DA}{\gamma_m h} \quad (2)$$

with the hypothesis that excipients in ASD will not alter the parameters of D, A, γ_m and h, the changing of the drug flux is proportional to the drug's thermodynamic activity:

$$J = B \cdot a_{aq} \quad (3)$$

where the B is a constant number. In the ASD solution, theoretically, the thermodynamic activity of the drug is associated with the apparent concentration and the activity coefficient of the drug in the aqueous solution. Thus, the thermodynamic activity of the drug increases proportionally to the drug concentration until the binodal point, which indicates the onset of the drug-rich phases (Equation (5)). Therefore, the drug's thermodynamic activity is obtained as:

$$\begin{cases} C < S \\ a_{aq} = \gamma_{aq} \cdot C \end{cases} \quad (4)$$

$$\begin{cases} C \geqslant S \\ a_{aq} = \gamma_{aq} \cdot S \end{cases} \quad (5)$$

where C is the drug's apparent concentration and S is the critical apparent solubility of the drug corresponding to the onset of LLPS, γ_{aq} is the activity coefficient of the drug in the aqueous donor solution. According to Equations (3)–(5), the drug influx across the membrane as a function of the drug's apparent concentration in the system may be illustrated by the blue line in Figure 4A. The flux initially increases with the increase in the drug's apparent concentration and reaches a plateau at a critical drug concentration where the drug-rich phase forms, indicating the maximum permeability in correlation to the critical apparent solubility of the drug in ASD. Based on this relation, other solubilization techniques by forming drug-containing micelles or drug-excipient complexes were introduced to further increase the critical apparent solubility of the drug in the ASD system [134]. Surprisingly, a much lower drug influx was observed, as schematically illustrated in the green line in Figure 4A. Indeed, the apparent solubility of the drug is not the only factor that determines the drug influx according to Equations (3)–(5). Other factors, such as the drug diffusion coefficient (D), activity coefficient in the membrane (γ_m) or in the donor solution (γ_{aq}), membrane area (A) or thickness (h) can also play important roles. For example, the solubilization technique of using surfactant might enhance the apparent solubility of the drug by significantly sacrificing its activity coefficient in ASD and during the membrane absorption. By contrast, the ASD formulation typically enhances the apparent solubility of the drug via the formation of the drug-rich metastable phase that does not compromise the activity coefficient of the drug. This contrasting effect is a likely explanation for the striking differences demonstrated in Figure 4A. However, more experimental evidence is still required to validate such an explanation.

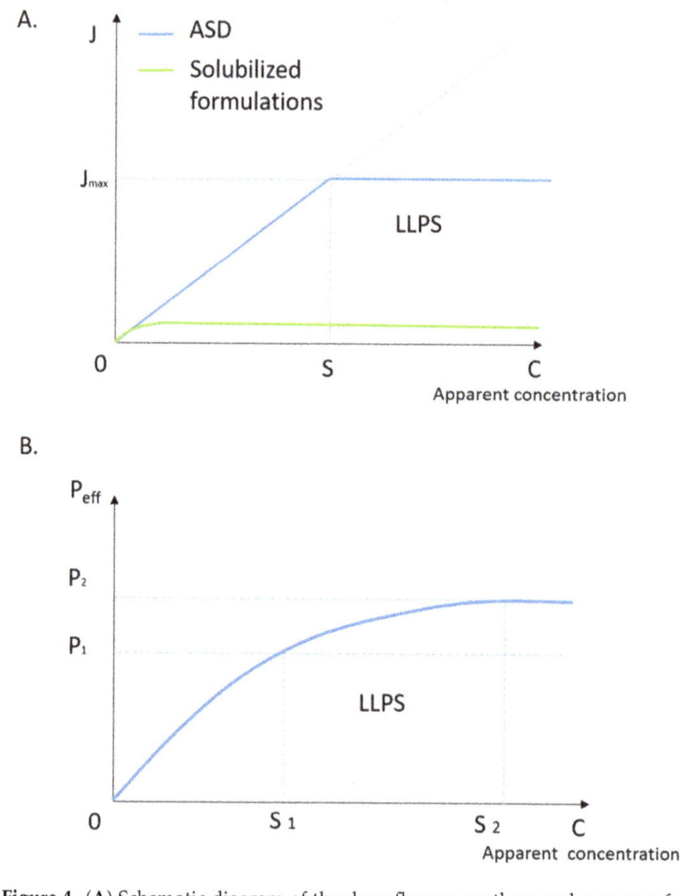

Figure 4. (**A**) Schematic diagram of the drug flux across the membrane as a function of apparent concentration. The blue line represents the theoretical drug flux across the membrane for ASD and the green line represents that of solubilizing formulations, S represents the concentration where the drug-rich phase forms. (**B**) The schematic diagram of permeability with the increasing of apparent concentration in the donor compartment with considering the effect of the drug-rich phase on the effective thickness of ABL. S1 represents the concentration at which the drug-rich phase forms, S2 represents the maximum effective concentration for the drug-rich phase in respect to the permeability enhancement.

3.2. The Roles of Drug-rich Phase in Membrane Transportation

The overall enhancement of drug permeability and bioavailability has been frequently reported with the presence of a drug-rich metastable phase during the dissolution of ASD formulations [56,123,135,136]. This metastable phase can temporarily enhance the drug's apparent solubility and, therefore, the drug influx [127,130]. Apart from the enhanced apparently solubility, drug-rich phases are suggested to further increase oral bioavailability via two other possible mechanisms. During the drug's oral absorption, with the free drug molecules diffusing across the epithelial membrane, the drug from the metastable phase can easily re-dissolve and maintain the overall drug concentration within the GI fluid. This is known as the "replenish" or "reservoir" mechanism [52,53]. Furthermore, drug nanoparticles or nanodrops could reduce the effective thickness of the aqueous boundary layer (ABL) next to the surface of the intestinal membrane and increase the drug concentration as a "shuttle" or "drifting" mechanism [54–57]. The ABL is an obstacle for permeation of the drugs, especially for the hydrophobic drugs for which the diffusivity across the ABL is lower than absorption through the membrane. The diffusivity of the ABL can be increased by enriching the drug's apparent concentration through the formation of the drug-rich phase [54,115,122,137,138]. These species can ultimately enhance the permeability (P_{ABL}) of the drug through the ABL and lead to the enhancement of the overall effective permeability (P_{eff}). The overall effective permeability (P_{eff}) is defined by both drug permeability through the membrane (P_m) and the apparent aqueous boundary layer (P_{ABL}) described as:

$$P_{eff} = \frac{1}{\frac{1}{P_m} + \frac{1}{P_{ABL}}} \quad (6)$$

Considering that the effect of the drug-rich phase enhances the drug permeability through the ABL, the schematic diagram of drug flux versus drug apparent concentration may be illustrated in Figure 4B. S_1 represents the concentration at the occurrence of the drug-rich phase and S_2 is the maximum effective concentration for the drug-rich phase in respect to the permeability enhancement. P_1 and P_2 are drug permeability at corresponding concentrations, respectively. The driving force of the increasing permeability from 0 to P_1 is achieved by the drug's apparent solubility. The slop is altered with the concentration increasing from S_1 to S_2 due to the change of permeability mechanism. The dominant factor for effective permeability (P_{eff}) enhancement is shifted from P_m to P_{ABL} as the apparent drug concentration reaches S_1 and beyond. Theoretically, with the presence of the drug-rich metastable phase and appropriate nanoparticle/nanodrop size, drug permeability will increase until the unbound amorphous drug is saturated at the surface of the membrane. The maximum permeability then reaches a new plateau at a high apparent drug concentration. Experimentally, Siddhi et al. demonstrated the dissolution performance and membrane mass transportation for atazanavir (ATZ) ASD using a high surface area apparatus [111]. Figure 5A illustrated the maximum drug concentration (C_{max}) and the area under the curve (AUC) values in the acceptor compartment in relation to the drug's apparent concentration in the donor compartment. A clear step-change on the drug membrane influx can be observed at the concentration value of the ATZ-rich metastable phase, indicating the potential benefits of such metastable phases during the dissolution of ASD. Aaron and Michael further evaluated this drug-rich phase-induced transport enhancement mechanism with a modified permeability model [139]. Accounting for the effects of the drug-rich phase on diffusivity across the ABL, the experimental behaviors of itraconazole ASD absorption in vivo was successfully described, as illustrated in Figure 5B. Similarly, a novel approach was developed by Freddy et al. for accurately measuring the thermodynamic activity of drugs and predicting their true flux with the presence of drug-rich nanodroplets [140].

3.3. The Importance of Polymeric Excipients for Drug Permeability Enhancement

It is worth highlighting that the drug permeability enhancement for ASD formulation heavily relies upon the properties of polymeric excipients. Through rational selection of polymers, additional benefits, in terms of the drug's oral bioavailability, could be achieved through the formation of the drug-rich metastable phase; however, a plateau may be reached when an excess amount was used. For example, in relation to the Eudragit® E, the equilibrium solubilities of bezfibrate, indomethacin, piroxicam, and warfarin were increased proportionally with the increases in the polymer concentration up to 2% (w/w) until a plateau is reached [141,142]. Besides, when a strong interaction is formed between drug and polymeric excipients, such as ionic interaction, the solubility-permeability trade-off occurs [143,144]. Similarly, the drug-polymer molecular complex can negatively impact the drug's membrane permeability. In the phenytoin-PVP system at polymer concentrations above 0.5 mg/mL, a strong interaction was observed, presenting a downfield shift of the phenytoin proton peak (NMR) [145]. Such interaction between drug and polymer can lead to the electron-withdrawing effect, resulting in a decrease in apparent permeability coefficients for corresponding drugs. In contrast, polymers stabilizing the supersaturated drug solution without significantly affecting the thermodynamic activity may result in a better permeability enhancement.

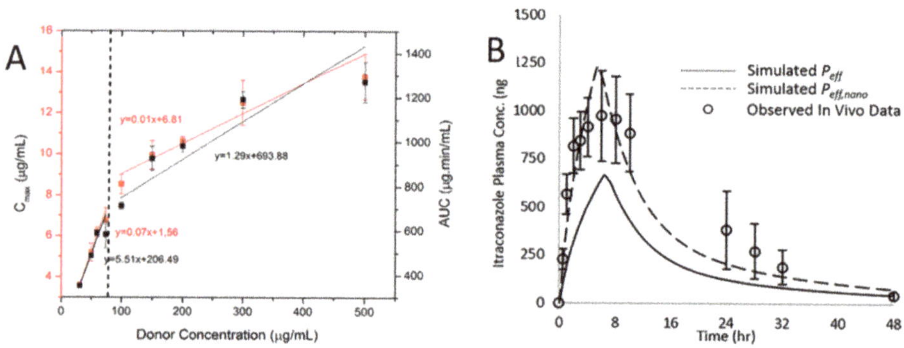

Figure 5. (**A**) The maximum Atazanavir (ATZ) concentration (Cmax) and the area under the curve (AUC) values plotted with respect to the initial donor concentration. The vertical line represents the drug's amorphous solubility. Reproduced with permission from [111], American Chemical Society, 2018. (**B**) A comparison of the itraconazole in vivo data and permeability models with or without modified with the effect of nanoparticles in aqueous solution "drifting" into the ABL. Reproduced with permission from [139], American Chemical Society, 2019.

Processing ASD into nanoparticles as means of formulation has also been frequently considered as a desirable strategy for increasing the therapeutical windows of highly potent compounds [61,146–150]. Examples of amorphous drug-polymer nanoparticle formulations exhibiting enhanced in vitro and in vivo performances are summarized in Table 3. Formulations were prepared using techniques such as antisolvent, ionic cross-linking, solvent displacement, spray drying, freeze-drying, and twin-screw extrusion, etc., with mean sizes ranging from 50 to 500 nm. Small fractions of surfactants were additionally added to further reduce the particle sizes and improve stability. Such a formulation strategy may be considered for similar species in the drug-rich phase generated by ASD, with added benefits such as targeted or site-specific drug release as well as penetration of the mucus layer within the body [151]. For example, the amorphous drug-polymer system is capable of controlled release of small molecules and large biological molecules to mucosal surfaces, such as those of the lung airways, GI tract, female reproductive tract, nose, and eye; therefore, it is of widespread interest in the field of nanomedicine [146,151–153]. Muco-penetration particles are aimed to slip through the mucus with the modification of the particle surface, especially surface charge and viscosity [147]. Due to the overall negative charge of the mucus, formulations with noninteracting polymers, high-charge density net-neutral particles, charge-shifting particles are designed for better penetration of the mucus layer [148,149,154]. On the other hand, the mucoadhesive drug delivery system (MDDS) featuring a longer adhesion time and subsequent penetration through the epithelial membrane is also a desirable formulation design. Polymers such as HPMC, thiolated polymers, polysaccharides, or maleimide modified polymers were reported for the MDDS [61,150,155–157].

Table 3. Examples of amorphous drug-polymer nanoparticle formulations reported to enhance in vivo drug absorption (from 2010).

Drugs	Excipients	Particle Size	Comments	References
heparin	thiolated chitosan; HPMCP	200–500nm (1360 nm for the special drug-excipient ratio)	A significant improvement of mucoadhesion was observed in rats; a 1.86-fold improvement of permeation-enhancing effect was observed in freshly excised carp intestine.	[156]
celecoxib	PVP K30; TPGS	less than 300 nm	The amorphous nanoparticles showed 4.6- and 5.7- times greater $AUC_{0–24\,h}$ and C_{max} in plasma, respectively, compared with the unprocessed form.	[158]
celecoxib	HPMC E5 and SDS (2:1, w/w)	mean 159 nm	The maximum drug concentration and $AUC_{0–24h}$ of amorphous celecoxib nanoparticles were observed to be 3-fold and 2-fold greater than those of the celecoxib capsules.	[159]
silibinin	HPMC	132.3 nm	The dissolution rate of silibinin nanoparticles was 48.2 times and 153.8 times higher than that of free silibinin in gastric juice and intestinal juice, respectively. The C_{max} and AUC values for silibinin nanoparticles were 15.3 times and 6.48 times greater than those in the free drug.	[160]
bezafibrate	PVP K30; cremophor ELP	less than 500 nm	Smooth-surfaced amorphous particles exhibited a 5.5-fold higher oral bioavailability compared with the plain bezafibrate powder.	[161]
valsartan	HPMC; poloxamer 407	less than 400 nm	Valsartan-HPMC-surfactant nanoparticles increased the drug release speed (up to 90% within 10 min), exhibited 7.2-fold greater maximum plasma concentration, and 4.6-fold higher $AUC_{0–24h}$.	[162]
cyclosporine A	HPMCP; including HP50 and HP55	50–60 nm	Significant sustain drug release; the bioavailability of pH-sensitive drug nanoparticles calculated by the $AUC_{0–72h}$ was 119.6% for HPMC HP55.	[163]
raloxifene	PVP	180 nm	The raloxifene loaded suspension nanoparticles were found to be 3.3-fold and 2.3-fold higher, in terms of AUC and C_{max} increment, respectively, in plasma than those of the drug powder.	[164]
anacetrapib	copovidone; vitamin E TPGS	/	Particles with sizes of less than 100 nm were observed to have approximately 2-fold higher average exposure in dogs compared with larger particles.	[123]
7-ethyl10-hydroxycamptothecin (SN-38)	disodium glycyrrhizin	69.41 nm	The amorphous formulation increased the drug solubility of the crystal drug 189-fold. The nanoparticles exhibited 4-fold greater bioavailability than that of drug suspension.	[165]
tadalafil	HPMC, VA64, and PVP K30	200 nm	The amorphous particles exhibit 8.5 times faster dissolution rates in the first minute of dissolution, 22-fold greater apparent solubility at 10 min, 3.67-fold greater in oral bioavailability than unprocessed tadalafil.	[166]
megestrol acetate	HPMC; PVPK30; Ryoto sugar ester L1695	less than 500 nm	The solid dispersion nanoparticles showed 4-fold and 5.5-fold higher $AUC_{0–24h}$ and C_{max} than those of raw drug powder.	[167]

Table 3. *Cont.*

Drugs	Excipients	Particle Size	Comments	References
sitagliptin	chitosan	210–618 nm	The chitosan-loaded amorphous nanoparticles showed a sustained release and mucoadhesive properties.	[157]
itraconazole	HPMCAS 716HP; HPMCAS 126HP	140–210 nm	The drug-rich nanoparticles have been demonstrated contribution to the ABL diffusion in proportion to their diffusion coefficients and drug loadings.	[115]
sirolimus	PVP K30; surfactants	mean 250 nm	The peak concentration and $AUC_{0\text{-}12h}$ of sirolimus were increased 18.3-fold and 15.2-fold, respectively.	[168]
enzalutamide	HPMCAS; PVPVA	42–43 nm	ASDs that formed drug nanoparticles during dissolution showed higher drug concentration for rat plasma exposure than samples that only yielded supersaturated solutions.	[53]

* HPMCP represents hydroxypropyl methylcellulose phthalate; TPGS represents d-α-tocopheryl polyethylene glycol 1000 succinate; Cmax represents the maximum drug concentration; AUC 0-t represents the area under the concentration-time curve over time (t).

4. Conclusions

The supersaturated ASD solution increases the oral bioavailability of drugs through the elevation of both the drug's apparent solubility and its permeability. We suggest that such enhancement is likely to be contributed by the excipient-assisted formation and stabilization of drug-rich metastable phases following the non-classical nucleation pathway. In this case, a high apparent solubility of a drug could be kept for an extended amount of time in the solution. Therefore, suitable thermodynamic and kinetic models featuring drug-excipient-water ternary phase behaviors may be useful for designing the ASD formulation with the ability of self-generating and kinetically stabilizing drug-rich phases. Furthermore, these drug-rich amorphous phases, might significantly increase the drug's oral absorption by other mechanisms such as (i) the "reservoir" mechanism, in which the drug is replenished when the drug concentration in the GI tract or the donor compartment is lower than its solubility; (ii) the increasing of the drug concentration on the surface of the membrane by a "shuttle" effect; (iii) potentially sliding into the mucus layer that is adjacent to the intestinal membrane. Designs of ASD formulations that enable the synergetic effects of these different mechanisms to enhance the drug oral absorption will be highly beneficial for future medicine development.

Author Contributions: Conceptualization, K.Q., Y.T. and H.D.; writing—original draft preparation, K.Q.; writing—review and editing, Y.T., H.D., K.Q., L.S., G.P.A., D.S.J.; visualization, K.Q., Y.T. and H.D.; supervision, Y.T.; project administration, Y.T.; funding acquisition, Y.T. All authors have read and agreed to the published version of the manuscript.

Funding: Royal Academy of Engineering Enterprise Fellowship (EF1819\8\12). Swiss National Science Foundation Early Postdoc. Mobility fellowship (P2ELP2_191667).

Institutional Review Board Statement: Not applicable.

Informed Consent Statement: Not applicable.

Data Availability Statement: Not applicable.

Acknowledgments: Y.T. would like to acknowledge the financial support from Royal Academy of Engineering Enterprise Fellowship (EF1819\8\12). H.D. would like to acknowledge the financial support from the Swiss National Science Foundation Early Postdoc. Mobility fellowship (P2ELP2_191667).

Conflicts of Interest: The authors declare no conflict of interest.

References

1. Waring, M.J.; Arrowsmith, J.; Leach, A.R.; Leeson, P.D.; Mandrell, S.; Owen, R.M.; Pairaudeau, G.; Pennie, W.D.; Pickett, S.D.; Wang, J.; et al. An Analysis of the Attrition of Drug Candidates from Four Major Pharmaceutical Companies. *Nat. Rev. Drug Discov.* **2015**, *14*, 475–486. [CrossRef] [PubMed]
2. Ting, J.M.; Porter, W.W.; Mecca, J.M.; Bates, F.S.; Reineke, T.M. Advances in Polymer Design for Enhancing Oral Drug Solubility and Delivery. *Bioconjugate Chem.* **2018**, *29*, 939–952. [CrossRef] [PubMed]
3. Lipinski, C.A.; Lombardo, F.; Dominy, B.W.; Feeney, P.J. Experimental and Computational Approaches to Estimate Solubility and Permeability in Drug Discovery and Development Settings. *Adv. Drug Deliv. Rev.* **2012**, *23*, 3–25. [CrossRef]
4. Dahan, A.; Miller, J.M.; Amidon, G.L. Prediction of Solubility and Permeability Class Membership: Provisional BCS Classification of the World's Top Oral Drugs. *AAPS J.* **2009**, *11*, 740–746. [CrossRef]
5. Loftsson, T.; Brewster, M.E. Pharmaceutical Applications of Cyclodextrins: Basic Science and Product Development. *J. Pharm. Pharmacol.* **2010**, *62*, 1607–1621. [CrossRef]
6. Takagi, T.; Ramachandran, C.; Bermejo, M.; Yamashita, S.; Yu, L.X.; Amidon, G.L. A Provisional Biopharmaceutical Classification of the Top 200 Oral Drug Products in the United States, Great Britain, Spain, and Japan. *Mol. Pharm.* **2006**. [CrossRef] [PubMed]
7. Thayer, A.M. Finding Solutions. *Chem. Eng. News* **2010**. [CrossRef]
8. Babu, N.J.; Nangia, A. Solubility Advantage of Amorphous Drugs and Pharmaceutical Cocrystals. *Cryst. Growth Des.* **2011**, *11*, 2662–2679. [CrossRef]
9. Schittny, A.; Huwyler, J.; Puchkov, M. Mechanisms of Increased Bioavailability through Amorphous Solid Dispersions: A Review. *Drug Deliv.* **2020**, *27*, 110–127. [CrossRef]
10. Kanaujia, P.; Poovizhi, P.; Ng, W.K.; Tan, R.B.H. Amorphous Formulations for Dissolution and Bioavailability Enhancement of Poorly Soluble APIs. *Powder Technol.* **2015**, *285*, 2–15. [CrossRef]

11. Buraphacheep, V.; Morakul, B. Nanocrystals for Enhancement of Oral Bioavailability of Poorly Water-Soluble Drugs. *Asian J. Pharm. Sci.* **2015**, *10*, 13–23. [CrossRef]
12. Gigliobianco, M.R.; Casadidio, C.; Censi, R.; Di Martino, P. Nanocrystals of Poorly Soluble Drugs: Drug Bioavailability and Physicochemical Stability. *Pharmaceutics* **2018**, *10*, 134. [CrossRef] [PubMed]
13. Porter, C.J.H.; Trevaskis, N.L.; Charman, W.N. Lipids and Lipid-Based Formulations: Optimizing the Oral Delivery of Lipophilic Drugs. *Nat. Rev. Drug Discov.* **2007**, *6*, 231–248. [CrossRef] [PubMed]
14. Lee, M.K. Liposomes for Enhanced Bioavailability of Water-Insoluble Drugs: In Vivo Evidence and Recent Approaches. *Pharmaceutics* **2020**, *12*, 264. [CrossRef] [PubMed]
15. Kang, B.K.; Lee, J.S.; Chon, S.K.; Jeong, S.Y.; Yuk, S.H.; Khang, G.; Lee, H.B.; Cho, S.H. Development of Self-Microemulsifying Drug Delivery Systems (SMEDDS) for Oral Bioavailability Enhancement of Simvastatin in Beagle Dogs. *Int. J. Pharm.* **2004**, *274*, 65–73. [CrossRef]
16. Mashinchian, O.; Johari-Ahar, M.; Ghaemi, B.; Rashidi, M.; Barar, J.; Omidi, Y. Impacts of Quantum Dots in Molecular Detection and Bioimaging of Cancer. *BioImpacts* **2014**, *4*, 149–166. [CrossRef]
17. Drugs@FDA: FDA-Approved Drugs. Available online: https://www.acessdata.fda.gov/scripts/cder/daf/index.cfm (accessed on 15 February 2021).
18. Hancock, B.C.; Parks, M. What Is the True Solubility Advantage for Amorphous Pharmaceuticals? *Pharm. Res.* **2000**. [CrossRef]
19. Xiong, X.; Xu, K.; Du, Q.; Zeng, X.; Xiao, Y.; Yang, H.; Li, H. Effects of Temperature and Solvent on the Solid-State Transformations of Pranlukast During Mechanical Milling. *J. Pharm. Sci.* **2017**. [CrossRef]
20. Huang, Y.; Dai, W.-G. Fundamental Aspects of Solid Dispersion Technology for Poorly Soluble Drugs. *Acta Pharm. Sin. B* **2014**, *4*, 18–25. [CrossRef]
21. LaFountaine, J.S.; McGinity, J.W.; Williams, R.O. Challenges and Strategies in Thermal Processing of Amorphous Solid Dispersions: A Review. *AAPS PharmSciTech* **2016**. [CrossRef]
22. Ricarte, R.G.; Van Zee, N.J.; Li, Z.; Johnson, L.M.; Lodge, T.P.; Hillmyer, M.A. Recent Advances in Understanding the Micro-and Nanoscale Phenomena of Amorphous Solid Dispersions. *Mol. Pharm.* **2019**, *16*, 4089–4103. [CrossRef]
23. Jackson, M.J.; Toth, S.J.; Kestur, U.S.; Huang, J.; Qian, F.; Hussain, M.A.; Simpson, G.J.; Taylor, L.S. Impact of Polymers on the Precipitation Behavior of Highly Supersaturated Aqueous Danazol Solutions. *Mol. Pharm.* **2014**, *11*, 3027–3038. [CrossRef]
24. Baghel, S.; Cathcart, H.; O'Reilly, N.J. Polymeric Amorphous Solid Dispersions: A Review of Amorphization, Crystallization, Stabilization, Solid-State Characterization, and Aqueous Solubilization of Biopharmaceutical Classification System Class II Drugs. *J. Pharm. Sci.* **2016**, *105*, 2527–2544. [CrossRef] [PubMed]
25. Vasconcelos, T.; Marques, S.; das Neves, J.; Sarmento, B. Amorphous Solid Dispersions: Rational Selection of a Manufacturing Process. *Adv. Drug Deliv. Rev.* **2016**, *100*, 85–101. [CrossRef] [PubMed]
26. Jermain, S.V.; Brough, C.; Williams, R.O. Amorphous Solid Dispersions and Nanocrystal Technologies for Poorly Water-Soluble Drug Delivery—An Update. *Int. J. Pharm.* **2018**, *535*, 379–392. [CrossRef] [PubMed]
27. Rams-Baron, M.; Jachowicz, R.; Boldyreva, E.; Zhou, D.; Jamroz, W.; Paluch, M. *Amorphous Drugs: Benefits and Challenges*; Springer International Publishing: Cham, Switzerland, 2018; ISBN 9783319720029.
28. Uttaro, E.; Pudipeddi, M.; Schweighardt, A.; Zhao, F. To Crush or Not to Crush: A Brief Review of Novel Tablets and Capsules Prepared from Nanocrystal and Amorphous Solid Dispersion Technologies. *Am. J. Health-Syst. Pharm.* **2021**, *78*, 389–394. [CrossRef]
29. Tian, Y.; Jacobs, E.; Jones, D.S.; McCoy, C.P.; Wu, H.; Andrews, G.P. The Design and Development of High Drug Loading Amorphous Solid Dispersion for Hot-Melt Extrusion Platform. *Int. J. Pharm.* **2020**, *586*, 119545. [CrossRef] [PubMed]
30. Kissi, E.O.; Khorami, K.; Rades, T. Determination of Stable Co-Amorphous Drug-Drug Ratios from the Eutectic Behavior of Crystalline Physical Mixtures. *Pharmaceutics* **2019**, *11*, 628. [CrossRef] [PubMed]
31. Shi, Q.; Moinuddin, S.M.; Cai, T. Advances in Coamorphous Drug Delivery Systems. *Acta Pharm. Sin. B* **2019**. [CrossRef]
32. Su, M.; Xia, Y.; Shen, Y.; Heng, W.; Wei, Y.; Zhang, L.; Gao, Y.; Zhang, J.; Qian, S. A Novel Drug-Drug Coamorphous System without Molecular Interactions: Improve the Physicochemical Properties of Tadalafil and Repaglinide. *RSC Adv.* **2019**. [CrossRef]
33. Shayanfar, A.; Jouyban, A. Drug-Drug Coamorphous Systems: Characterization and Physicochemical Properties of Coamorphous Atorvastatin with Carvedilol and Glibenclamide. *J. Pharm. Innov.* **2013**. [CrossRef]
34. Balk, A.; Wiest, J.; Widmer, T.; Galli, B.; Holzgrabe, U.; Meinel, L. Transformation of Acidic Poorly Water Soluble Drugs into Ionic Liquids. *Eur. J. Pharm. Biopharm.* **2015**. [CrossRef] [PubMed]
35. Stoimenovski, J.; MacFarlane, D.R.R. Enhanced Membrane Transport of Pharmaceutically Active Protic Ionic Liquids. *Chem. Commun.* **2011**, *47*, 11429. [CrossRef]
36. Yang, D.; Liu, C.; Piao, H.; Quan, P.; Fang, L. Enhanced Drug Loading in the Drug-in-Adhesive Transdermal Patch Utilizing a Drug-Ionic Liquid Strategy: Insight into the Role of Ionic Hydrogen Bonding. *Mol. Pharm.* **2021**, *18*, 1157–1166. [CrossRef] [PubMed]
37. Löbmann, K.; Grohganz, H.; Laitinen, R.; Strachan, C.; Rades, T. Amino Acids as Co-Amorphous Stabilizers for Poorly Water Soluble Drugs—Part 1: Preparation, Stability and Dissolution Enhancement. *Eur. J. Pharm. Biopharm.* **2013**. [CrossRef]
38. Wiest, J.; Saedtler, M.; Balk, A.; Merget, B.; Widmer, T.; Bruhn, H.; Raccuglia, M.; Walid, E.; Picard, F.; Stopper, H.; et al. Mapping the Pharmaceutical Design Space by Amorphous Ionic Liquid Strategies. *J. Control. Release* **2017**. [CrossRef] [PubMed]
39. Wilkes, J.S. A Short History of Ionic Liquids—From Molten Salts to Neoteric Solvents. *Green Chem.* **2002**, *4*, 73–80. [CrossRef]

40. Hough, W.L.; Smiglak, M.; Rodríguez, H.; Swatloski, R.P.; Spear, S.K.; Daly, D.T.; Pernak, J.; Grisel, J.E.; Carliss, R.D.; Soutullo, M.D.; et al. The Third Evolution of Ionic Liquids: Active Pharmaceutical Ingredients. *New J. Chem.* **2007**. [CrossRef]
41. Egorova, K.S.; Gordeev, E.G.; Ananikov, V.P. Biological Activity of Ionic Liquids and Their Application in Pharmaceutics and Medicine. *Chem. Rev.* **2017**, *117*, 7132–7189. [CrossRef]
42. Shamshina, J.L.; Barber, P.S.; Rogers, R.D. Ionic Liquids in Drug Delivery. *Expert Opin. Drug Deliv.* **2013**, *10*, 1367–1381. [CrossRef]
43. Lovejoy, K.S.; Corley, C.A.; Cope, E.K.; Valentine, M.C.; Leid, J.G.; Purdy, G.M.; Wilkes, J.S.; Koppisch, A.T.; Del Sesto, R.E. Utilization of Metal Halide Species Ambiguity to Develop Amorphous, Stabilized Pharmaceutical Agents as Ionic Liquids. *Cryst. Growth Des.* **2012**. [CrossRef]
44. Choi, S.Y.; Rodríguez, H.; Gunaratne, H.Q.N.; Puga, A.V.; Gilpin, D.; McGrath, S.; Vyle, J.S.; Tunney, M.M.; Rogers, R.D.; McNally, T. Dual Functional Ionic Liquids as Antimicrobials and Plasticisers for Medical Grade PVCs. *RSC Adv.* **2014**. [CrossRef]
45. Araújo, J.M.M.; Florindo, C.; Pereiro, A.B.; Vieira, N.S.M.; Matias, A.A.; Duarte, C.M.M.; Rebelo, L.P.N.; Marrucho, I.M. Cholinium-Based Ionic Liquids with Pharmaceutically Active Anions. *RSC Adv.* **2014**. [CrossRef]
46. Egorova, K.S.; Ananikov, V.P. Fundamental Importance of Ionic Interactions in the Liquid Phase: A Review of Recent Studies of Ionic Liquids in Biomedical and Pharmaceutical Applications. *J. Mol. Liq.* **2018**, *272*, 271–300. [CrossRef]
47. Verreck, G.; Chun, I.; Peeters, J.; Rosenblatt, J.; Brewster, M.E. Preparation and Characterization of Nanofibers Containing Amorphous Drug Dispersions Generated by Electrostatic Spinning. *Pharm. Res.* **2003**. [CrossRef] [PubMed]
48. Young, T.J.; Mawson, S.; Johnston, K.P.; Henriksen, I.B.; Pace, G.W.; Mishra, A.K. Rapid Expansion from Supercritical to Aqueous Solution to Produce Submicron Suspensions of Water-Insoluble Drugs. *Biotechnol. Prog.* **2000**. [CrossRef] [PubMed]
49. Amstad, E.; Gopinadhan, M.; Holtze, C.; Osuji, C.O.; Brenner, M.P.; Spaepen, F.; Weitz, D.A. Production of Amorphous Nanoparticles by Supersonic Spray-Drying with a Microfluidic Nebulator. *Science* **2015**. [CrossRef]
50. Cheow, W.S.; Kiew, T.Y.; Hadinoto, K. Amorphous Nanodrugs Prepared by Complexation with Polysaccharides: Carrageenan versus Dextran Sulfate. *Carbohydr. Polym.* **2015**. [CrossRef]
51. Jog, R.; Burgess, D.J. Pharmaceutical Amorphous Nanoparticles. *J. Pharm. Sci.* **2017**, *106*, 39–65. [CrossRef]
52. Indulkar, A.S.; Gao, Y.; Raina, S.A.; Zhang, G.G.Z.; Taylor, L.S. Exploiting the Phenomenon of Liquid-Liquid Phase Separation for Enhanced and Sustained Membrane Transport of a Poorly Water-Soluble Drug. *Mol. Pharm.* **2016**, *13*, 2059–2069. [CrossRef] [PubMed]
53. Wilson, V.; Lou, X.; Osterling, D.J.; Stolarik, D.F.; Jenkins, G.; Gao, W.; Zhang, G.G.Z.; Taylor, L.S. Relationship between Amorphous Solid Dispersion In Vivo Absorption and In Vitro Dissolution: Phase Behavior during Dissolution, Speciation, and Membrane Mass Transport. *J. Control. Release* **2018**, *292*, 172–182. [CrossRef]
54. Stewart, A.M.; Grass, M.E.; Mudie, D.M.; Morgen, M.M.; Friesen, D.T.; Vodak, D.T. Development of a Biorelevant, Material-Sparing Membrane Flux Test for Rapid Screening of Bioavailability-Enhancing Drug Product Formulations. *Mol. Pharm.* **2017**, *14*, 2032–2046. [CrossRef] [PubMed]
55. Roos, C.; Dahlgren, D.; Berg, S.; Westergren, J.; Abrahamsson, B.; Tannergren, C.; Sjögren, E.; Lennernäs, H. In Vivo Mechanisms of Intestinal Drug Absorption from Aprepitant Nanoformulations. *Mol. Pharm.* **2017**, *14*, 4233–4242. [CrossRef] [PubMed]
56. Sugano, K. Possible Reduction of Effective Thickness of Intestinal Unstirred Water Layer by Particle Drifting Effect. *Int. J. Pharm.* **2010**, *387*, 103–109. [CrossRef]
57. Amidon, G.E.; Higuchi, W.I.; Ho, N.F.H. Theoretical and Experimental Studies of Transport of Micelle-solubilized Solutes. *J. Pharm. Sci.* **1982**, *71*, 77–84. [CrossRef] [PubMed]
58. Denninger, A.; Westedt, U.; Rosenberg, J.; Wagner, K.G. A Rational Design of a Biphasic Dissolution Setup—Modelling of Biorelevant Kinetics for a Ritonavir Hot-Melt Extruded Amorphous Solid Dispersion. *Pharmaceutics* **2020**, *12*, 237. [CrossRef] [PubMed]
59. Paisana, M.C.; Lino, P.R.; Nunes, P.D.; Pinto, J.F.; Henriques, J.; Paiva, A.M. Laser Diffraction as a Powerful Tool for Amorphous Solid Dispersion Screening and Dissolution Understanding. *Eur. J. Pharm. Sci.* **2021**, *163*. [CrossRef]
60. Ueda, K.; Taylor, L.S. Partitioning of Surfactant into Drug-Rich Nanodroplets and Its Impact on Drug Thermodynamic Activity and Droplet Size. *J. Control. Release* **2021**, *330*, 229–243. [CrossRef]
61. Mašková, E.; Kubová, K.; Raimi-Abraham, B.T.; Vllasaliu, D.; Vohlídalová, E.; Turánek, J.; Mašek, J. Hypromellose—A Traditional Pharmaceutical Excipient with Modern Applications in Oral and Oromucosal Drug Delivery. *J. Control. Release* **2020**, *324*, 695–727. [CrossRef]
62. Davey, R.J.; Schroeder, S.L.M.M.; Ter Horst, J.H. Nucleation of Organic Crystals—A Molecular Perspective. *Angew. Chem. Int. Ed.* **2013**, *52*, 2167–2179. [CrossRef]
63. De Yoreo, J. A Perspective on Multistep Pathways of Nucleation. *ACS Symp. Ser.* **2020**, *1358*, 1–17. [CrossRef]
64. De Yoreo, J.J.; Gilbert, P.U.P.A.; Sommerdijk, N.A.J.M.; Penn, R.L.; Whitelam, S.; Joester, D.; Zhang, H.; Rimer, J.D.; Navrotsky, A.; Banfield, J.F.; et al. Crystallization by Particle Attachment in Synthetic, Biogenic, and Geologic Environments. *Science* **2015**, *349*, aaa6760. [CrossRef] [PubMed]
65. Tian, Y.; Qian, K.; Jacobs, E.; Amstad, E.; Jones, D.S.; Stella, L.; Andrews, G.P. The Investigation of Flory-Huggins Interaction Parameters for Amorphous Solid Dispersion across the Entire Temperature and Composition Range. *Pharmaceutics* **2019**, *11*, 420. [CrossRef] [PubMed]
66. Ilevbare, G.A.; Taylor, L.S. Liquid-Liquid Phase Separation in Highly Supersaturated Aqueous Solutions of Poorly Water-Soluble Drugs: Implications for Solubility Enhancing Formulations. *Cryst. Growth Des.* **2013**, *13*, 1497–1509. [CrossRef]

67. Xu, S.; Zhang, H.; Qiao, B.; Wang, Y. Review of Liquid-Liquid Phase Separation in Crystallization: From Fundamentals to Application. *Cryst. Growth Des.* **2021**. [CrossRef]
68. Du, H.; Amstad, E. Water: How Does It Influence the $CaCO_3$ Formation? *Angewandte Chemie—International Edition* **2020**, *59*, 1798–1816. [CrossRef]
69. Xu, Y.; Tijssen, K.C.H.; Bomans, P.H.H.; Akiva, A.; Friedrich, H.; Kentgens, A.P.M.; Sommerdijk, N.A.J.M. Microscopic Structure of the Polymer-Induced Liquid Precursor for Calcium Carbonate. *Nat. Commun.* **2018**, *9*, 1–12. [CrossRef]
70. Li, N.; Taylor, L.S. Tailoring Supersaturation from Amorphous Solid Dispersions. *J. Control. Release* **2018**, *279*, 114–125. [CrossRef]
71. Raina, S.A.; Alonzo, D.E.; Zhang, G.G.Z.; Gao, Y.; Taylor, L.S. Using Environment-Sensitive Fluorescent Probes to Characterize Liquid-Liquid Phase Separation in Supersaturated Solutions of Poorly Water Soluble Compounds. *Pharm. Res.* **2015**, *32*, 3660–3673. [CrossRef]
72. Indulkar, A.S.; Box, K.J.; Taylor, R.; Ruiz, R.; Taylor, L.S. PH-Dependent Liquid-Liquid Phase Separation of Highly Supersaturated Solutions of Weakly Basic Drugs. *Mol. Pharm.* **2015**, *12*, 2365–2377. [CrossRef]
73. Lafferrère, L.; Hoff, C.; Veesler, S. Study of Liquid-Liquid Demixing from Drug Solution. *J. Cryst. Growth* **2004**, *269*, 550–557. [CrossRef]
74. Baird, J.A.; Van Eerdenbrugh, B.; Taylor, L.S. A Classification System to Assess the Crystallization Tendency of Organic Molecules from Undercooled Melts. *J. Pharm. Sci.* **2010**, *99*, 3787–3806. [CrossRef]
75. Tian, Y.; Jones, D.S.D.S.; Andrews, G.P.G.P. An Investigation into the Role of Polymeric Carriers on Crystal Growth within Amorphous Solid Dispersion Systems. *Mol. Pharm.* **2015**, *12*, 1180–1192. [CrossRef]
76. Lauer, M.E.; Grassmann, O.; Siam, M.; Tardio, J.; Jacob, L.; Page, S.; Kindt, J.H.; Engel, A.; Alsenz, J. Atomic Force Microscopy-Based Screening of Drug-Excipient Miscibility and Stability of Solid Dispersions. *Pharm. Res.* **2011**, *28*, 572–584. [CrossRef]
77. Li, N.; Taylor, L.S. Nanoscale Infrared, Thermal, and Mechanical Characterization of Telaprevir-Polymer Miscibility in Amorphous Solid Dispersions Prepared by Solvent Evaporation. *Mol. Pharm.* **2016**, *13*, 1123–1136. [CrossRef] [PubMed]
78. Guo, Y.; Shalaev, E.; Smith, S. Physical Stability of Pharmaceutical Formulations: Solid-State Characterization of Amorphous Dispersions. *Trac—Trends Anal. Chem.* **2013**, *49*, 137–144. [CrossRef]
79. Guan, J.; Liu, Q.; Jin, L.; Xu, H.; Wu, H.; Zhang, X.; Mao, S. Synergistic Effect of Soluplus and Hyaluronic Acid on the Supersaturation Maintenance of Lovastatin: The Facilitated in Vitro-in Vivo Performance and Improved Physical Stability. *Carbohydr. Polym.* **2019**, *222*, 114978. [CrossRef] [PubMed]
80. Sarode, A.L.; Sandhu, H.; Shah, N.; Malick, W.; Zia, H. Hot Melt Extrusion for Amorphous Solid Dispersions: Temperature and Moisture Activated Drug-Polymer Interactions for Enhanced Stability. *Mol. Pharm.* **2013**, *10*, 3665–3675. [CrossRef]
81. Lin, X.; Hu, Y.; Liu, L.; Su, L.; Li, N.; Yu, J.; Tang, B.; Yang, Z. Physical Stability of Amorphous Solid Dispersions: A Physicochemical Perspective with Thermodynamic, Kinetic and Environmental Aspects. *Pharm. Res.* **2018**, *35*. [CrossRef] [PubMed]
82. Mehta, M.; Suryanarayanan, R. Accelerated Physical Stability Testing of Amorphous Dispersions. *Mol. Pharm.* **2016**, *13*, 2661–2666. [CrossRef]
83. Li, N.; Gilpin, C.J.J.; Taylor, L.S.S. Understanding the Impact of Water on the Miscibility and Microstructure of Amorphous Solid Dispersions: An AFM-LCR and TEM-EDX Study. *Mol. Pharm.* **2017**, *14*, 1691–1705. [CrossRef] [PubMed]
84. Dohrn, S.; Luebbert, C.; Lehmkemper, K.; Kyeremateng, S.O.; Degenhardt, M.; Sadowski, G. Solvent Influence on the Phase Behavior and Glass Transition of Amorphous Solid Dispersions. *Eur. J. Pharm. Biopharm.* **2021**, *158*, 132–142. [CrossRef]
85. Meere, M.; Pontrelli, G.; McGinty, S. Modelling Phase Separation in Amorphous Solid Dispersions. *Acta Biomater.* **2019**, *94*, 410–424. [CrossRef] [PubMed]
86. Duarte, Í.; Santos, J.L.; Pinto, J.F.; Temtem, M. Screening Methodologies for the Development of Spray-Dried Amorphous Solid Dispersions. *Pharm. Res.* **2015**, *32*, 222–237. [CrossRef] [PubMed]
87. Luebbert, C.; Huxoll, F.; Sadowski, G.; Van Den Mooter, G.; Grohganz, H. Amorphous-Amorphous Phase Separation in API/Polymer Formulations. *Molecules* **2017**, *22*, 296. [CrossRef] [PubMed]
88. Taylor, L.S.; Li, N.; Cape, J.L.; Mankani, B.R.; Zemlyanov, D.Y.; Shepard, K.B.; Morgen, M.M. Water-Induced Phase Separation of Spray-Dried Amorphous Solid Dispersions. *Mol. Pharm.* **2020**, *17*, 4004–4017. [CrossRef]
89. Knopp, M.M.; Wendelboe, J.; Holm, R.; Rades, T. Effect of Amorphous Phase Separation and Crystallization on the in Vitro and in Vivo Performance of an Amorphous Solid Dispersion. *Eur. J. Pharm. Biopharm.* **2018**, *130*, 290–295. [CrossRef]
90. Meng, F.; Ferreira, R.; Zhang, F. Effect of Surfactant Level on Properties of Celecoxib Amorphous Solid Dispersions. *J. Drug Deliv. Sci. Technol.* **2019**, *49*, 301–307. [CrossRef]
91. Ueda, K.; Higashi, K.; Yamamoto, K.; Moribe, K. Inhibitory Effect of Hydroxypropyl Methylcellulose Acetate Succinate on Drug Recrystallization from a Supersaturated Solution Assessed Using Nuclear Magnetic Resonance Measurements. *Mol. Pharm.* **2013**, *10*, 3801–3811. [CrossRef]
92. Haiss, W.; Thanh, N.T.K.; Aveyard, J.; Fernig, D.G. Determination of Size and Concentration of Gold Nanoparticles from UV-Vis Spectra. *Anal. Chem.* **2007**, *79*, 4215–4221. [CrossRef] [PubMed]
93. Purohit, H.S.; Taylor, L.S. Phase Behavior of Ritonavir Amorphous Solid Dispersions during Hydration and Dissolution. *Pharm. Res.* **2017**, *34*, 2842–2861. [CrossRef]
94. Almeida, E.; Sousa, L.; Reutzel-Edens, S.M.; Stephenson, G.A.; Taylor, L.S. Assessment of the Amorphous "Solubility" of a Group of Diverse Drugs Using New Experimental and Theoretical Approaches. *Mol. Pharm.* **2015**, *12*, 484–495. [CrossRef]

95. Ilevbare, G.A.; Liu, H.; Pereira, J.; Edgar, K.J.; Taylor, L.S. Influence of Additives on the Properties of Nanodroplets Formed in Highly Supersaturated Aqueous Solutions of Ritonavir. *Mol. Pharm.* **2013**, *10*, 3392–3403. [CrossRef]
96. Zhao, Z.; Katai, H.; Higashi, K.; Ueda, K.; Kawakami, K.; Moribe, K. Cryo-TEM and AFM Observation of the Time-Dependent Evolution of Amorphous Probucol Nanoparticles Formed by the Aqueous Dispersion of Ternary Solid Dispersions. *Mol. Pharm.* **2019**, *16*, 2184–2198. [CrossRef]
97. Wang, S.; Liu, C.; Chen, Y.; Zhu, A.; Qian, F. Aggregation of Hydroxypropyl Methylcellulose Acetate Succinate under Its Dissolving PH and the Impact on Drug Supersaturation. *Mol. Pharm.* **2018**, *15*, 4643–4653. [CrossRef] [PubMed]
98. Mosquera-Giraldo, L.I.; Taylor, L.S. Glass-Liquid Phase Separation in Highly Supersaturated Aqueous Solutions of Telaprevir. *Mol. Pharm.* **2015**, *12*, 496–503. [CrossRef] [PubMed]
99. Zhu, C.; Liang, S.; Song, E.; Zhou, Y.; Wang, W.; Shan, F.; Shi, Y.; Hao, C.; Yin, K.; Zhang, T.; et al. In-Situ Liquid Cell Transmission Electron Microscopy Investigation on Oriented Attachment of Gold Nanoparticles. *Nat. Commun.* **2018**. [CrossRef] [PubMed]
100. Weatherall, E.; Willmott, G.R. Applications of Tunable Resistive Pulse Sensing. *Analyst* **2015**, *140*, 3318–3334. [CrossRef]
101. Sivakumaran, M.; Platt, M. Tunable Resistive Pulse Sensing: Potential Applications in Nanomedicine. *Nanomedicine* **2016**, *11*, 2197–2214. [CrossRef]
102. Bootz, A.; Vogel, V.; Schubert, D.; Kreuter, J. Comparison of Scanning Electron Microscopy, Dynamic Light Scattering and Analytical Ultracentrifugation for the Sizing of Poly(Butyl Cyanoacrylate) Nanoparticles. *Eur. J. Pharm. Biopharm.* **2004**. [CrossRef]
103. Gioria, S.; Caputo, F.; Urbán, P.; Maguire, C.M.; Bremer-Hoffmann, S.; Prina-Mello, A.; Calzolai, L.; Mehn, D. Are Existing Standard Methods Suitable for the Evaluation of Nanomedicines: Some Case Studies. *Nanomedicine* **2018**, *13*, 539–554. [CrossRef] [PubMed]
104. Bonnett, P.E.; Carpenter, K.J.; Dawson, S.; Davey, R.J. Solution Crystallisation via a Submerged Liquid-Liquid Phase Boundary: Oiling Out. *Chem. Commun.* **2003**, *3*, 698–699. [CrossRef]
105. Ricarte, R.G.; Li, Z.; Johnson, L.M.; Ting, J.M.; Reineke, T.M.; Bates, F.S.; Hillmyer, M.A.; Lodge, T.P. Direct Observation of Nanostructures during Aqueous Dissolution of Polymer/Drug Particles. *Macromolecules* **2017**, *50*, 3143–3152. [CrossRef]
106. Egami, K.; Higashi, K.; Yamamoto, K.; Moribe, K. Crystallization of Probucol in Nanoparticles Revealed by AFM Analysis in Aqueous Solution. *Mol. Pharm.* **2015**, *12*, 2972–2980. [CrossRef] [PubMed]
107. Ueda, K.; Higashi, K.; Moribe, K. Direct NMR Monitoring of Phase Separation Behavior of Highly Supersaturated Nifedipine Solution Stabilized with Hypromellose Derivatives. *Mol. Pharm.* **2017**, *14*, 2314–2322. [CrossRef]
108. Ueda, K.; Taylor, L.S. Polymer Type Impacts Amorphous Solubility and Drug-Rich Phase Colloidal Stability: A Mechanistic Study Using Nuclear Magnetic Resonance Spectroscopy. *Mol. Pharm.* **2020**, *17*, 1352–1362. [CrossRef] [PubMed]
109. Jha, P.K.; Larson, R.G. Assessing the Efficiency of Polymeric Excipients by Atomistic Molecular Dynamics Simulations. *Mol. Pharm.* **2014**, *11*, 1676–1686. [CrossRef]
110. Indulkar, A.S.; Lou, X.; Zhang, G.G.Z.; Taylor, L.S. Insights into the Dissolution Mechanism of Ritonavir-Copovidone Amorphous Solid Dispersions: Importance of Congruent Release for Enhanced Performance. *Mol. Pharm.* **2019**, *16*, 1327–1339. [CrossRef] [PubMed]
111. Hate, S.S.; Reutzel-Edens, S.M.; Taylor, L.S. Insight into Amorphous Solid Dispersion Performance by Coupled Dissolution and Membrane Mass Transfer Measurements. *Mol. Pharm.* **2019**, *16*, 448–461. [CrossRef]
112. Jackson, M.J.; Kestur, U.S.; Hussain, M.A.; Taylor, L.S. Dissolution of Danazol Amorphous Solid Dispersions: Supersaturation and Phase Behavior as a Function of Drug Loading and Polymer Type. *Mol. Pharm.* **2016**, *13*, 223–231. [CrossRef]
113. Saboo, S.; Mugheirbi, N.A.; Zemlyanov, D.Y.; Kestur, U.S.; Taylor, L.S. Congruent Release of Drug and Polymer: A "Sweet Spot" in the Dissolution of Amorphous Solid Dispersions. *J. Control. Release* **2019**, *298*, 68–82. [CrossRef] [PubMed]
114. Ueda, K.; Higashi, K.; Moribe, K. Mechanistic Elucidation of Formation of Drug-Rich Amorphous Nanodroplets by Dissolution of the Solid Dispersion Formulation. *Int. J. Pharm.* **2019**, *561*, 82–92. [CrossRef] [PubMed]
115. Stewart, A.M.; Grass, M.E.; Brodeur, T.J.; Goodwin, A.K.; Morgen, M.M.; Friesen, D.T.; Vodak, D.T. Impact of Drug-Rich Colloids of Itraconazole and HPMCAS on Membrane Flux in Vitro and Oral Bioavailability in Rats. *Mol. Pharm.* **2017**, *14*, 2437–2449. [CrossRef]
116. Mosquera-Giraldo, L.I.; Li, N.; Wilson, V.R.; Nichols, B.L.B.; Edgar, K.J.; Taylor, L.S. Influence of Polymer and Drug Loading on the Release Profile and Membrane Transport of Telaprevir. *Mol. Pharm.* **2018**, *15*, 1700–1713. [CrossRef]
117. Li, Z.; Johnson, L.M.; Ricarte, R.G.; Yao, L.J.; Hillmyer, M.A.; Bates, F.S.; Lodge, T.P. Enhanced Performance of Blended Polymer Excipients in Delivering a Hydrophobic Drug through the Synergistic Action of Micelles and HPMCAS. *Langmuir* **2017**, *33*, 2837–2848. [CrossRef] [PubMed]
118. Dalsin, M.C.; Tale, S.; Reineke, T.M. Solution-State Polymer Assemblies Influence BCS Class II Drug Dissolution and Supersaturation Maintenance. *Biomacromolecules* **2014**, *15*, 500–511. [CrossRef]
119. Li, Z.; Lenk, T.I.; Yao, L.J.; Bates, F.S.; Lodge, T.P. Maintaining Hydrophobic Drug Supersaturation in a Micelle Corona Reservoir. *Macromolecules* **2018**, *51*, 540–551. [CrossRef]
120. Li, Z.; Van Zee, N.J.; Bates, F.S.; Lodge, T.P. Polymer Nanogels as Reservoirs to Inhibit Hydrophobic Drug Crystallization. *ACS Nano* **2019**, *13*, 1232–1243. [CrossRef]

121. Indulkar, A.S.; Waters, J.E.; Mo, H.; Gao, Y.; Raina, S.A.; Zhang, G.G.Z.; Taylor, L.S. Origin of Nanodroplet Formation Upon Dissolution of an Amorphous Solid Dispersion: A Mechanistic Isotope Scrambling Study. *J. Pharm. Sci.* **2017**, *106*, 1998–2008. [CrossRef]
122. Harmon, P.; Galipeau, K.; Xu, W.; Brown, C.; Wuelfing, W.P. Mechanism of Dissolution-Induced Nanoparticle Formation from a Copovidone-Based Amorphous Solid Dispersion. *Mol. Pharm.* **2016**, *13*, 1467–1481. [CrossRef]
123. Kesisoglou, F.; Wang, M.; Galipeau, K.; Harmon, P.; Okoh, G.; Xu, W. Effect of Amorphous Nanoparticle Size on Bioavailability of Anacetrapib in Dogs. *J. Pharm. Sci.* **2019**, *108*, 2917–2925. [CrossRef]
124. Miller, J.M.; Beig, A.; Carr, R.A.; Spence, J.K.; Dahan, A. A Win-Win Solution in Oral Delivery of Lipophilic Drugs: Supersaturation via Amorphous Solid Dispersions Increases Apparent Solubility without Sacrifice of Intestinal Membrane Permeability. *Mol. Pharm.* **2012**, *9*, 2009–2016. [CrossRef] [PubMed]
125. Beig, A.; Miller, J.M.; Lindley, D.; Carr, R.A.; Zocharski, P.; Agbaria, R.; Dahan, A. Head-To-Head Comparison of Different Solubility-Enabling Formulations of Etoposide and Their Consequent Solubility-Permeability Interplay. *J. Pharm. Sci.* **2015**, *104*, 2941–2947. [CrossRef]
126. Dahan, A.; Beig, A.; Lindley, D.; Miller, J.M. The Solubility-Permeability Interplay and Oral Drug Formulation Design: Two Heads Are Better than One. *Adv. Drug Deliv. Rev.* **2016**, *101*, 99–107. [CrossRef]
127. Dahan, A.; Beig, A.; Ioffe-Dahan, V.; Agbaria, R.; Miller, J.M. The Twofold Advantage of the Amorphous Form as an Oral Drug Delivery Practice for Lipophilic Compounds: Increased Apparent Solubility and Drug Flux through the Intestinal Membrane. *AAPS J.* **2013**, *15*, 347–353. [CrossRef]
128. Dahan, A.; Miller, J.M. The Solubility-Permeability Interplay and Its Implications in Formulation Design and Development for Poorly Soluble Drugs. *AAPS J.* **2012**, *14*, 244–251. [CrossRef] [PubMed]
129. Miller, J.M.; Beig, A.; Carr, R.A.; Webster, G.K.; Dahan, A. The Solubility-Permeability Interplay When Using Cosolvents for Solubilization: Revising the Way We Use Solubility-Enabling Formulations. *Mol. Pharm.* **2012**, *9*, 581–590. [CrossRef] [PubMed]
130. Beig, A.; Fine-Shamir, N.; Lindley, D.; Miller, J.M.; Dahan, A. Advantageous Solubility-Permeability Interplay When Using Amorphous Solid Dispersion (ASD) Formulation for the BCS Class IV P-Gp Substrate Rifaximin: Simultaneous Increase of Both the Solubility and the Permeability. *AAPS J.* **2017**, *19*, 806–813. [CrossRef]
131. Dahan, A.; Miller, J.M.; Hoffman, A.; Amidon, G.E.; Amidon, G.L. The Solubility-Permeability Interplay in Using Cyclodextrins as Pharmaceutical Solubilizers: Mechanistic Modeling and Application to Progesterone. *J. Pharm. Sci.* **2010**. [CrossRef]
132. Miller, J.M.; Beig, A.; Krieg, B.J.; Carr, R.A.; Borchardt, T.B.; Amidon, G.E.; Amidon, G.L.; Dahan, A. The Solubility-Permeability Interplay: Mechanistic Modeling and Predictive Application of the Impact of Micellar Solubilization on Intestinal Permeation. *Mol. Pharm.* **2011**, *8*, 1848–1856. [CrossRef] [PubMed]
133. Raina, S.A.; Zhang, G.G.Z.; Alonzo, D.E.; Wu, J.; Zhu, D.; Catron, N.D.; Gao, Y.; Taylor, L.S. Impact of Solubilizing Additives on Supersaturation and Membrane Transport of Drugs. *Pharm. Res.* **2015**, *32*, 3350–3364. [CrossRef]
134. Loftsson, T.; Brewster, M.E. Cyclodextrins as Functional Excipients: Methods to Enhance Complexation Efficienc. *J. Pharm. Sci.* **2012**, *101*, 3019–3032. [CrossRef]
135. Kesisoglou, F.; Wu, Y. Understanding the Effect of API Properties on Bioavailability through Absorption Modeling. *AAPS J.* **2008**, *10*, 516–525. [CrossRef]
136. Wu, Y.; Loper, A.; Landis, E.; Hettrick, L.; Novak, L.; Lynn, K.; Chen, C.; Thompson, K.; Higgins, R.; Batra, U.; et al. The Role of Biopharmaceutics in the Development of a Clinical Nanoparticle Formulation of MK-0869: A Beagle Dog Model Predicts Improved Bioavailability and Diminished Food Effect on Absorption in Human. *Int. J. Pharm.* **2004**, *285*, 135–146. [CrossRef] [PubMed]
137. Wilson, F.A.; Dietschy, J.M. Characterization of Bile Acid Absorption across the Unstirred Water Layer and Brush Border of the Rat Jejunum. *J. Clin. Investig.* **1972**, *51*, 3015–3025. [CrossRef]
138. Loftsson, T. Drug Permeation through Biomembranes: Cyclodextrins and the Unstirred Water Layer. *Pharmazie* **2012**, *67*, 363–370. [CrossRef]
139. Stewart, A.M.; Grass, M.E. Practical Approach to Modeling the Impact of Amorphous Drug Nanoparticles on the Oral Absorption of Poorly Soluble Drugs. *Mol. Pharm.* **2020**, *17*, 180–189. [CrossRef] [PubMed]
140. Arce, F.A.; Setiawan, N.; Campbell, H.R.; Lu, X.; Nethercott, M.J.; Bummer, P.; Su, Y.; Marsac, P.J. Toward Developing Discriminating Dissolution Methods for Formulations Containing Nanoparticulates in Solution: The Impact of Particle Drift and Drug Activity in Solution. *Mol. Pharm.* **2020**, *17*, 4125–4140. [CrossRef] [PubMed]
141. Fine-Shamir, N.; Dahan, A. Methacrylate-Copolymer Eudragit EPO as a Solubility-Enabling Excipient for Anionic Drugs: Investigation of Drug Solubility, Intestinal Permeability, and Their Interplay. *Mol. Pharm.* **2019**, *16*, 2884–2891. [CrossRef]
142. Saal, W.; Ross, A.; Wyttenbach, N.; Alsenz, J.; Kuentz, M. A Systematic Study of Molecular Interactions of Anionic Drugs with a Dimethylaminoethyl Methacrylate Copolymer Regarding Solubility Enhancement. *Mol. Pharm.* **2017**, *14*, 1243–1250. [CrossRef]
143. Higashi, K.; Yamamoto, K.K.K.; Pandey, M.K.K.; Mroue, K.H.H.; Moribe, K.; Yamamoto, K.K.K.; Ramamoorthy, A. Insights into Atomic-Level Interaction between Mefenamic Acid and Eudragit EPO in a Supersaturated Solution by High-Resolution Magic-Angle Spinning NMR Spectroscopy. *Mol. Pharm.* **2014**, *11*, 351–357. [CrossRef]
144. Raghavan, S.L.; Kiepfer, B.; Davis, A.F.; Kazarian, S.G.; Hadgraft, J. Membrane Transport of Hydrocortisone Acetate from Supersaturated Solutions; the Role of Polymers. *Int. J. Pharm.* **2001**, *221*, 95–105. [CrossRef]

145. Otsuka, N.; Ueda, K.; Ohyagi, N.; Shimizu, K.; Katakawa, K.; Kumamoto, T.; Higashi, K.; Yamamoto, K.; Moribe, K. An Insight into Different Stabilization Mechanisms of Phenytoin Derivatives Supersaturation by HPMC and PVP. *J. Pharm. Sci.* **2015**, *104*, 2574–2582. [CrossRef]
146. Yu, M.; Wu, J.; Shi, J.; Farokhzad, O.C. Nanotechnology for Protein Delivery: Overview and Perspectives. *J. Control. Release* **2016**, *240*, 24–37. [CrossRef] [PubMed]
147. Engin, A.B.; Nikitovic, D.; Neagu, M.; Henrich-Noack, P.; Docea, A.O.; Shtilman, M.I.; Golokhvast, K.; Tsatsakis, A.M. Mechanistic Understanding of Nanoparticles' Interactions with Extracellular Matrix: The Cell and Immune System. *Part. Fibre Toxicol.* **2017**, *14*, 1–16. [CrossRef]
148. Crater, J.S.; Carrier, R.L. Barrier Properties of Gastrointestinal Mucus to Nanoparticle Transport. *Macromol. Biosci.* **2010**. [CrossRef]
149. Maisel, K.; Ensign, L.; Reddy, M.; Cone, R.; Hanes, J. Effect of Surface Chemistry on Nanoparticle Interaction with Gastrointestinal Mucus and Distribution in the Gastrointestinal Tract Following Oral and Rectal Administration in the Mouse. *J. Control. Release* **2015**, *197*, 48–57. [CrossRef] [PubMed]
150. Sahatsapan, N.; Rojanarata, T.; Ngawhirunpat, T.; Opanasopit, P. 6-Maleimidohexanoic Acid-Grafted Chitosan: A New Generation Mucoadhesive Polymer. *Carbohydr. Polym.* **2018**, *202*, 258–264. [CrossRef] [PubMed]
151. Lai, S.K.; Wang, Y.Y.; Hanes, J. Mucus-Penetrating Nanoparticles for Drug and Gene Delivery to Mucosal Tissues. *Adv. Drug Deliv. Rev.* **2009**, *61*, 158–171. [CrossRef] [PubMed]
152. Yu, M.; Xu, L.; Tian, F.; Su, Q.; Zheng, N.; Yang, Y.; Wang, J.; Wang, A.; Zhu, C.; Guo, S.; et al. Rapid Transport of Deformation-Tuned Nanoparticles across Biological Hydrogels and Cellular Barriers. *Nat. Commun.* **2018**, *9*, 1–11. [CrossRef]
153. Abe, O.; Abe, R.; Enomoto, K.; Kikuchi, K.; Koyama, H.; Masuda, H.; Nomura, Y.; Sakai, K.; Sugimachi, K.; Tominaga, T.; et al. A Comparison of Spray Drying and Milling in the Production of Amorphous Dispersions of Sulfathiazole/Polyvinylpyrrolidone and Sulfadimidine/Polyvinylpyrrolidone. *J. Pharm. Sci.* **2011**, *8*, 1131–1138. [CrossRef]
154. Taipaleenmäki, E.; Städler, B. Recent Advancements in Using Polymers for Intestinal Mucoadhesion and Mucopenetration. *Macromol. Biosci.* **2020**, *20*, 1–9. [CrossRef] [PubMed]
155. Ways, T.M.M.; Lau, W.M.; Khutoryanskiy, V.V. Chitosan and Its Derivatives for Application in Mucoadhesive Drug Delivery Systems. *Polymers* **2018**, *10*, 267. [CrossRef] [PubMed]
156. Fan, B.; Xing, Y.; Zheng, Y.; Sun, C.; Liang, G. PH-Responsive Thiolated Chitosan Nanoparticles for Oral Low-Molecular Weight Heparin Delivery: In Vitro and in Vivo Evaluation. *Drug Deliv.* **2016**, *23*, 238–247. [CrossRef]
157. Sreeharsha, N.; Ramnarayanan, C.; Al-Dhubiab, B.E.; Nair, A.B.; Hiremath, J.G.; Venugopala, K.N.; Satish, R.T.; Attimarad, M.; Shariff, A. Mucoadhesive Particles: A Novel, Prolonged-Release Nanocarrier of Sitagliptin for the Treatment of Diabetics. *BioMed Res. Int.* **2019**, *2019*, 3950942. [CrossRef]
158. Ha, E.S.; Choo, G.H.; Baek, I.H.; Kim, M.S. Formulation, Characterization, and in Vivo Evaluation of Celecoxib-PVP Solid Dispersion Nanoparticles Using Supercritical Antisolvent Process. *Molecules* **2014**, *19*, 20325–20339. [CrossRef]
159. Liu, Y.; Sun, C.; Hao, Y.; Jiang, T.; Zheng, L.; Wang, S. Mechanism of Dissolution Enhancement and Bioavailability of Poorly Water Soluble Celecoxib by Preparing Stable Amorphous Nanoparticles. *J. Pharm. Pharm. Sci.* **2010**, *13*, 589–606. [CrossRef]
160. Wu, W.; Zu, Y.; Wang, L.; Wang, L.; Li, Y.; Liu, Y.; Wu, M.; Zhao, X.; Zhang, X. Preparation, Characterization and Antitumor Activity Evaluation of Silibinin Nanoparticles for Oral Delivery through Liquid Antisolvent Precipitation. *RSC Adv.* **2017**, *7*, 54379–54390. [CrossRef]
161. Sun, R.; Shen, C.; Shafique, S.; Mustapha, O.; Hussain, T.; Khan, I.U.; Mehmood, Y.; Anwer, K.; Shahzad, Y.; Yousaf, A.M. Electrosprayed Polymeric Nanospheres for Enhanced Solubility, Dissolution Rate, Oral Bioavailability and Antihyperlipidemic Activity of Bezafibrate. *Int. J. Nanomed.* **2020**, *15*, 705–715. [CrossRef]
162. Kim, M.S.; Baek, I.H. Fabrication and Evaluation of Valsartan-Polymer- Surfactant Composite Nanoparticles by Using the Supercritical Antisolvent Process. *Int. J. Nanomed.* **2014**. [CrossRef] [PubMed]
163. Wang, X.Q.; Dai, J.D.; Chen, Z.; Zhang, T.; Xia, G.M.; Nagai, T.; Zhang, Q. Bioavailability and Pharmacokinetics of Cyclosporine A-Loaded PH-Sensitive Nanoparticles for Oral Administration. *J. Control. Release* **2004**, *97*, 421–429. [CrossRef]
164. Tran, T.H.; Poudel, B.K.; Marasini, N.; Chi, S.C.; Choi, H.G.; Yong, C.S.; Kim, J.O. Preparation and Evaluation of Raloxifene-Loaded Solid Dispersion Nanoparticle by Spray-Drying Technique without an Organic Solvent. *Int. J. Pharm.* **2013**, *443*, 50–57. [CrossRef]
165. Sun, X.; Zhu, D.; Cai, Y.; Shi, G.; Gao, M.; Zheng, M. One-Step Mechanochemical Preparation and Prominent Antitumor Activity of SN-38 Self-Micelle Solid Dispersion. *Int. J. Nanomed.* **2019**, *14*, 2115–2126. [CrossRef] [PubMed]
166. Rao, Q.; Qiu, Z.; Huang, D.; Lu, T.; Zhang, Z.J.; Luo, D.; Pan, P.; Zhang, L.; Liu, Y.; Guan, S.; et al. Enhancement of the Apparent Solubility and Bioavailability of Tadalafil Nanoparticles via Antisolvent Precipitation. *Eur. J. Pharm. Sci.* **2019**, *128*, 222–231. [CrossRef] [PubMed]
167. Ha, E.S.; Kim, J.S.; Baek, I.H.; Yoo, J.W.; Jung, Y.; Moon, H.R.; Kim, M.S. Development of Megestrol Acetate Solid Dispersion Nanoparticles for Enhanced Oral Delivery by Using a Supercritical Antisolvent Process. *Drug Des. Dev. Ther.* **2015**, *9*, 4269–4277. [CrossRef]
168. Kim, M.S.; Kim, J.S.; Park, H.J.; Cho, W.K.; Cha, K.H.; Hwang, S.J. Enhanced Bioavailability of Sirolimus via Preparation of Solid Dispersion Nanoparticles Using a Supercritical Antisolvent Process. *Int. J. Nanomed.* **2011**, *6*, 2997–3009. [CrossRef]

pharmaceutics

MDPI

Review

Therapeutic Applications of Solid Dispersions for Drugs and New Molecules: In Vitro and In Vivo Activities

Verônica da Silva Oliveira [1], Amanda Silva de Almeida [2], Ingrid da Silva Albuquerque [2], Fernanda Ílary Costa Duarte [1], Bárbara Cristina Silva Holanda Queiroz [1], Attilio Converti [3] and Ádley Antonini Neves de Lima [1,2,*]

1 Postgraduate Program in Pharmaceutical Sciences, Health Sciences Center, Federal University of Rio Grande do Norte, Natal-RN 59012-570, Brazil; veronicasoliver47@gmail.com (V.d.S.O.); fernandailary@gmail.com (F.Í.C.D.); barbaracqueiroz@gmail.com (B.C.S.H.Q.)
2 Department of Pharmacy, Universidade Federal do Rio Grande do Norte, Natal-RN 59012-570, Brazil; amandasdealmeida@gmail.com (A.S.d.A.); ingridmariaalbuquerque@gmail.com (I.d.S.A.)
3 Department of Civil, Chemical and Environmental Engineering, Pole of Chemical Engineering, Genoa University, I-16145 Genoa, Italy; converti@unige.it
* Correspondence: adleylima13@gmail.com; Tel.: +55-8499-928-8864

Received: 17 September 2020; Accepted: 27 September 2020; Published: 30 September 2020

check for updates

Abstract: This review aims to provide an overview of studies that address the use, in therapeutic applications, of solid dispersions (SDs) with biological activities in vitro and/or in vivo mainly made up of polymeric matrices, as well as to evaluate the bioactive activity of their constituents. This bibliographic survey shows that the development of solid dispersions provides benefits in the physicochemical properties of bioactive compounds, which lead to an increase in their biological potential. However, despite the reports found on solid dispersions, there is still a need for biological assay-based studies, mainly in vivo, to assist in the investigation and to devise new applications. Therefore, studies based on such an approach are of great importance to enhance and extend the use of solid dispersions in the most diverse therapeutic applications.

Keywords: solid dispersions; drug delivery; biological applications

1. Introduction

Many drug candidates have low aqueous solubility, which can make their oral absorption inadequate. According to the literature, approximately 40% of marketed drugs are poorly soluble in water, as are, according to the Biopharmaceutical Classification System, about 40–90% of new drug candidates [1–4].

The slow dissolution rate and low solubility of some drugs lead to unpredictable bioavailability, non-reproducible clinical response or treatment inefficiency due to low therapeutic plasma levels; therefore, their use must be optimized using formulation strategies capable of improving their administration [5].

Several technological strategies have been developed and employed to circumvent this situation such as complexation with cyclodextrins, particle size reduction by micronization, crystal development, nanotechnology [5], salt formation and solid dispersions (SDs) [6,7]. However, some of these strategies have shown disadvantages including the development of active forms in vivo, high execution cost and considerable levels of toxicity. Some of these limits can be overcome by employing solid dispersions, a viable, well-established and widely-used strategy to increase the dissolution rate and solubility of poorly water-soluble drugs [4,6,7].

SDs can be defined as molecular mixtures of drugs that are not soluble in hydrophilic carriers, which exhibit a drug release profile driven by the polymer properties [8]. The so-called first-generation crystalline SDs, which use crystalline carriers such as urea and sugars including sucrose, dextrose and galactose, have the disadvantage of high thermodynamic stability that prevents a quick release of drugs [9]. In second-generation SDs drugs are dispersed in usually polymeric amorphous carriers [8], while the carriers used in third-generation SDs are surfactant or a mixture of amorphous polymers and surfactants [10].

The use of hydrophilic polymeric matrices to develop SDs for the release of commercial drugs or new drugs candidates has been shown to be a promising alternative able to improve their pharmacokinetic properties [7].

This review deals with SDs, mainly those belonging to the second-generation class that use polymers as carriers. It describes the main in vitro and in vivo activities (Figure 1) of SDs having different compositions and various preparation methods, intending to innovate and enhance the release of poorly water-soluble drugs, as well as improving the biological activities of the loaded bioactive compounds.

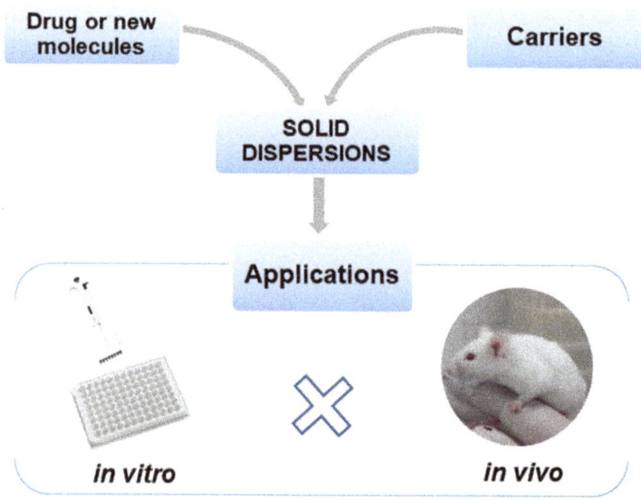

Figure 1. Representation of the application of solid dispersions in biological assays.

2. Methods

This literature search was conducted through the ScienceDirect and PubMed specialized databases, using different combinations of the following keywords: "amorphous solid dispersions" or "in vitro solid dispersions" or "in vivo solid dispersions". Studies performed in the last 10 years in vitro and in vivo were included, in which the biological action and use of solid dispersions to improve the biological performance of substances were investigated.

3. In Vitro Study of Solid Dispersions in Polymeric Matrices

This section deals with solid dispersions (SDs) prepared to improve the properties and release of poorly soluble drugs and drug candidates of natural or synthetic origin. The preparation and use of SDs have been reported in several in vitro studies, which have been numerically quantified and classified in Figure 2 based on the biological activities of their active compounds, while the main information on these studies is summarized in Table 1.

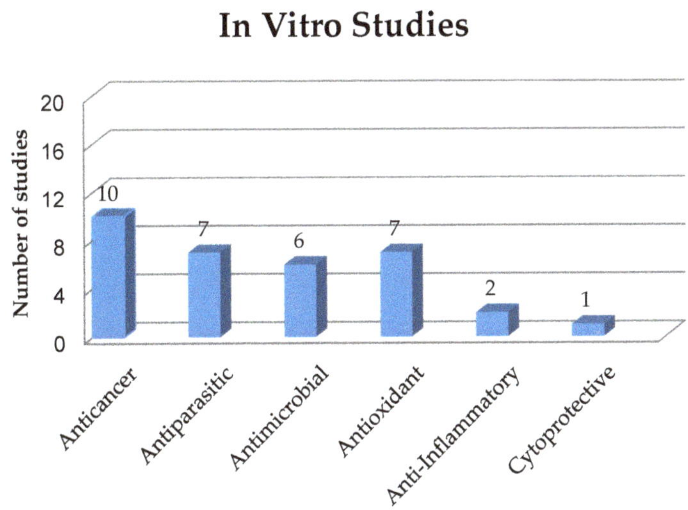

Figure 2. Quantification and classification of in vitro studies on solid dispersions published in the period from 2009 to 2020.

3.1. Anticancer Activities of Solid Dispersions

To improve the solubility and bioavailability of niclosamide, an anthelmintic with promising anticancer activity, a SD of its mixture with the octenylsuccinate hydroxypropyl phytoglycogen (OHPP) biopolymer was prepared by spray drying. In vitro tests by the 3-(4,5-dimethylthiazol-2-yl)-2,5-diphenyltetrazolium bromide (MTT) assay provided half inhibitory concentration (IC50 values) of SD against HeLa (cervical carcinoma), PC-3 (prostate cancer) and A549 (lung cancer) cells of 0.92, 0.68 and 1.18 μg/mL, while those of niclosamide alone were 1.82, 5.93 and 1.46 μg/mL, respectively. These results demonstrated that niclosamide-OHPP SD showed higher cytotoxicity than DMSO-assisted niclosamide solutions, mostly on PC-3 cell line, although OHPP, as the solubilizer, did not contribute to cytotoxicity. In addition, the analysis on OHPP alone is missed [11].

Although paclitaxel (PTX) is a natural compound with powerful antitumor activity against lung, breast and ovarian cancers, its low water solubility poses a challenge for the manufacture and administration of its formulations [12]. There are currently two commercial forms, Taxol® and Abraxane®, which; however, have a high rate of hypersensitivity and side effects.

Considering these drawbacks, Xie and Yao [13] incorporated PTX with OHPP in the form of solid dispersion (PTX-OHPP SD) by spray drying, whose in vitro antitumor activity was tested by the MTT method against HeLa, PC-3 and A549 cells and compared to that of PTX alone. Cytotoxicity was dose dependent and OHPP alone was not cytotoxic to cells, with an inhibition rate lower than 15% at the highest concentration (93.75 μg/mL). The IC50 values of DMSO-assisted PTX solutions against the above tumor cells were 0.02, 0.08 and 0.03 μg/mL, respectively, while PTX-OHPP SD demonstrated the greatest efficacy against cancer cells, with values of 0.008, 0.017 and 0.002 μg/mL, respectively.

In another study conducted by Choi et al. [14], the goal was to develop PTX-containing SDs. Several polymers were tested to increase the drug solubility and, subsequently, to select the best polymer to prepare SDs by the solvent evaporation method. Solubility and dissolution tests allowed to select a SD composed of D-α-tocopheryl polyethylene glycol 1000 succinate (TPGS), polyvinylpyrrolidone/vinyl acetate (PVP/VA), Aerosil®200 and PTX (SD4) and another containing PTX, TPGS and Aerosil®200 (SD9), whose antitumor activities were assessed, at drug concentrations of 0.1, 1, 5, 10 and 20 μg/mL, against breast cancer cells (BT-474, MCF-7 and SK-BR-3) and non-tumor cells by the MTT assay. Whereas both SDs were nontoxic to normal cells, at a dose of 20 μg/mL the cytotoxicity of SD4 and SD9 was 3.0- and 2.8-fold higher than that of PTX alone against BT-474 cells, 4.0- and 4.1-fold against

MCF-7 cells and 5.6- and 5.0-fold against SK-BR-3 cells. These results show greater cytotoxicity of SD4 and SD9 compared to PTX alone at the same concentration of 20 μg/mL.

In order to improve the solubility and absorption of chrysin, a potent inhibitor of breast cancer resistance protein, Lee et al. [15] have developed SD formulations with hydrophilic carriers, including polyoxyethylene (4) lauryl ether (Brij®L4), using the solvent evaporation method. To evaluate the cytotoxicity of chrysin alone and SD in the concentration range of 5–160 μM, the MTT assay against HT29 (human colorectal carcinoma) cells was used. SD displayed a 50% cytotoxic concentration (CC50) of 26.3 μM, while pure chrysin showed low activity even at the highest concentration tested.

Despite its wide variety of bioactive properties, Curcumin (CM), the main bioactive polyphenolic compound in the rhizomes of *Curcuma longa* L., has low solubility in water and; therefore, a limited bioavailability, which reduces its applicability. Based on these limitations, SDs of curcumin were prepared and tested for cytotoxic activity against tumor cells, namely MCF-7, NCIH460 (non-small cell lung carcinoma), HeLa and HepG2 (carcinoma hepatocellular) cells and non-tumor (PLP2) cells [16]. SDs showed cytotoxicity to all tumor cell lines tested, with concentrations causing 50% cell growth inhibition (GI50) of 207 ± 4, 208 ± 3, 259 ± 8 and 240 ± 7 μg/mL against NCIH460, HeLa, HepG2 and MCF-7, respectively, whereas no cytotoxicity was detected to non-tumor cells, with GI50 > 400 μg/mL. The improved CM activity was attributed to its association with copolymer surfactants such as Poloxamer 407 used to prepare SDs, which increased the dissolution rate and bioavailability of this hydrophobic drug.

In another study on CM, SD of zinc(II)-curcumin complex was developed using polyvinylpyrrolidone (PVP K30) [17]. Metal-curcumin complexes are known to exert anticancer activity, but they have low bioavailability and poor solubility in water, which hinder oral formulations. To evaluate the in vitro activity, the MTT assay was used in hepatocellular carcinoma cell lines (HepG2 and SK-HEP1), using different concentrations of SDs containing pure CM (CM-SD) or Zn(II)-curcumin complex (ZnCM-SD). Treatment with CM-SD significantly reduced the viability of HepG2 and SK-HEP1 cells within 48 h, with IC50 of 17.73 and 57.31 nM, respectively, while cells treated with ZnCM-SD exhibited after the same time IC50 of 38.52 and 10.41 nM, respectively, in a dose-dependent manner.

Xiong et al. [18] prepared telaprevir-containing SDs by the co-grinding technique using hydroxypropyl methylcellulose (HPMC), PVP K30 and polyethylene glycol (PEG) 6000 as polymers. SDs were characterized and their cytotoxicity was evaluated in vitro by the MTT method against HepG2 and normal liver cells (LO2) at drug concentrations between 0.1 and 30 μM. Although the viability of HepG2 cells gradually decreased as the drug concentration increased, it was almost the same in the presence of SDs and pure drug; furthermore, the efficacy of telaprevir was maintained even after the addition of polymers and no systems developed were toxic to normal liver cells.

Considering that uterine carcinoma still affects several women worldwide and is responsible for the highest rate of female malignancy [19], a study developed by Jiang et al. [20] sought to evaluate the anticancer activity of *Angelica gigas* Nakai, a medicinal herb containing coumarins, essential oils and polyacetylene that has several biological activities. However, its main components have a hydrophobic character that affects its application in formulations. To overcome this difficulty, these authors developed a SD of the above herb using the Soluplus® polymer and subjected it to shear by hot-melt extrusion. The MTT method was used for the in vitro antiproliferative activity and the developed systems were tested at concentrations of 50, 100 and 200 μg/mL against HeLa cells and normal cells (HEK 293 lineage). All formulations reduced cell viability, but the hot-melt extruded powder led to the lowest HeLa cells viability after 24 h. When these cells were treated for 48 h with the SD, there were a considerable reduction in viability and a higher cytotoxicity compared to the others, with only 17.37% of viability being observed at the concentration of 200 μg/mL. On the other hand, none of the systems tested showed cytotoxicity in 24 or 48 h to HEK 293 cells.

The research developed by Guo et al. [21] aimed at obtaining a SD using a methacrylic acid polymer Eudragit S-100 in association with berberine hydrochloride (HB), which is an isoquinoline-derived alkaloid found widely in the root and stem bark of numerous medicinal plants. Previous in vitro studies did in fact show that berberine decreases the viability of breast cancer cells, lung and hepatocellular carcinoma [22,23]. However, the development of formulations using HB is hampered by the low bioavailability of this compound that causes irritation to the gastrointestinal tract after oral administration. In this context, a SD made up of HB and Eudragit S-100 was prepared by the solvent evaporation method and its antitumor activity evaluated by the MTT method against SW480, HCT116 and Caco-2 colon cancer cells, over time periods of 12, 24 and 48 h, at HB equivalent concentrations from 0 to 100 μM [21]. The IC50 values obtained after 24 h treatment with HB alone against the above cells were 88.64, 67.83 and 89.78 μM, respectively, while those of SD were 52.96, 42.81 and 37.68 μM, respectively. After 48 h treatment, both showed better results, with pure HB IC50 values of 43.10, 34.13 and 32.82 μM and SD IC50 values of 28.15, 22.06 and 15.39 μM against the same cells in the above order. These results showed that cell viability of these strains suffered a dose and time dependent reduction in the presence either of HB or SD.

Kumar et al. [24] developed binary and ternary SDs incorporating IIIM-290, a synthetic chromone alkaloid acting as a potent inhibitor of CDK2/A and CDK9/T1 kinases, whose poor water solubility causes it to be used in high doses. SDs were prepared by solvent casting and evaporation method using PVP K30, PEG-Polypropylene glycol(PPG)-PEG or xanthan gum as polymers. In in vitro cytotoxicity tests against Ehrlich ascites carcinoma cells, IIIM-290 alone and the best drug-containing SD prepared with PVP K30 showed IC50 values of 1.10 ± 0.05 and 1.35 ± 0.02 μg/mL, respectively.

The results summarized in this section demonstrated that the use of SDs not only allowed to increase the solubility and dissolution of the bioactive substances, but also to improve their in vitro antitumor activity, as evidenced mainly by a IC50 reduction; therefore, it can be deduced that this improvement derives from the increase in the solubility of active compounds when present in the dispersed systems. Furthermore, it is important to note that SDs were obtained for a large number of active compounds and that toxicity tests were performed on different cell types, which demonstrates their potential as systems for the release of lipophilic compounds to fight a variety of cancers.

3.2. *Antiparasitic Activity of Solid Dispersions*

3.2.1. Antichagasic Activity of Solid Dispersions

Chagas disease is an infection that affects around eight million people around the world, mainly in Latin America, which is treated with nifurtimox and benznidazole (BNZ). Although the latter is preferable because of fewer adverse effects than the former, it has low solubility, which limits its effectiveness in treating the disease [25].

One of the first studies developed to obtain BNZ-containing SDs was conducted by Lima et al. [26] with the aim of improving the solubility of this drug. In this work, the use of PEG 6000 and PVP K30 as hydrophilic polymer carriers led to an increase in the solubility of BNZ in aqueous medium. The interaction of BNZ with polymers was evaluated by physicochemical characterization and in vitro dissolution tests. The dissolution profile of all SDs was significantly better than that of BNZ alone, thus demonstrating the solubilizing effect of PEG and PVP. Another work published by the same group [27] aimed to develop SDs with HPMC and β-cyclodextrin inclusion complexes to increase the drug solubility and dissolution.

In a study by Fonseca-Berzal et al. [28], BNZ-containing SDs were prepared by the freeze-drying technique using sodium deoxycholate (NaDC) as surfactant and low-substituted hydroxypropyl cellulose (HPC) as polymer, whose trypanocidal activity was tested in vitro against axenic cultures of *Trypanosoma cruzi*. The results showed greater activity of the 1:3 BNZ:NaDC solid dispersion, with IC50 values of 33.92 ± 6.41 μM for epimastigote and 0.40 ± 0.05 μM for amastigote, compared to pure BNZ (IC50 of 52.95 ± 2.22 and 1.8 ± 0.66 μM, respectively).

To potentiate the effect of the same drug, Simonazzi et al. [25] prepared a BNZ-containing SD using Poloxamer 407 (P407) as polymer, whose in vitro activity was tested by the MTT method against

T. cruzi (Tulahuén lineage) in the epimastigote phase at a concentration of 4 × 107 cells/mL. SD showed higher antiparasitic activity (IC50 of 21.68 ± 1.6 μM) than BNZ alone (32.47 ± 4.9 μM), likely because P407 significantly improved BNZ solubility and consequently increased the rate of its release compared to the other preparations. According to the authors, this SD could be of great help in the treatment of Chagas disease, as it would allow for more effective treatment with fewer adverse effects.

In a study carried out by Eloy et al. [29], a SD has been developed with ursolic acid, a compound that helps to prevent mass loss and muscle atrophy, inhibits fat accumulation and possesses trypanocidal activity, acting as an important ally in treatment of Chagas disease. SD was prepared by the solvent method using ursolic acid, polyoxylglyceride (Gelucire 50/13) as carrier and silica dioxide as drying agent. In in vitro tests against the trypomastigote form obtained through infection of Swiss albino mice by *T. cruzi* (strain Y), SD showed an IC50 value (219.2 μM) 44.7% lower than that of the physical mixture. These data suggest that the surfactant interacted intermolecularly with ursolic acid in the SD, enhancing its trypanocidal activity.

3.2.2. Antischistosomal Activity of Solid Dispersions

Praziquantel (PZQ) is a drug widely used in the treatment of parasitic diseases, especially schistosomiasis, popularly known as "water belly" in Brazil, which is caused by contact of the parasite *Schistosoma mansoni* with the skin of the individual. Despite being the only drug in the treatment of flattened worms, PZQ has several limitations due to its low water solubility, which implies the need to administer it in high doses to patients [30,31].

Solid dispersions were developed by Perissutti et al. [30] to increase the solubility and bioavailability of this drug and then reduce its doses. SDs with PZQ were prepared using different carriers; however, the best results were obtained using PVP (Kollidon K30), PVP/VA, Kollidon-CL-M and sodium starch glycolate as polymers. Adult schistosomes were removed from female mice infected by the infectious form of *S. mansoni* (cercaria) and incubated. A ratio of 1:1 (*w*/*w*) Kollidon-CL-M:PZQ showed better performance in terms of solubility (642.54 mg/mL) and IC50 value (0.102 μg/mL) than pure PZQ (solubility of 140.30 mg/mL and IC50 of 0.165 μg/mL). According to the authors, PZQ-loaded SDs with Kollidon-CL-M could be a valid candidate for the treatment of schistosomiasis.

In another study performed by Albertini et al. [31] with the aim to improve the antiparasitic activity of PZQ, a SD with PVP K30 was prepared by the mechanochemical activation process and spray congealing technique. For in vitro testing of SDs, newly transformed schistosomula, obtained by mechanical transformation from *S. mansoni* and adult forms of the parasite were used. Adult *S. mansoni* individuals were removed from the hepatic portal and mesenteric veins of infected mice. The IC50 values of the physical mixture, pure PZQ and SD against schistosomula were 3.16, 2.58 and 2.4 μg/mL, respectively, while those of physical mixture and SD against adult parasites were 0.48 and <0.23 μg/mL. These results suggest that the SD contributed to the increase in PZQ solubility thanks to a 70% reduction of its crystallinity, thus reducing its dosage to a level suitable even for pediatric use.

3.2.3. Antimalarial Activity of Solid Dispersions

Malaria is an ancient disease dating back to the fifth century BC, but today it is one of the main public health problems in the world. It is estimated to affect around 200 million people worldwide, particularly individuals living in subtropical and tropical areas, causing 500,000 deaths each year [32]. It is caused by parasites belonging to the genus *Plasmodium*, of which the most aggressive species is *P. falciparum* that multiplies in the bloodstream, whereas *P. vivax* is milder affecting less than 1% of red blood cells and hardly leads individuals to death [33].

Artemether is a drug used in combination with lumefantrine to treat malaria. In the study proposed by Fule et al. [34], SDs were developed to improve the dissolution rate of this drug, as well as to increase its solubility in water. Solid dispersions were prepared by hot-melt extrusion using a combination of several polymers, such as Soluplus, PEG 400, Lutrol F127 and Lutrol F68 and sodium lauryl sulphate as surfactant. In in vitro studies, 15 artemether dispersions formulations were tested for antimalarial activity against the protozoan *P. falciparum* 3D7 (chloroquine-sensitive cell line) in schizont form and against ITG cells (chloroquine-resistant cell line). On average, the IC50 values of SDs, ranging from 0.054 to 0.081 ng/mL, were 39 times less than that of pure artemether and about 70 times less than that of chloroquine, the standard antimalarial drug. Among the formulations, the one formed by 1.0:3.8:0.2 drug, Soluplus and PEG 400, was the most promising with an IC50 value of 0.054 ng/mL.

Lumefantrine is a medicine used in association with artemether in a 6:1 ratio to treat malaria caused by *P. vivax* strains, either sensitive or resistant to chloroquine, as well as cerebral malaria caused by *P. falciparum*. However, it has low oral bioavailability due to its low aqueous solubility, which also affects its dissolution rate. To circumvent this problem and enhance its therapeutic potential, Fule et al. [35] prepared solid dispersions of lumefantrine by hot-melt extrusion using Soluplus, Kollidon VA64 and Plasdone S630 as polymers, which were tested in vitro for their antiparasitic activity against *P. falciparum* 3D7 and ITG cells. SDs showed IC50 values of 0.084–0.213 ng/mL, while pure lumefantrine and chloroquine showed values of 18.2 and 3.8 ng/mL, respectively, which demonstrates the remarkable increase in the antiparasitic activity of this drug induced by the SDs environment.

From the studies cited in this section aimed at antiparasitic activity, it can be concluded that SDs are a valid alternative in the treatment of some parasitic diseases such as Chagas disease, schistosomiasis and malaria. Solid dispersions containing drugs already used for these diseases have in fact shown an increase in their solubility and dissolution rate, as well as a consequent improvement in the activity against parasites, showing IC50 values lower than those of pure drugs. It should also be remembered that a dose reduction can reduce the incidence of toxicity during pharmacotherapy.

3.3. *Antimicrobial Activity of Solid Dispersions*

In a study conducted by Crucitti et al. [36], the antimicrobial activity of SDs of abietic acid and chitosan prepared in different drug:polymer proportions was checked against *Staphylococcus epidermidis*. Minimum inhibitory concentrations (MICs) of abietic acid, chitosan, 1:1 drug:chitosan mixture and SD containing this mixture were 0.8, 0.5, 1.0 and 0.25 mg/mL, respectively. According to the authors, not only the amorphous state of the drug in SD but also the synergistic effect between the two components may have improved the antimicrobial activity of the bioactive compound.

The low solubility of griseofulvin, which is one of the antimicrobials generally used in the treatment of dermatophytosis caused by the fungus *Trichophyton rubrum*, makes oral administration of its formulation particularly difficult, requiring high doses to reach therapeutic levels. To get around this problem, Al-Obaidi et al. [37] have successfully increased the solubility of this drug developing forms of solvates, also known as pseudopolymorphs, consisting of drug crystals with solvents, stabilized within polymeric matrices. To obtain solid dispersions, different solvated (chloroform) and unsolvated (methanol and acetone) materials were used, including microcrystalline cellulose, PVP K30 and hydroxypropyl methylcellulose acetate succinate (HPMCAS). Griseofulvin solubility in the HPMCAS-based SD (200 μg/mL) was almost seven-fold that of the pure drug. The antifungal potential of griseofulvin and this SD was tested in vitro against *T. rubrum* NCPF 935. Biofilm production and fungal proteolytic activity were significantly reduced in the presence of SDs incorporating either solvated or unsolvated materials when compared to the control. These results demonstrated the role of polymeric dispersion in increasing the antifungal activity of this drug.

Gatifloxacin is an antibiotic used to treat various eye infections such as conjunctivitis, keratitis and keratoconjunctivitis caused by bacteria [38]. The low solubility of this drug promoted the preparation of polymeric micelles incorporating it into Pluronic F127 as polymer by the solvent diffusion method [39]. Among the dispersions obtained, the one prepared at a 0.25:2.52 (*w/w*) drug:polymer ratio showed the best results in terms of particle size, stability and solubility. In the agar diffusion test conducted using *Staphylococcus aureus* as the target microorganism, this dispersion yielded an inhibition zone diameter (10.33 ± 0.57 mm) 18.3% larger than that produced by commercial gatifloxacin.

Among the numerous attempts to exploit the biological potential of CM, the results of studies aimed at the preparation of CM-containing SDs with antibacterial and pesticidal activity are summarized here. Hernandez-Patlan et al. [40] prepared CM-SDs using boric acid and PVP by the solvent evaporation method. In in vitro tests, 1% boric acid, 1% SD containing 10:90 (*w*/*w*) CM:PVP and a 50:50 (*w*/*w*) mixture of them were tested against *Salmonella enteritidis* in media simulating culture, intestine and proventriculus. Boric acid led to $\log_{10}$ CFU/mL values of 7.78 ± 00 in culture, 3.81 ± 0.05 in proventriculus and even 0.00 ± 0.00 in intestine, while 1% CM:PVP of 7.73 ± 00, 3.91 ± 0.08 and 5.00 ± 0.55 and their mixture of 7.65 ± 00, 3.71 ± 0.05 and 3.94 ± 0.23, respectively. In summary, the promising results shown by the SD containing CM and PVP were even improved in the presence of boric acid, thanks to its ability to penetrate the cell membrane of the target microorganism.

Solid dispersions were also prepared to enhance the efficacy of CM in antimicrobial photodynamic therapy. Bearing in mind that CM as a photosensitizer has low solubility and phototoxicity limited to Gram-negative bacteria, Hegge et al. [41] prepared SDs using concentrated solutions of hydroxypropyl-β-cyclodextrin/CM by solvent evaporation and then HPMC by lyophilization. The antibacterial activity was tested against *Escherichia coli* (ATCC 25922) (1 × 106 CFU/mL) at concentrations of 10 and 25 μM of CM in SDs exposed to irradiation. The dispersion that showed the best results in dissolution tests was then used in in vitro tests. Since it displayed phototoxicity at both concentrations (10 and 25 μM) at a radiation dose of 5 J/cm^2, one can infer that the dispersion increased the bactericidal power of CM against *E. coli*, so it could play an important role in photodynamic therapy.

Another study was conducted by Alves et al. [42] with the aim of developing a formulation with antimicrobial activity using CM and silver nanoparticles, as both have activity against bacteria, fungi and viruses. For this purpose, CM-containing SDs were prepared by co-precipitation using CM and P407 at 1:1, 1:2 and 2:1 (*w*/*w*) ratios and then silver nanoparticles were added. Nanoparticles were synthesized and stabilized using polyvinyl alcohol and PVP as polymers and sodium citrate. The antibacterial activity was determined in in vitro tests and expressed as MIC against *E. coli* ATCC 25922, *Pseudomonas aeruginosa* ATCC 9721 and *S. aureus* ATCC 10390. The SD composed of P407/CM/nanoparticles/PVP showed better MIC values against *P. aeruginosa* (50 μg/mL) and *E. coli* (25 μg/mL) compared to those reported in the literature (175 and 163 μg/mL, respectively) [43]. All formulations were less effective against *S. aureus* (Gram-positive) and the best performance was observed for the P407/CM/nanoparticles/citrate system. The different SD activity against Gram-positive and Gram-negative bacteria was probably due to the thinner cell wall of the latter, which may have facilitated SD internalization and cell wall disruption.

In this context, DSs with drugs and antimicrobials showed good results including increased solubility, low MIC values and reduction of bacterial colonies. Although there are currently many antimicrobial drug options on the market, they are still being used indiscriminately leading to the emergence of resistant bacteria as well as other adverse effects. Therefore, SDs not only can be an option to improve their solubility, but also require lower dosages, making antimicrobial treatment safer and more effective.

3.4. Antioxidant Activity of Solid Dispersions

Quercetin is a flavonoid that has, among other activities, the ability to capture free radicals. In the study by Costa et al. [44], a solid dispersion of quercetin with PVP K25 was developed with the aim of increasing its solubility and antioxidant activity determined by the 2,2-diphenyl-1-picryl-hydrazyl-hydrate (DPPH) radical assay and expressed as IC50. IC50 values of 0.61 ± 0.03 and 1.00 ± 0.02 μg/mL were obtained for SD dissolved in water and for quercetin dissolved in methanol, respectively, which demonstrates that SD enhanced quercetin antioxidant activity.

The SDs prepared by Crucitti et al. [36] using 1:1 abietic acid:chitosan already discussed in the previous section were also tested for their antioxidant activity with the same method. As expected, the antioxidant activity of chitosan (IC50 of approximately 11 mg/mL) was much lower than that of abietic acid (IC50 of 1.65 mg/mL), but the one of the 1:1 AB/chitosan SD was even stronger (IC50 of 0.61 mg/mL). There are at least two factors that can explain the particularly high antioxidant activity of SDs. First, the amorphous state may have influenced the solubility and availability of the drug compared to its crystalline nature. Secondly, the natural tendency of abietic acid to react with oxygen forming peroxides may have been reduced by its stabilization due to interaction with chitosan.

The previously mentioned CM-SD prepared by Sá et al. [16] using P407 was investigated for possible effect on the activities of acetylcholinesterase (AChE), butyrylcholinesterase (BChE), glutathione S-transferase (GST) and two isoforms (A and B) of monoamine oxidase (MAO). The SD was able to inhibit the activities of AChE, BChE and GST in aqueous medium, which suggests that the hydrophobic nature of CM was overcome by the SD amorphous state. In addition, it inhibited both MAO isoforms at a concentration of 100 μM, while pure CM had no effect at the same concentration, suggesting that CM interaction with P407 in SD was responsible for enzyme inhibition. It is known that the reaction catalyzed by MAO produces hydrogen peroxide, a source of hydroxyl radicals, so MAO inhibitors, as well as this SD, may be usefully exploited to treat pathologies associated with oxidative stress.

Coenzyme Q10 has two main functions, namely the transport of mitochondrial electrons to form phosphate for muscle contraction and the antioxidant activity of its reduced form (ubiquinol). However, its bioavailability is low and variable because of its poor solubility and high molecular weight. A study performed by Ge et al. [45] aimed to investigate the therapeutic potential of water-soluble nanoscale Coenzyme Q10-SDs with mannitol prepared by high-pressure homogenization. To assess the antioxidant activity, intracellular reactive oxygen species (ROS) were detected using myocardial cells divided into three groups treated with SDs with different particle sizes, namely 408, 300 and 146 nm and a control. Myocardial cells were subjected to hypoxia/reoxygenation to detect the free-radical scavenging ability of substances. In vitro tests showed that the ROS level in the control group increased by 96%, while those in the three other groups were reduced by 48.3–87.3%, which proved the strong antioxidant activity of SDs.

To increase water solubility of usnic acid, a secondary metabolite of lichen with antioxidant activity, Fitriani et al. [46] developed SDs with PVP K30 by spray-drying and freeze-drying and determined their antioxidant activity by the DPPH assay. Since the drug and the developed systems showed IC50 values in the range 50–100 ppm, they can be classified as moderate antioxidants, with the lowest value (63.87 μg/mL) detected for the SD prepared by freeze-drying in the proportion 1:2 (usnic acid:polymer).

Luteolin is a compound belonging to the class of biflavonoids, which is found in several plants and has a rich arsenal of therapeutic activities; however, its low solubility and instability in water pose a challenge for its use in pharmaceutical formulations. Therefore, a study promoted the production of luteolin-containing SDs by solvent evaporation, microwave irradiation or fusion using PEG 4000 as polymer [47]. After dissolving the drug (control) and the three different SDs in dimethyl sulfoxide up to a concentration of 5 mg/mL, their antioxidant activity was assessed by the DPPH assay. The antioxidant activity of the drug was 61.12 ± 4.11% and those of the above SDs 61.12 ± 4.11, 94.14 ± 6.11, 88.55 ± 3.98 and 76.23 ± 5.12%, respectively, suggesting that this property increased due to improved solubilization of luteolin in the presence of PEG 4000 as carrier.

In summary, the antioxidant activity of SDs was quantified by different methods including the activities of some enzymes, but with a certain prevalence of the DPPH assay. In the above studies aimed at therapeutic purposes, the best results have always been obtained with SDs, probably thanks to the increase in the solubility of drugs in the administration systems.

3.5. Anti-Inflammatory Activity of Solid Dispersions

α,β-Amyrin (ABAM) is a natural blend of pentacyclic triterpenes that has shown numerous pharmacological properties, including the anti-inflammatory activity [7]. To increase this activity, ABAM-SDs, prepared using PVP K30, PEG 6000 and HPMC by physical mixture, kneading and rotary evaporation, were submitted to the in vitro anti-inflammatory test based on nitric oxide (NO) production by lipopolysaccharide (LPS)-stimulated macrophages J774. The results showed that ABAM inhibited NO production at concentration of 20 μg/mL, although the effect of SDs was stronger than those of the control and the drug alone. Specifically, the ABAM-PVP SD prepared by the kneading method had the greatest effect with 62.49% inhibition, while ABAM showed a value of 44.55%. According to the authors, SDs prepared with hydrophilic polymers were effective in increasing the solubility of ABAM and allowed to enhance its anti-inflammatory activity.

In the cited study of Sá et al. [16], CM-SD was tested for anti-inflammatory activity via inhibition of NO production by LPS-stimulated murine macrophages (RAW 264.7). The SD was able to improve the affinity of water for CM which, in turn, exerted cytotoxic effects in the aqueous medium, with an IC50 value > 400 μg/mL. This result suggests possible SD administration in vivo, avoiding the use of toxic solvents to increase its clinical efficacy.

These studies as a whole demonstrated that the use of SDs containing low water-soluble active ingredients, both natural and synthetic, prepared with different methods, is an effective means to increase their anti-inflammatory activity, which can be evaluated in vitro by quantifying the inhibition of NO production by LPS-stimulated macrophages.

3.6. Cytoprotective Activity on Liver Cells

SDs of CM prepared using HPMC as the polymer have been reported by Shin et al. [48], who studied their possible hepatoprotective action in HepG2 cells treated with tert-butyl hydroperoxide (t-BHP) as a hepatotoxic agent. The survival rate of cells, which in the presence of t-BHP was about 40%, significantly increased after treatment with DSs-CM at concentrations of 10, 20 and 40 μg/mL compared to the treatment with CM alone at equivalent concentrations (2, 4 and 8 μg/mL, respectively). Additionally, the extracellular activity of lactate dehydrogenase, an enzyme released by HepG2 cells under the stress conditions induced by t-BHP, was reduced by treatment with CM-containing SDs more than by that with CM alone, indicating a stronger cytoprotective activity. Treatment with SDs containing 10 and 20 μg/mL CM was also able to inhibit, more effectively than CM alone, the activation of caspase-7, -8 and poly (ADP-ribose) polymerase, responsible for t-BHP-induced apoptosis in HepG2 cells.

In summary, given the hepatoprotective activity of CM-containing SDs, their use can be considered a valid alternative to other therapies, considering that the administration of drugs for long periods and in large doses can cause liver damage, especially in elderly patients. In this sense, further in vitro and in vivo studies on DSs of CM and other bioactive compounds are needed, given the hepatoprotective potential associated with this drug delivery system.

Table 1. In vitro studies on solid dispersions.

Carrier Type	Substance	Cell Type	Activity	Improved Characteristics	Reference
OHPP	Niclosamide	PC-3, HeLa, A549	Anticancer	SDs showed higher cytotoxicity to target cells (lower IC50) than the niclosamide solution.	[11]
OHPP	Paclitaxel	PC-3, HeLa, A549	Anticancer	SDs showed significantly higher cytotoxicity to target cells (lower IC50) than the paclitaxel solution.	[13]
PVP/VA TPGS	Paclitaxel	BT-474, MCF-7, SK-BR-3	Anticancer	SDs showed higher cytotoxicity against cancer cells compared to the pure drug.	[14]
Brij®L4	Chrysin	HT29	Anticancer	The higher solubility of chrysin in SDs compared to water solution increased cytotoxicity.	[15]
Poloxamer 407	Curcumin (CM)	NCIH460, HeLa, HepG2, MCF-7 and PLP2	Anticancer	SDs showed cytotoxicity against all tumor cell lines tested, but no toxic effects on non-tumor cells.	[16]
		AChE, BChE, GST, MAO A-B	Enzyme inhibitory /Antioxidant	SD was able to inhibit the activities of AChE, BChE and GST in aqueous medium.	
		LPS-stimulated murine macrophages (RAW 264.7)	Anti-inflammatory	IC50 (inhibitory concentration of 50% NO production by macrophages) > 400 µg/mL.	
PVP K30	Zn(II)-curcumin complex	HepG2, SK-HEP1	Anticancer	SD of Zn(II)-curcumin complex had a potent anticancer effect.	[17]
HPMC, PVP K30, PEG 6000	Telaprevir	HepG2	Anticancer	The antitumor activity was dose dependent and even with the addition of the polymer the drug maintained its efficacy.	[18]
Soluplus®	Angelica gigas Nakai	HeLa, HEK 293	Anticancer	SD at the concentration of 200 µg/mL showed a significant decrease (to only 17.37%) in cell viability. There was no toxicity to normal cells.	[20]
Eudragit S-100	Berberine hydrochloride (HB)	SW480, HCT116, Caco-2	Anticancer	The release of HB from SDs was effective and cell viability was reduced in a dose and time dependent manner.	[21]
PVP K30	IIIM-290	Ehrlich ascites carcinoma cells	Cytotoxic	Despite the reduced amount of IIIM-290 in SD, the IC50 value of SD was lower than that of IIIM-290 alone.	[24]
Poloxamer 407	Benznidazole	Epimastigotes of *Trypanosoma cruzi*	Antichagasic	SDs enhanced drug solubility, release kinetics and parasitic activity	[25]
Low-substituted HPC	Benznidazole	Epimastigotes and intracellular amastigotes of *T. cruzi* (CL-B5)	Antichagasic	SDs had higher antiparasitic activity against amastigotes than epimastigotes.	[28]
Gelucire 50/13	Ursolic acid	Trypomastigotes of *T. cruzi* Y	Antichagasic	Increased antiparasitic activity.	[29]

Table 1. *Cont.*

Carrier Type	Substance	Cell Type	Activity	Improved Characteristics	Reference
PVP K30,PVP/VA, Kollidon-CL-M, sodium starch glycolate	Praziquantel	Adult Schistosomes of *Schistosoma mansoni*	Antischistosomal	Increased solubility, better bioavailability and stronger antiparasitic activity.	[30]
PVP K30	Praziquantel	Newly transformed schistosomula of *S. mansoni* and adults	Antischistosomal	Increased solubility, reduced dosage especially for children and increased antiparasitic activity.	[31]
Soluplus, PEG 400, Lutrol F127 and Lutrol F68	Artemether	Schizonts of *Plasmodium falciparum* 3D7	Antimalarial	Increased dissolution rate, amorphous form, increased solubility and, mainly, increased antimalarial activity.	[34]
Soluplus, Kollidon VA64, Plasdone S630	Lumefantrine	ITG cells	Antimalarial	Increased antiparasitic activity.	[35]
Chitosan	Abietic acid	*Staphylococcus epidermidis*	Antimicrobial	SD exhibited better MIC values against *S. epidermidis* than chitosan and abietic acid alone.	[36]
		DPPH radical scavenging	Antioxidant	SD had higher antioxidant power (IC50 of 0.61 mg/mL) than abietic acid alone (IC50 of 11 mg/mL).	
PVP K30 and HPMCAS	Griseofulvin	Dermatophytes of *Trichophyton rubrum* NCPF 935	Antimicrobial	SDs significantly reduced biofilm formation when compared to the control.	[37]
Pluronic F127	Gatifloxacin	*Staphylococcus aureus*	Antimicrobial	The gatifloxacin/Pluronic F127 system exhibited antimicrobial efficacy when compared to commercialized eye drops.	[39]
PVP K30	Curcumin	*Salmonella enteritidis*	Antimicrobial	SD had a strong antimicrobial effect on *S. enteritidis*, while CM alone did not show antimicrobial activity in vitro.	[40]
HPMC	Curcumin	*Escherichia coli*	Antimicrobial	SD used to prepare phototoxic supersaturated solutions showed significant bactericidal activity against *E. coli*.	[41]
Polaxamer 407	Curcumin	*E. coli, Pseudomonas aeruginosa* and *S. aureus*	Antimicrobial	The association between SD and silver nanoparticles increased CM antimicrobial and antioxidant activities.	[42]
		DPPH radical scavenging	Antioxidant		
PVP K25	Quercetin	DPPH radical scavenging	Antioxidant	Increased quercetin antioxidant activity in SD (0.61 ± 0.03 ≤ IC50 ≤ 1.00 ± 0.02 μg/mL).	[44]
Mannitol	Coenzyme Q10	Intracellular ROS level	Antioxidant	The SD with the smallest particle size showed the greatest absorption of UVB radiation as well as the highest antioxidant activity in vitro.	[45]

Table 1. *Cont.*

Carrier Type	Substance	Cell Type	Activity	Improved Characteristics	Reference
PVP K30	Usnic acid	DPPH radical scavenging	Antioxidant	Increased usnic acid solubility and antioxidant activity.	[46]
PEG 4000	Luteolin	DPPH	Antioxidant	Polymers increased luteolin solubility and antioxidant activity.	[47]
PVP K30, PEG 6000 and HPMC	α,β-Amyrin	LPS-stimulated macrophages J774	Anti-inflammatory	SDs enhanced the anti-inflammatory activity of α,β-amyrin.	[7]
HPMC	Curcumin	HepG2	Cytoprotective	SDs showed better cytoprotective activity than pure CM and inhibited cell death induced by *t*-BHP.	[48]

OHPP, Octenylsuccinate hydroxypropyl phytoglycogen; IC50, Half inhibitory concentration; PVP/VA, Polyvinylpyrrolidone/vinyl acetate; TPGS, D-α-toco-pheryl polyethylene glycol-1000-succinate; PVP, Polyvinylpyrrolidone; SD, Solid dispersion; AChE: Acetylcholinesterase; BChE, Butyrylcholinesterase; CM, Curcumin; GST, Glutathione S-transferase; MAO, Monoamine oxidase; LPS, Lipopolysaccharide; NO, Nitric oxide; PEG, Polyethylene glycol; HPMC, Hydroxypropyl methylcellulose; ITG (chloroquine-resistant cell line); DPPH, 2,2-diphenyl-1-picryl-hydrazyl-hydrate; HPMCAS, Hydroxypropyl methylcellulose acetate succinate; MIC, Minimum inhibitory concentration; ROS, Reactive oxygen species; t-BHP, *tert*-Butylhydroperoxide.

4. In Vivo Studies on Solid Dispersions in Polymeric Matrices

As already mentioned, solid dispersions (SDs) have been used as a strategy in pharmaceutical technology to circumvent some limitations presented by drugs and new bioactive compounds such as low solubility and bioavailability. In this sense, this section addresses in vivo studies on SDs with different biological activities, as shown in Figure 3 and quantitatively expressed in Figure 4. The main information about these studies is summarized in Table 2.

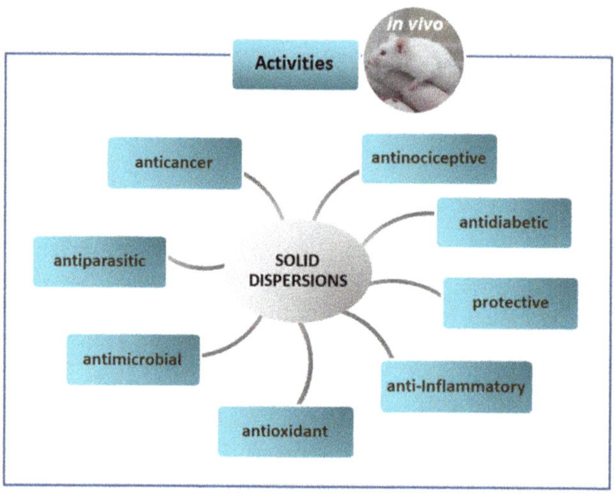

Figure 3. In vivo activities of solid dispersions.

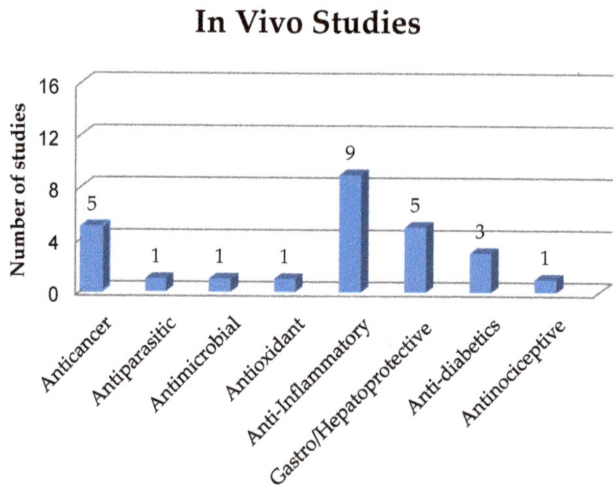

Figure 4. Quantification and classification of in vivo studies on solid dispersions published in the period from 2009 to 2020.

4.1. *Anticancer Activity of Solid Dispersions*

As previously mentioned, the low solubility of many active compounds complicates oral formulations requiring high doses. Therefore, SDs have been prepared to increase their solubility and, consequently, their therapeutic effects. Most SDs that were successful in in vitro experiments described previously in Section 3 were also assessed in vivo assays, whose results are summarized and critically discussed here together with those of other studies.

The IIIM-290-containing SDs prepared by Kumar et al. [24] with the highest solubility were assessed for in vivo anticancer activity on mice with Ehrlich ascites carcinoma. The animals were divided into four groups to receive treatment and two controls. Group I received only IIIM-290 (50 mg/kg), while groups II, III and IV received SD consisting of PVP K30 and drug at doses of 25, 50 and 75 mg/kg, respectively. While the group that received the drug alone showed a tumor inhibition rate of only 24.75%, the SD reduced tumor growth by 31–43% and ensured better performance even at a half dosage of drug.

To improve the solubility and permeability of 9-nitrocamptothecin, a topoisomerase I inhibitor and potent chemotherapy drug to treat pancreatic cancer, a SD was developed using Soluplus by lyophilization and its antitumor activity was in vivo assessed in mice with sarcoma 180 [49]. SD showed higher tumor growth inhibitory rate (approximately 80%) than the drug alone due to improved oral bioavailability.

(−)-Oleocanthal, a phenolic compound derived from olive oil with anti-inflammatory, antitumor and anti-Alzheimer activities, exerts an irritating action in the oropharyngeal region that limits its use in oral formulations. To improve its taste and dissolution, Qusa et al. [50] developed by the hot-melting method a SD containing 1:7 oleocanthal:xylitol, whose antitumor activity was tested in vivo against human triple-negative breast cancer cells (MDA-MB-231) xenografted in mice. SD administration resulted in a two-day delay and 49% tumor growth rate when compared with the group treated only with placebo. Additionally, in the recurrence test the SD-treated group showed recurrence-free survival in two days, with only two of the five animals suffering from recurrent but smaller tumors, while in the control group a 100% tumor reappearance rate was observed.

Stimulated by the promising results obtained in vitro, Wu et al. [17] also tested in vivo the antitumor effect of Zn(II)-curcumin (CM) SDs with PVP K30. Hepatic xenografts of H22 and HepG2 cells were performed in male BALB/c mice and female B-NDG mice, respectively. BALB/c mice were divided into three groups, in which each received a dose of PVP K30 (700 mg/kg), or CM-SD or Zn-CM-SD (equivalent doses of 100 mg/kg CM and ZnCM), while B-NDG mice were treated with the same formulations except CM-SD at the same doses. Using H22 cells, while the untreated group had tumors weighing over 3 g and volume of 2000 mm^3, tumors in animals treated with Zn-CM-SD weighed less than 2 g and had a volume of less than 1000 mm^3. In addition, greater extension of necrotic tissue but no reduction in animal body weight was reported.

Selaginella doederleinii Hieron is a widely distributed medicinal herb in southern China that has traditionally been used as a folk medicine for cancer, especially lung cancer. However, its total biflavonoid extract (TBESD) has low water solubility, gastrointestinal permeability and oral absorption that limit its use in therapeutic formulations [51,52].

To overcome these difficulties, Chen et al. [53] prepared a solid dispersion of TBESD with the polymer PVP K30 (TBESD-SD) by the solvent evaporation method. To assess the in vivo antitumor effect, mice with xenograft of A549 cells derived from human alveolar basal epithelial cell adenocarcinoma were treated with doses of 200 mg/kg TBSED and TBESD-SD and 2 mg/kg doxorubicin. Tumor growth inhibition rates in the TBESD, TBESD-SD and pure drug groups were 29.48, 46 and 58.44%, respectively and the respective reduction rates of tumor angiogenesis 24.30, 52.20 and 59.91%. However, the drug group, unlike the other two, showed significant body weight loss. Therefore, the dispersion developed in this study proved to be a promising formulation for the treatment of tumors.

In summary, it can be concluded, despite the limited number of studies concerning the in vivo antitumor activity, that the use of SDs allowed a higher rate of tumor inhibition than the isolated compounds, also corroborating the results obtained in vitro in other studies.

4.2. *Antiparasitic Activity of Solid Dispersions*

The Benznidazole (BNZ)-containing SDs developed by Fonseca-Berzal et al. [28] using sodium deoxycholate (NaDC) as surfactant and low-substituted hydroxypropyl cellulose (HPC) as polymer were also tested in vivo for trypanocidal activity. For this purpose, parasites of *Trypanosoma cruzi* (strain Y) isolated from human individuals in the acute phase of the disease were used in their trypomastigote form to infect mice, which were administered at a 25 mg/kg/day oral dosage. The percentage reductions in the area under the parasitic curve were 91.93, 96.65 and 95.27% administering SD with 1:3 BNZ:NaDC, SD with 1:3 BNZ-HPC and BNZ alone, with the best anti-interference performance obtained for the second formulation, while the survival rate of in the control group was only 44.44%. None of the mice treated with SDs or BNZ alone died, which indicates no toxic effects at the established dosage. Although these SDs have shown great potential in in vivo tests, further studies are needed because BNZ has a reduced pharmacological activity in the chronic phase of Chagas disease and has several side effects.

4.3. *Antimicrobial Activity of Solid Dispersions*

In the study developed by Hernandez-Patlan et al. [40], the antibacterial activity of CM-loaded SDs prepared using PVP K30 and boric acid was evaluated against *Salmonella enteritidis*. In vivo tests were performed to determine the effects of (a) SD with 10/90 (*w*/*w*) CM/PVP, (b) boric acid and (c) 50/50 (*w*/*w*) CM/PVP plus boric acid administered orally to broiler chickens treated with 10^7 CFU/mL *S. enteritidis* each. While boric acid had no significant effect, the CM/PVP plus boric acid triad significantly reduced the number of colonies in the intestine of animals compared to CM/PVP, revealing a synergistic effect. Since such a reduction occurred only in the above CM/PVP proportion in SD, it can be concluded that this effect was the result of the increase in CM solubility. Although this in vivo antimicrobial activity of CM-SDs would allow for equivalent or even greater therapeutic effects than the drug alone in lower doses, further studies involving other species of target microorganisms and animals are needed.

4.4. *Antioxidant Activity of Solid Dispersions*

Some studies have revealed that zinc plays an important role in maintaining redox balance as well as protecting against oxidative stress and damage induced by ethanol, a compound capable of reducing the defenses of the gastric mucosa and inducing gastric ulcers [54,55]. Taurine is an amino acid that participates in numerous physiological reactions and can exert a cytoprotective effect against lesions of the stomach mucosa [56] thanks to its antioxidant activity [57].

In this sense, SDs of the taurine-zinc complex with the PVP K30 were prepared by Yu et al. [58] in the 1:6 (*w*/*w*) proportion by spray drying. In the in vivo study, the gastric mucosa of rats with ethanol-induced lesions was examined. The treatment with SD taurine-zinc (200 mg/kg) led to a 46% increase in the superoxide dismutase (SOD) activity and a 3.9-fold increase in the level of glutathione in the gastric mucosa compared to the model group (28.40 U/mg protein and 0.81 ± 0.56 mg/g protein, respectively), which suggests a direct relationship between endogenous level of this antioxidant and SOD activity to be exploited for better therapeutic response. Unlike in vitro activity, we found in the databases only this study published on in vivo antioxidant activity of SDs, which highlights the need for further research in this field.

4.5. *Anti-Inflammatory Activity of Solid Dispersions*

In an anti-inflammatory study conducted by Gou et al. [59], a solid dispersion of triacetylated andrographolide (TA) with Kollidon (VA64) had its in vivo anti-inflammatory activity assessed in mice ulcerative colitis model through parameters such as length, weight and levels of inflammatory cytokines (IL-2, IL-6) and SOD activity in the colon. Reduced length and increased weight of the colon indicated severe edema in the model group and in the group administered with TA suspension. Significant differences were also observed in these parameters between the model/crude TA group and the groups treated with TA-SD at different doses. While high a dose (200 mg/kg/day) of TA-SD remarkably reduced colon weight, low (50 mg/kg/day), or medium (100 mg/kg/day) doses showed a therapeutic effect against ulcerative colitis comparable to that of andrographolide pills (commercial anti-inflammatory drug). In addition, a reduction in the levels of IL-2 and IL-6 cytokines was observed in both the TA-SD and pill groups compared to the model group and the crude TA group. Finally, SOD activity significantly decreased in the model group compared to the normal one, while administration of TA-SD at medium and high doses caused an increase in SOD activity comparable to that induced by andrographolide pills. These results taken together demonstrated the anti-inflammatory effect of TA-SD.

Different solid dispersions of curcumin with PVP K30, Poloxamer 188 or 2-hydroxypropyl-β-cyclodextrin (HP-β-CD) were prepared by the solvent evaporation method and their anti-inflammatory activity was tested in mouse ear edema model [60]. The histological cuts of the ears revealed a decrease in the number of inflammatory cells infiltrated when mice were submitted to treatment with curcumin and CM-PVP. The results of ELISA assay showed a stronger anti-inflammatory activity of SDs compared to pure CM, revealed by a reduction in the expression of matrix metallopeptidase 9 as well as the levels of IL-1β and IL-6 cytokines.

Still regarding curcumin, Teixeira et al. [61], with the aim of improving its solubility, prepared atomized ternary SDs using Gelucire®50/13-Aerosil®, which provided important results in terms of anti-inflammatory activity in vivo. Administration of SDs containing 40% CM in a dose of 100 mg/kg showed a greater capacity to reduce rat paw edema than indomethacin (5 mg/kg). This anti-inflammatory activity, which appeared to be dose-dependent in the 10–100 mg/kg range, was ascribed to the increase in gastrointestinal absorption of SDs observed in the study.

In another research, Chuah et al. [62] investigated the bioavailability and bio-efficacy of CM in a solid dispersion prepared using HPMC, lecithin and isomalt by hot-melt extrusion for possible application in food products. To test the bio-efficacy of CM-SD on male Sprague-Dawley rats, they used the in vivo LPS-induced lung inflammation model and a dose of 5 mg/kg in 0.25% carboxymethylcellulose as vehicle administered for 21 days by oral gavage. Administration of LPS resulted in acute inflammatory lung injury and septic shock. Because the bioavailability of CM was increased by more than 10 times in solid dispersion, its efficacy was tested at a concentration 10 times lower than that of unformulated CM. While the treatment with CM alone at 50 mg/kg for 21 days reduced the leukocyte count marginally, that with CM-SD at only 5 mg of CM/kg significantly reduced the counts of both bronchoalveolar lavage fluid leukocytes and neutrophils. Damages in the lung tissues were less pronounced in CM-SD treated rats compared to those treated with LPS or CM alone. CM-SD, but not unformulated CM, was also able to downregulate the levels of IL-1, IL-6 and IL-8 inflammatory cytokines significantly. In summary, despite a 10 times lower dose, CM-SD was responsible for greater anti-inflammatory activity than CM alone. This study, which is the first report on CM-SDs suitable for food, demonstrated that the use of CM in solid dispersion in significantly low doses does not affect its efficacy and may improve color, taste and smell of foods.

Aceclofenac, chemically [2-(2′,6′-dichlorophenyl)amino] phenylacetoxyacetic acid, is used as a non-steroidal anti-inflammatory drug in the symptomatic treatment of pain and inflammation. However, like other drugs of this class, its oral administration is associated with gastrointestinal side effects like gastric ulceration, gastrointestinal bleeding and liver and kidney trouble. For these reasons, aceclofenac is increasingly administered by topical route. In this context, Carbopol gels offer a good alternative to oil-based topical formulations [63] and crospovidone has been used as a carrier of SDs of various drugs to improve their aqueous solubility [64]. In the study performed by Jana et al. [65], an attempt was made to develop Carbopol 940 gel for topical application containing SDs of aceclofenac using crospovidone as carrier to improve the skin permeation profile of this drug. The aceclofenac-crospovidone (1:4) SD was prepared by the solvent evaporation technique and tested in vivo for anti-inflammatory activity on male Sprague-Dawley rats using carrageenan-induced rat-paw edema model. The results obtained applying 1 g of optimized gel on the skin (1 cm^2) back of rats showed that SDs increased the rate of permeation of the drug, increased the intensity of response and gave results comparable with those of a marketed gel without leading to any skin irritation.

Another non-steroidal anti-inflammatory drug widely used in the treatment of mild-to-moderate pain and fever is ibuprofen ([2RS]-1[4-(2-methyl propyl) phenyl] propionic acid). Ofokansi et al. [66] tested in vivo the anti-inflammatory activity of ibuprofen-loaded PEG 8000 SDs prepared by the fusion method. The paw edema test was applied on Wister rats by orally administering ibuprofen-SD doses equivalent to 20 mg/kg body weight, while the reference group received 20 mg/kg of pure drug and the control one 1% (*w*/*v*) PEG 8000 suspension used as vehicle. Thirty minutes after treatment, paw edema was induced by injection of egg albumin. All SDs showed good anti-inflammatory properties being able to inhibit up to 90% of edema after 6 h versus only 77% of inhibition observed with the drug alone. This study clearly demonstrated that the use of hydrophilic carriers such as PEG 8000 can improve the drug pharmacological activity.

To understand and improve the dissolution behavior of poorly water-soluble drugs from SDs, flurbiprofen, a phenylalkanoic acid derivative belonging to the non-steroidal anti-inflammatory drugs, was used as a model [67]. SDs were prepared by the fusion method using urea and mannitol as hydrophilic carriers. Microspheres were also prepared by the solvent evaporation method using Eudragit L-100 (EL100) and Eudragit RS PO (ERS) as rate-controlling polymers. Flurbiprofen:urea-SDs and microspheres were used to develop controlled-release formulations by mixing them in different proportions. In in vivo anti-inflammatory tests, the pure drug and formulations were administered orally at a dose of 11.69 mg/kg body weight. The formulation that showed the best inhibition of rat paw edema up to 16 h was identified as a suitable product for controlled drug delivery.

Still within this context, meloxicam is a potent nonsteroidal anti-inflammatory drug belonging to the selective cyclo-oxygenase II inhibitors with the advantage of being safer in the gastrointestinal tract; however, it is poorly water-soluble and highly permeable (class II). A study by Al-Remawi et al. [68] aimed to improve meloxicam solubility and maximizing its pharmacological activity by preparing, by rotary evaporation and freeze drying, binary SDs with paracetamol, a highly water-soluble and poorly permeable drug. The anti-inflammatory effect was also investigated by measuring the changes in tail volume of edema induced by carrageenan injection. The formulation prepared by lyophilization using ethanol as a solvent and a meloxicam:paracetamol ratio of 1:10 (*w*/*w*) showed a decrease of more than 50% in the volume of carrageenan-induced tail edema compared to the physical mixture.

Chelerythrine (CHE), a quaternary benzo[c]phenanthridine alkaloid widely distributed in the Papaveraceae and Rutaceae families of plants, is a medicine officially listed in the Chinese Pharmacopoeia. Although this drug has shown a wide range of biological activities, its use in formulations is hampered by its poor water solubility. With the aim of improving CHE solubility and bioavailability, Li et al. [69] prepared solid dispersions with PVP K30 by the solvent evaporation method. Their anti-inflammatory activity was assessed in LPS-induced endotoxic shock experiments after single dose administration of dispersions (CHE-SDs) to mice in comparison with the CHE solution. CHE-SDs (10 mg/kg) significantly improved the bioavailability and anti-inflammatory activity of CHE by reducing the levels of TNF-α, IL-6 and NO in mice serum.

In summary, most in vivo studies on the anti-inflammatory activity used the paw edema test in rats, which revealed a significant increase in the biological activity of drugs in SDs compared to pure drugs, reducing lesion size and inflammatory cytokines production. These results corroborate those of tests often carried out by the same authors on solubility, bioavailability and physicochemical properties of drugs, which are directly related to the pharmacokinetic characteristics of SDs.

4.6. Gastro and Hepatoprotective Activity of Solid Dispersions

Among the activities and effects reported in the literature for SDs, the gastroprotective and hepatoprotective ones were also found.

Regarding the gastroprotective effect, a dispersion of the zinc taurine complex with PVP K30, already mentioned in the section on antioxidant activity, was also tested in vivo for its gastroprotective effect in experiments conducted on rats with ethanol-induced lesions in the gastric mucosa [58]. The gastric mucosa pretreated with taurine-zinc SD had an almost normal appearance, while that of the untreated group exhibited severe hemorrhagic ulcers. In addition, administration of SD reduced the ulcer index from 124.63 ± 26.72 in the untreated group to no more than 13.25 ± 4.77 in that treated with 200 mg/kg of SD. The gastroprotective effect of SDs was confirmed by a reduction in serum NO levels and a significant increase in gastric prostaglandin E2 compared to the untreated group.

Instead, Shin et al. [48] tested in vivo the hepatoprotective activity of CM-containing SDs with HPMC against tert-butyl hydroperoxide (t-BHP)-treated mice by administering 80 mg/kg of CM (positive control) or an equivalent dose of SD (400 mg/kg). Levels of alanine aminotransferase (ALT) and aspartate aminotransferase (AST), which indicate liver damage in response to t-BHP, were reduced by SD, while CM alone had no statistically significant effect. In the same study, the hepatoprotective effects of dispersion in liver tissues of mice were also evaluated. The liver section of the CM-administered group showed a slightly reduced necrotic region and decreased cellular infiltration, whereas these effects were much more marked in the SD-administered groups (200 and 400 mg/kg), with consequent improvement of the histological damage caused to the liver.

Solid dispersions of nobiletin and hydroxypropyl cellulose were also developed with the aim of improving drug solubility and bioavailability, as well as detecting possible hepatoprotective properties. Nobiletin is a polymethoxylated flavone with pharmacological activities such as anti-inflammatory, anti-apoptotic, anti-dementia, anti-oxidative and anti-tumor activities. The method used by Onoue et al. [70] to prepare nobiletin-SD was wet-milling followed by freeze drying. The evaluation of hepatoprotective activity was conducted in a rat model with acute liver damage induced by carbon tetrachloride (CCl_4). Administration of 2 mg/kg of SD led to a significant decrease in liver damage, evidenced by 60 and 43% reductions in ALT and AST levels and to greater hepatoprotective efficacy compared to the crystalline drug.

Silymarin is an extract of *Silybum marianum* seeds, used in the treatment of liver diseases, which however has low solubility in water. SDs and spherical agglomerates of silymarin were developed by Balata and Shamrool [71] to investigate their therapeutic effects on liver and kidney diseases. SDs with PVP K30 were prepared by the solvent evaporation method in 1:2, 1:4 and 1:8 (*w*/*w*) extract:polymer proportions. In vivo liver tests were conducted on albino rats with CCl_4-induced hepatotoxicity. The group treated with the SD with the lowest extract content significantly reduced the serum activities of ALT, AST and ALP as well as the level of total bilirubin, but increased the activity of alkaline phosphatase compared to the untreated group. In summary, this SD improved biomarker rates and had a significantly better therapeutic effect than the commercial extract. Furthermore, the results of histopathological studies confirmed the SD potential to reverse the chemically induced liver toxicity.

Another study with silymarin developed by Hwang et al. [72] aimed to evaluate a new system of SDs prepared using PVP K30 and Tween 80 by spray drying. Adhesion of the former ingredient to the outer surface of the drug led to a 650-fold increase in silymarin solubility. The hepatoprotective activity was determined on male rats and the tested substances were administered before and after exposure to CCl_4. The level of AST, which had been increased by exposition to CCl_4, was significantly reduced by the administration of silymarin-containing SD, while both the raw powder and commercial product did not show any appreciable hepatoprotective effect. These results agree with those of the histopathological study.

In summary, with regard to gastro and hepatoprotection, an evident increase in the biological activity of drugs was observed when they were incorporated into polymers, which reduced the level of damage more than the isolated drugs, thus demonstrating that SD have advantages that go beyond the increase in aqueous solubility and improvement of physicochemical properties of pharmacologically active molecules.

4.7. Antidiabetic Activity of Solid Dispersions

Diabetes mellitus is an endocrine-metabolic disorder characterized by chronic hyperglycemia, which can cause microvascular damage (retinopathy, nephropathy and neuropathy) and macrovascular damage (ischemic heart disease, stroke and peripheral vascular disease) [5,73].

Repaglinide, a hypoglycemic drug indicated for the treatment of patients with type 2 diabetes mellitus, was used by Zawar and Bari [74] to develop dispersions with Poloxamer 188 in the 1:3 (*w/w*) drug:polymer proportion by the conventional fusion or microwave method. In vivo experiments conducted on Wistar rats showed that the SD prepared by the microwave method reduced the serum glucose level determined by the glucose-oxidase method to 88 ± 0.73 mg/dL within 2 h, while the pure drug and the control led to glucose levels of 94 ± 2.63 and 120 ± 2.1 mg/dL, respectively. This study has shown that SDs can be an effective way to increase the solubility and bioavailability of repaglinide, leading to an enhancement of its antihyperglycemic activity and that the microwave method can be a green and effective alternative to obtain them.

Among the substances used in the treatment of non-insulin dependent diabetes mellitus is glimepiride, one of the sulfonylurea antidiabetics responsible for stimulating insulin secretion. These drugs have low solubility and high permeability and their use in the treatment of diabetes has been impaired by side effects [75]. In this sense, glimepiride-SDs were prepared by the solvent evaporation method using mixtures of Soluplus1 and PEG 4000 as polymers in different proportions and their antidiabetic activity in vivo was determined by changes in blood glucose concentrations in albino rats [5]. Glycemia was reduced by as much as 90.2–91.0% after 3 h in rats treated with the most effective SDs compared to the pure drug (38%) and commercial formulation (Amaryl1) (85%) and glucose in blood remained within the normoglycemic level for 9–12 h.

Pioglitazone is an antidiabetic drug that acts reducing glucose production in the liver and increasing insulin-dependent glucose availability. However, limitations related to its low water solubility and dissolution have led Pokharkar et al. [76] to prepare solid dispersions using different polymers such as HPMC E3, PVP K17, PVP K30, Soluplus and surfactants including Tween 80, Poloxamer 407 and Labrasol. The PVP K17-SD showed a better dissolution profile in vitro compared to pure pioglitazone, therefore it was chosen to assess hypoglycemic activity in vivo. This dispersion exerted an anti-hyperglycemic effect that resulted within 10 h in a serum glucose level (270.50 ± 4.47 mg/dL) approximately 20% lower than that observed using pure pioglitazone or a commercial drug, corresponding to a 33.85% reduction in concentration versus about 17%. The results of this study showed that SD was effective in improving the physicochemical properties of pioglitazone and may contribute to medical treatments as a hypoglycemic drug.

4.8. Antinociceptive Activity of Solid Dispersions

Neuropathic pain is a state of chronic pain caused by lesions or diseases of the somatosensory nervous system [77], for the treatment of which different drugs are used, which; however, have shown inconsistent results. To develop alternative therapies, among the pharmacological activities attributed to hecogenin acetate, the analgesic one has proven interesting in the treatment of chronic pain [78]. In this context Moreira et al. [77] developed solid dispersions with hydrophilic polymers, namely PVP K30, PEG 6000, HPMC, Soluplus® and PVP/VA 64, to increase the solubility of this drug and improve its dissolution profile. In this study, the anti-hyperalgesic activity of SDs was evaluated in vivo in mice selected as a neuropathic pain model. To assess the hyperalgesia of formulations, after induction of damage in mice by crushing the sciatic nerve, hecogenin acetate or its SD prepared with HPMC were administered and the parameters of mechanical and thermal hyperalgesia as well as grip strength were evaluated. The thermal hyperalgesia test on groups treated with SD showed that the incorporation of the drug in the HPMC matrix significantly improved its antinociceptive activity, while mice treated with hecogenin acetate and SD-HPMC only showed a significant effect on nociceptive behavior after mechanical and thermal stimulus testing.

Table 2. In vivo studies on solid dispersions.

Carrier Type	Substance	Animal	Dose	Activity	Improved Characteristics	References
PVP K30	IIIM-290	Swiss male mice (18–23 g)	25, 50 and 75 mg/kg	Anticancer	SD was able to reduce the IIIM-290 dose by at least 1.5-fold thanks to its efficiency in Ehrlich solid tumor model.	[24]
Soluplus®	9-Nitrocamptothecin	Male Sprague-Dawley rats (20 ± 2 g)	4 mg/kg	Anticancer	SD showed higher tumor growth inhibitory rate than the pure compound due to improved oral bioavailability	[49]
(+)-Xylitol	(−)-Oleocanthal	Athymic nude mice	10 mg/kg	Anticancer	Treatment with SD showed prevention, lower growth rate and less recurrence of tumor.	[50]
PVP K30	Zinc(II)-curcumin complex	B-NDG, BALB/c mice	100 mg/kg	Anticancer	SD reduced tumor size and weight in animals.	[17]
PVP K30	*Selaginella doederleinii* Hieron	BALB/c mice	200 mg/kg	Anticancer	SD reduced tumor size as well as the level of tumor angiogenesis.	[53]
Low substituted HPC	Benznidazole	Female NMRI mice (25 ± 2 g)	25 mg/kg/day	Antichagasic	The best SD showed a 96.65% trypanocidal activity, expressed as percentage reduction in the area under the parasitic curve.	[28]
PVP K30	Curcumin (CM)	Male Cobb-Vantress broiler chickens	1 g/kg of feed	Antimicrobial	The synergistic effect of 0.05% CM/PVP SD with 0.05% boric acid reduced colonization of *Salmonella enteritidis* in crop and ceca-cecal tonsils.	[40]
PVP K30	Taurine-zinc complex	Female Sprague-Dawley rats (240–260 g)	100 and 200 mg/kg/day	Antioxidant Gastroprotective	SDs protected rat gastric mucosa from ethanol-induced injury and increased SOD activity and glutathione level.	[58]
Kollidon (VA64)	Triacetylated andrographolide (TA)	Male Kunming mice	50, 100 and 200 mg/kg/day	Anti-inflammatory	TA-SD prepared with VA64 significantly improved the drug activity against ulcerative colitis.	[59]
PVP K30, Poloxamer 188	Curcumin	Female CD-1 mice	100 mg/kg oral doses	Anti-inflammatory	CM-SD prepared with PVP decreased matrix metallo-peptidase 9 expression and levels of IL-1β and IL-6 cytokines.	[60]
Gelucire®50/13-Aerosil®	Curcumin	Rat	10 to 100 mg/kg	Anti-inflammatory	A CM-SD dose of 100 mg/kg was more effective than 5 mg/kg indomethacin in reducing edema.	[61]
HPMC, lecithin and isomalt	Curcumin	Male Sprague-Dawley rats	5 mg/kg	Anti-inflammatory	A CM-SD dose of 5 mg/kg had greater anti-inflammatory activity than 50 mg/kg curcumin alone.	[62]
Crospovidone	Aceclofenac	Male Sprague-Dawley rats	1 g/cm^2 (topical)	Anti-inflammatory	The enhanced drug permeation increased the intensity of the anti-inflammatory response.	[65]
PEG 8000	Ibuprofen	Wistar rats	20 mg/kg	Anti-inflammatory	All SDs showed better anti-inflammatory activity than the pure drug, allowing up to 90% edema inhibition after 6 h.	[66]
Urea and mannitol	Flurbiprofen	Rat	11.69 mg/kg	Anti-inflammatory	SD showed better inhibition of rat paw edema up to 16 h.	[67]

Table 2. *Cont.*

Carrier Type	Substance	Animal	Dose	Activity	Improved Characteristics	References
Paracetamol	Meloxicam	Rat	-	Anti-inflammatory	SDs reduced by more than 50% the volume of carrageenan-induced tail edema compared to the physical mixture.	[68]
PVP K30	Chelerythrine (CHE)	Mice	10 mg/kg	Anti-inflammatory	SD enhanced CHE anti-inflammatory effect by reducing the levels of TNF-α, IL-6 and NO in mice serum.	[69]
HPMC	Curcumin	Male BABL/c mice	200 and 400 mg/kg	Hepatoprotective	The best SD increased the hepatoprotective efficacy of CM.	[48]
HPC	Nobiletin	Male Sprague-Dawley rats(220 g)	2 mg of drug/kg	Hepatoprotective	SD was more effective than the crystalline drug in rats with acute liver injury.	[70]
PVP K30	Silymarin	Adult male albino rats (150–200 g)	25 mg/kg	Hepatoprotective	The best SD improved biomarker rates and had a significantly better hepatoprotective effect than the commercial extract.	[71]
PVP K30	Silymarin	Male Sprague-Dawley rats (190–210 g)	50 mg/kg	Hepatoprotective	SD improved drug solubility and hepatoprotective activity, reducing the AST levels.	[72]
Poloxamer 188	Repaglinide	Wistar rats (150–250 g)	(1 mg of drug)	Antihyperglycemic	SD obtained by the microwave method improved the drug anti-hyperglycemic activity.	[74]
Soluplus1 and PEG 4000	Glimepiride	Albino Wistar rats (200–250 g)	0.0285 mg of drug/kg	Anti-diabetic	SD reduced the glucose level in rats more than the pure drug and a commercial product.	[5]
PVP K17	Pioglitazone	Male Swiss albino mice (25–30 g)	30 mg/kg SD	Antihyperglycemic	SD reduced the mean glucose level in mice more than the pure drug and a commercial product.	[76]
HPMC	Hecogenin acetate	Male Swiss mice (28–35 g)	40 mg/kg	Antinociceptive	Both the drug alone and its SD with HPMC-reduced mechanical and thermal hyperalgesia induced by crushing of the sciatic nerve in mice.	[77]

PVP, Polyvinylpyrrolidone; SD, Solid dispersion; HPC, Hydroxypropyl cellulose; CM, Curcumin; CM/PVP SD, curcumin/polyvinylpyrrolidone solid dispersion; SOD, Superoxide dismutase; PEG, Polyethylene glycol; TNF-α, tumor necrosis factor alpha; IL, interleukin; NO, Nitric oxide; HPMC, Hydroxypropyl methylcellulose; AST, Aspartate aminotransferase.

5. Critical Analysis

In this review, the main studies on the biological activities of solid dispersions (SDs) have been examined to make a comparison of the results of in vitro and in vivo tests carried out in the last 10 years. In vitro studies have demonstrated the ability of SDs incorporating various types of active compounds to enhance their antitumor, antiparasitic, antimicrobial, antioxidant, anti-inflammatory, or cytoprotective activities, while additional activities, such as the gastroprotective, hepatoprotective, antidiabetic, or antinociceptive ones, have been highlighted by in vivo studies.

Most of the articles published and cited in this review, for a total of 33, refer to in vitro studies particularly addressed to the antitumor activity of active compounds (cell viability or cytotoxicity), while the in vivo ones were relatively fewer in number, equal to 26 and mostly devoted to their anti-inflammatory activity.

Although SDs have already been studied and cited in the literature, the number of studies published on them with this focus is still relatively small, considering the great potential of these formulations in pharmaceutical technology and the possibility of preparing both amorphous and crystalline ones for the most diverse applications. In this review, different processes, carriers and drugs to prepare SDs were also discussed, a variety that can be of great value in the field of pharmaceutical technology to develop new SDs for the release of a large number of different drugs.

The results of biological activity studies have shown that SDs, as a tool of drug release, do not constitute a limiting factor for the execution of in vitro and in vivo tests; indeed, they stand out as a promising system in which the active principle and the vector interact allowing, in most cases, an increase in the pharmacological potential of the former as a consequence of changes in the physicochemical properties of the constituents. Similarly, SDs can enhance the biological properties of the incorporated bioactive compounds, making them more effective against a wide variety of pathogens such as bacteria and protozoa.

As a final remark, SDs could play a role in various biological mechanisms, thus representing a safe and effective alternative for the development and improvement of drugs targeting a wide range of pharmacological treatments.

6. Conclusions

Solid dispersions are a technological strategy to improve the pharmacological potential of natural or synthetic bioactive molecules, thanks to an increase in their solubility and bioavailability, thus leading to possible enhancement of their biological activities. In this context, this review sought to summarize and critically examine studies conducted with this approach both in vitro and in vivo, which have provided evidence of a significant therapeutic potential that solid dispersions could offer in the context of the most diverse biological activities.

As poorly water-soluble active ingredients incorporated in solid dispersions exhibited improved physicochemical and pharmacological properties, these systems can be targeted in the pharmaceutical industry as possible therapeutic alternatives for certain diseases, with greater selectivity, safety and efficacy than the existing treatments.

Author Contributions: All authors collaborated in data collection, analysis and writing of the review. All authors have read and agreed to the published version of the manuscript.

Funding: This study was financed in part by the Coordenação de Aperfeiçoamento de Pessoal de Nível Superior -Brasil (CAPES) -Finance Code 001.

Acknowledgments: The authors thank CNPq and CAPES for their support and Federal University of Rio Grande do Norte.

Conflicts of Interest: The authors declare no conflict of interest.

Abbreviations

AChE, Acetylcholinesterase; ALT, Alanine aminotransferase; AST, Aspartate aminotransferase; BChE, Butyrylcholinesterase; BNZ, Benznidazole; CC50, 50% Cytotoxic concentration; CFU, Colony forming units; CHE, Chelerythrine; CM, Curcumin; DPPH, 2,2-diphenyl-1-picryl-hydrazyl-hydrate; GI50, Concentration causing 50% cell growth inhibition; GST, Glutathione S-transferase; HB, Berberine hydrochloride; HPMC, Hydroxypropyl methylcellulose; HPMCAS, Hydroxypropyl methylcellulose acetate succinate; IC50, Half inhibitory concentration; LPS, Lipopolysaccharide; HPC, Hydroxypropyl cellulose; MAO, Monoamine oxidase; MIC, Minimum inhibitory concentration; MTT, 3-(4,5-dimethylthiazol-2-yl)-2,5-diphenyltetrazolium bromide; NaDC, Sodium deoxycholate; NO, Nitric oxide; OHPP, Octenylsuccinate hydroxypropyl phytoglycogen; P407, Poloxamer 407; PEG, Polyethylene glycol; PPG, Polypropylene glycol; PTX, Paclitaxel; PVP, Polyvinylpyrrolidone; PVP/VA, Polyvinylpyrrolidone/vinyl acetate; PZQ, Praziquantel; ROS, Reactive oxygen species; SD, Solid dispersion; SOD, Superoxide dismutase; TA, Triacetylated andrographolide; TBESD, Total biflavonoids extract of *Selaginella doederleinii*; t-BHP, *tert*-Butylhydroperoxide; TPGS, D-α-toco-pheryl polyethylene glycol-1000-succinate.

References

1. Kambayashi, A.; Kiyota, T.; Fujiwara, M.; Dressman, J.B. PBPK modeling coupled with biorelevant dissolution to forecast the oral performance of amorphous solid dispersion formulations. *Eur. J. Pharm. Sci.* **2019**, *135*, 83–90. [CrossRef] [PubMed]
2. Jermain, S.V.; Brough, C.; Williams, R.O. Amorphous solid dispersions and nanocrystal technologies for poorly water-soluble drug delivery—An update. *Int. J. Pharm.* **2018**, *535*, 379–392. [CrossRef] [PubMed]
3. Loftsson, T.; Brewster, M.E. Pharmaceutical applications of cyclodextrins: Basic science and product development. *J. Pharm. Pharmacol.* **2010**, *62*, 1607–1621. [CrossRef] [PubMed]
4. Worku, Z.A.; Aarts, J.; Singh, A.; Mooter, G.V.D. Drug–Polymer Miscibility across a Spray Dryer: A Case Study of Naproxen and Miconazole Solid Dispersions. *Mol. Pharm.* **2014**, *11*, 1094–1101. [CrossRef]
5. Reginald-Opara, J.N.; Attama, A.; Ofokansi, K.; Umeyor, C.; Kenechukwu, F. Molecular interaction between glimepiride and Soluplus®-PEG 4000 hybrid based solid dispersions: Characterisation and anti-diabetic studies. *Int. J. Pharm.* **2015**, *496*, 741–750. [CrossRef]
6. Danda, L.J.D.A.; Batista, L.D.M.; Melo, V.C.S.; Sobrinho, J.L.S.; Soares, M.F.D.L.R. Combining amorphous solid dispersions for improved kinetic solubility of posaconazole simultaneously released from soluble PVP/VA64 and an insoluble ammonio methacrylate copolymer. *Eur. J. Pharm. Sci.* **2019**, *133*, 79–85. [CrossRef]
7. Júnior, W.F.D.S.; Pinheiro, J.G.D.O.; De Menezes, D.L.B.; Silva, N.E.D.S.E.; De Almeida, P.D.O.; Lima, E.S.; Veiga-Junior, V.F.; De Azevedo, E.P.; Lima, E.S. Development, Physicochemical Characterization and In Vitro Anti-Inflammatory Activity of Solid Dispersions of α,β Amyrin Isolated from Protium Oilresin. *Molecules* **2017**, *22*, 1512. [CrossRef]
8. Vasconcelos, T.; Sarmento, B.; Costa, P.C. Solid dispersions as strategy to improve oral bioavailability of poor water soluble drugs. *Drug Discov. Today* **2007**, *12*, 1068–1075. [CrossRef]
9. Leuner, C. Improving drug solubility for oral delivery using solid dispersions. *Eur. J. Pharm. Biopharm.* **2000**, *50*, 47–60. [CrossRef]
10. Zhang, Z.; Chen, Y.; Deng, J.; Jia, X.-B.; Zhou, J.; Lv, H. Solid dispersion of berberine–phospholipid complex/TPGS 1000/SiO2: Preparation, characterization and in vivo studies. *Int. J. Pharm.* **2014**, *465*, 306–316. [CrossRef]
11. Xie, Y.; Yao, Y. Octenylsuccinate hydroxypropyl phytoglycogen enhances the solubility and in-vitro antitumor efficacy of niclosamide. *Int. J. Pharm.* **2018**, *535*, 157–163. [CrossRef] [PubMed]
12. Steed, H.; Sawyer, M.B. Pharmacology, pharmacokinetics and pharmacogenomics of paclitaxel. *Pharmacogenomics* **2007**, *8*, 803–815. [CrossRef] [PubMed]
13. Xie, Y.; Yao, Y. Incorporation with Dendrimer-Like Biopolymer Leads to Improved Soluble Amount and In Vitro Anticancer Efficacy of Paclitaxel. *J. Pharm. Sci.* **2019**, *108*, 1984–1990. [CrossRef] [PubMed]
14. Choi, J.-S.; Cho, N.H.; Kim, D.-H.; Park, J.-S. Comparison of paclitaxel solid dispersion and polymeric micelles for improved oral bioavailability and in vitro anti-cancer effects. *Mater. Sci. Eng. C* **2019**, *100*, 247–259. [CrossRef] [PubMed]

15. Lee, S.H.; Lee, Y.-S.; Song, J.G.; Han, H.-K. Improved In vivo Effect of Chrysin as an Absorption Enhancer Via the Preparation of Ternary Solid Dispersion with Brij®L4 and Aminoclay. *Curr. Drug Deliv.* **2018**, *16*, 86–92. [CrossRef] [PubMed]
16. De Sá, I.S.; Peron, A.P.; Leimann, F.V.; Bressan, G.N.; Krum, B.N.; Fachinetto, R.; Pinela, J.; Calhelha, R.C.; Barreiro, M.F.; Ferreira, I.C.; et al. In vitro and in vivo evaluation of enzymatic and antioxidant activity, cytotoxicity and genotoxicity of curcumin-loaded solid dispersions. *Food Chem. Toxicol.* **2019**, *125*, 29–37. [CrossRef]
17. Wu, R.; Mei, X.; Ye, Y.; Xue, T.; Wang, J.; Sun, W.; Lin, C.; Xue, R.; Zhang, J.; Xu, D. Zn(II)-curcumin solid dispersion impairs hepatocellular carcinoma growth and enhances chemotherapy by modulating gut microbiota-mediated zinc homeostasis. *Pharmacol. Res.* **2019**, *150*, 104454. [CrossRef]
18. Xiong, X.; Zhang, M.; Hou, Q.; Tang, P.; Suo, Z.; Zhu, Y.; Li, H. Solid dispersions of telaprevir with improved solubility prepared by co-milling: Formulation, physicochemical characterization and cytotoxicity evaluation. *Mater. Sci. Eng. C* **2019**, *105*, 110012. [CrossRef]
19. Lewinska, A.; Adamczyk, J.; Pajak, J.; Stoklosa, S.; Kubis, B.; Pastuszek, P.; Slota, E.; Wnuk, M. Curcumin-mediated decrease in the expression of nucleolar organizer regions in cervical cancer (HeLa) cells. *Mutat. Res. Toxicol. Environ. Mutagen.* **2014**, *771*, 43–52. [CrossRef]
20. Jiang, Y.; Piao, J.; Cho, H.-J.; Kang, W.-S.; Kim, H. Improvement in antiproliferative activity of Angelica gigas Nakai by solid dispersion formation via hot-melt extrusion and induction of cell cycle arrest and apoptosis in HeLa cells. *Biosci. Biotechnol. Biochem.* **2015**, *79*, 1635–1643. [CrossRef]
21. Guo, S.; Wang, G.; Wu, T.; Bai, F.; Xu, J.; Zhang, X. Solid dispersion of berberine hydrochloride and Eudragit® S100: Formulation, physicochemical characterization and cytotoxicity evaluation. *J. Drug Deliv. Sci. Technol.* **2017**, *40*, 21–27. [CrossRef]
22. Wang, L.; Li, H.; Wang, S.; Liu, R.; Wu, Z.; Wang, C.; Wang, Y.; Chen, M. Enhancing the Antitumor Activity of Berberine Hydrochloride by Solid Lipid Nanoparticle Encapsulation. *AAPS PharmSciTech* **2014**, *15*, 834–844. [CrossRef] [PubMed]
23. Patil, J.B.; Kim, J.; Jayaprakasha, G.K. Berberine induces apoptosis in breast cancer cells (MCF-7) through mitochondrial-dependent pathway. *Eur. J. Pharmacol.* **2010**, *645*, 70–78. [CrossRef] [PubMed]
24. Kumar, V.; Mintoo, M.J.; Mondhe, D.M.; Bharate, S.B.; Vishwakarma, R.A.; Bharate, S.S. Binary and ternary solid dispersions of an anticancer preclinical lead, IIIM-290: In vitro and in vivo studies. *Int. J. Pharm.* **2019**, *570*, 118683. [CrossRef]
25. Simonazzi, A.; Davies, C.; Cid, A.G.; Gonzo, E.; Parada, L.; Bermúdez, J.M. Preparation and Characterization of Poloxamer 407 Solid Dispersions as an Alternative Strategy to Improve Benznidazole Bioperformance. *J. Pharm. Sci.* **2018**, *107*, 2829–2836. [CrossRef] [PubMed]
26. Lima, Á.A.; Soares-Sobrinho, J.L.; Silva, J.L.; Corrêa-Júnior, R.A.; Lyra, M.A.; Santos, F.L.; Oliveira, B.G.; Hernandes, M.Z.; Rolim, L.A.; Rolim-Neto, P.J. The Use of Solid Dispersion Systems in Hydrophilic Carriers to Increase Benznidazole Solubility. *J. Pharm. Sci.* **2011**, *100*, 2443–2451. [CrossRef]
27. Lima, Á.A.N.; Sobrinho, J.L.S.; De Lyra, M.A.M.; Dos Santos, F.L.A.; Figueirêdo, C.B.M.; Neto, P.J.R. Evaluation of in vitro dissolution of benznidazole and binary mixtures: Solid dispersions with hydroxypropylmethylcellulose and β-cyclodextrin inclusion complexes. *Int. J. Pharm. Pharm. Sci.* **2015**, *7*, 371–375.
28. Fonseca-Berzal, C.; Palmeiro-Roldán, R.; Escario, J.A.; Torrado, S.; Arán, V.J.; Torrado-Santiago, S.; Gómez-Barrio, A. Novel solid dispersions of benznidazole: Preparation, dissolution profile and biological evaluation as alternative antichagasic drug delivery system. *Exp. Parasitol.* **2015**, *149*, 84–91. [CrossRef]
29. Eloy, J.O.; Saraiva, J.; De Albuquerque, S.; Marchetti, J.M. Solid Dispersion of Ursolic Acid in Gelucire 50/13: A Strategy to Enhance Drug Release and Trypanocidal Activity. *AAPS PharmSciTech* **2012**, *13*, 1436–1445. [CrossRef]
30. Perissutti, B.; Passerini, N.; Trastullo, R.; Keiser, J.; Zanolla, D.; Zingone, G.; Voinovich, D.; Albertini, B. An explorative analysis of process and formulation variables affecting comilling in a vibrational mill: The case of praziquantel. *Int. J. Pharm.* **2017**, *533*, 402–412. [CrossRef]
31. Albertini, B.; Perissutti, B.; Bertoni, S.; Zanolla, D.; Franceschinis, E.; Voinovich, D.; Lombardo, F.; Keiser, J.; Passerini, N. Combining Mechanochemistry and Spray Congealing for New Praziquantel Pediatric Formulations in Schistosomiasis Treatment. *Int. J. Mol. Sci.* **2019**, *20*, 1233. [CrossRef] [PubMed]
32. White, J.; Rathod, P.K. Indispensable malaria genes. *Science* **2018**, *360*, 490–491. [CrossRef] [PubMed]

33. Abreu, T.P.; Silva, L.D.S.; Takiya, C.M.; Souza, M.C.; Henriques, M.D.G.M.O.; Pinheiro, A.A.S.; Caruso-Neves, C. Mice Rescued from Severe Malaria Are Protected against Renal Injury during a Second Kidney Insult. *PLoS ONE* **2014**, *9*, e93634. [CrossRef] [PubMed]
34. Fule, R.A.; Meer, T.A.; Sav, A.R.; Amin, P. Artemether-Soluplus Hot-Melt Extrudate Solid Dispersion Systems for Solubility and Dissolution Rate Enhancement with Amorphous State Characteristics. *J. Pharm.* **2013**, *2013*, 1–15. [CrossRef]
35. Fule, R.; Meer, T.; Sav, A.; Amin, P. Solubility and dissolution rate enhancement of lumefantrine using hot melt extrusion technology with physicochemical characterisation. *J. Pharm. Investig.* **2013**, *43*, 305–321. [CrossRef]
36. Crucitti, V.C.; Migneco, L.M.; Piozzi, A.; Taresco, V.; Garnett, M.; Argent, R.H.; Francolini, I. Intermolecular interaction and solid state characterization of abietic acid/chitosan solid dispersions possessing antimicrobial and antioxidant properties. *Eur. J. Pharm. Biopharm.* **2018**, *125*, 114–123. [CrossRef]
37. Al-Obaidi, H.; Kowalczyk, R.M.; Kalgudi, R.; Zariwala, M.G. Griseofulvin solvate solid dispersions with synergistic effect against fungal biofilms. *Colloids Surfaces B Biointerfaces* **2019**, *184*, 110540. [CrossRef]
38. Gorle, A.P.; Gattani, S.G. Development and evaluation of ocular drug delivery system. *Pharm. Dev. Technol.* **2010**, *15*, 46–52. [CrossRef]
39. Kanoujia, J.; Kushwaha, P.S.; Saraf, S.A. Evaluation of gatifloxacin pluronic micelles and development of its formulation for ocular delivery. *Drug Deliv. Transl. Res.* **2014**, *4*, 334–343. [CrossRef]
40. Hernandez-Patlan, D.; Solis-Cruz, B.; Pontin, K.P.; Latorre, J.D.; Baxter, M.F.A.; Hernandez-Velasco, X.; Merino-Guzman, R.; Méndez-Albores, A.; Hargis, B.M.; Lopez-Arellano, R.; et al. Evaluation of a Solid Dispersion of Curcumin With Polyvinylpyrrolidone and Boric Acid Against Salmonella Enteritidis Infection and Intestinal Permeability in Broiler Chickens: A Pilot Study. *Front. Microbiol.* **2018**, *9*, 1289. [CrossRef]
41. Hegge, A.B.; Vukićević, M.; Bruzell, E.; Kristensen, S.; Tønnesen, H. Solid dispersions for preparation of phototoxic supersaturated solutions for antimicrobial photodynamic therapy (aPDT). *Eur. J. Pharm. Biopharm.* **2013**, *83*, 95–105. [CrossRef] [PubMed]
42. Alves, T.; Chaud, M.; Grotto, D.; Jozala, A.F.; Pandit, R.; Rai, M.; Dos Santos, C.A. Association of Silver Nanoparticles and Curcumin Solid Dispersion: Antimicrobial and Antioxidant Properties. *AAPS PharmSciTech* **2017**, *19*, 225–231. [CrossRef] [PubMed]
43. Gunes, H.; Gulen, D.; Mutlu, R.; Gumus, A.; Tas, T.; Topkaya, A.E. Antibacterial effects of curcumin. *Toxicol. Ind. Health* **2013**, *32*, 246–250. [CrossRef] [PubMed]
44. Costa, A.R.D.M.; Marquiafável, F.S.; Vaz, M.M.D.O.L.L.; Rocha, B.A.; Bueno, P.C.P.; Amaral, P.L.M.; Barud, H.D.S.; Berreta-Silva, A.A. Quercetin-PVP K25 solid dispersions. *J. Therm. Anal. Calorim.* **2010**, *104*, 273–278. [CrossRef]
45. Ge, Y.; Zhao, X.; Wang, S.; Zu, Y.; Zhao, T.; Sang, M.; Sui, X.; Wang, K. 10 (CoQ10) Prepared by Heat Melt and High-pressure Homogenization Method. *Curr. Nanosci.* **2014**, *10*, 292–296. [CrossRef]
46. Fitriani, L.; Rismawati, E.; Umar, S.; Zaini, E. Solid dispersion of usnic acid-PVP K30 and evaluation of antioxidant activity. *Rasayan J. Chem.* **2018**, *11*, 1643–1648. [CrossRef]
47. Alshehri, S.; Imam, S.S.; Altamimi, M.A.; Hussain, A.; Shakeel, F.; Elzayat, E.M.; Mohsin, K.; Ibrahim, M.; Alanazi, F. Enhanced Dissolution of Luteolin by Solid Dispersion Prepared by Different Methods: Physicochemical Characterization and Antioxidant Activity. *ACS Omega* **2020**, *5*, 6461–6471. [CrossRef]
48. Shin, M.-S.; Yu, J.S.; Lee, J.; Ji, Y.S.; Joung, H.J.; Han, Y.-M.; Yoo, H.H.; Kang, K.; Shin, Y. A Hydroxypropyl Methylcellulose-Based Solid Dispersion of Curcumin with Enhanced Bioavailability and its Hepatoprotective Activity. *Biomolecules* **2019**, *9*, 281. [CrossRef]
49. Lian, X.; Dong, J.; Zhang, J.; Teng, Y.; Lin, Q.; Fu, Y.; Gong, T. Soluplus® based 9-nitrocamptothecin solid dispersion for peroral administration: Preparation, characterization, in vitro and in vivo evaluation. *Int. J. Pharm.* **2014**, *477*, 399–407. [CrossRef]
50. Qusa, M.H.; Siddique, A.B.; Nazzal, S.; El Sayed, K. Novel olive oil phenolic (−)-oleocanthal (+)-xylitol-based solid dispersion formulations with potent oral anti-breast cancer activities. *Int. J. Pharm.* **2019**, *569*, 118596. [CrossRef]
51. Yang, S.; Shi, P.; Huang, X.; Zhao, M.; Li, S.; Wu, Y.; Lin, X.-H.; Yao, H. Pharmacokinetics, Tissue Distribution and Protein Binding Studies of Chrysocauloflavone I in Rats. *Planta Med.* **2015**, *82*, 217–223. [CrossRef] [PubMed]

52. Chen, B.; Wang, X.; Zou, Y.; Chen, W.; Wang, G.; Yao, W.; Shi, P.; Li, S.; Lin, S.; Lin, X.-H.; et al. Simultaneous quantification of five biflavonoids in rat plasma by LC-ESI–MS/MS and its application to a comparatively pharmacokinetic study of Selaginella doederleinii Hieron extract in rats. *J. Pharm. Biomed. Anal.* **2018**, *149*, 80–88. [CrossRef] [PubMed]
53. Chen, B.; Wang, X.; Zhang, Y.; Huang, K.; Liu, H.; Xu, D.; Li, S.; Liu, Q.; Huang, J.; Yao, H.; et al. Improved solubility, dissolution rate and oral bioavailability of main biflavonoids from Selaginella doederleinii extract by amorphous solid dispersion. *Drug Deliv.* **2020**, *27*, 309–322. [CrossRef] [PubMed]
54. Zhang, W.-H.; Wu, X.-J.; Niu, J.-X.; Yan, H.; Wang, X.-Z.; Yin, X.-D.; Pang, Y. Serum zinc status and Helicobacter pylori infection in gastric disease patients. *Asian Pac. J. Cancer Prev.* **2012**, *13*, 5043–5046. [CrossRef] [PubMed]
55. Zhong, W.; Zhao, Y.; Sun, X.; Song, Z.; McClain, C.J.; Zhou, Z. Dietary Zinc Deficiency Exaggerates Ethanol-Induced Liver Injury in Mice: Involvement of Intrahepatic and Extrahepatic Factors. *PLoS ONE* **2013**, *8*, e76522. [CrossRef] [PubMed]
56. Ma, N.; Sasaki, T.; Sakata-Haga, H.; Ohta, K.-I.; Gao, M.; Kawanishi, S.; Fukui, Y. Protective effect of taurine against nitrosative stress in the stomach of rat with water immersion restraint stress. In *Advances in Experimental Medicine and Biology*; Springer Science and Business Media LLC: Berlin/Heidelberg, Germany, 2009; Volume 643, pp. 273–283.
57. Marcinkiewicz, J.; Kontny, E. Taurine and inflammatory diseases. *Amino Acids* **2012**, *46*, 7–20. [CrossRef]
58. Yu, C.; Mei, X.-T.; Zheng, Y.-P.; Xu, D. Gastroprotective effect of taurine zinc solid dispersions against absolute ethanol-induced gastric lesions is mediated by enhancement of antioxidant activity and endogenous PGE2 production and attenuation of NO production. *Eur. J. Pharmacol.* **2014**, *740*, 329–336. [CrossRef]
59. Gou, J.; Fei, S.; Xue, B.; Zhang, J.; Zhang, Y.; Wang, X.; Zhang, Y.; Yin, T.; He, H.; Tang, X. Triacetylated andrographolide solid dispersions: Preparation, stability study and in vivo anti-inflammation in mice ulcerative colitis model. *J. Drug Deliv. Sci. Technol.* **2019**, *51*, 91–100. [CrossRef]
60. He, Y.; Liu, H.; Bian, W.; Liu, Y.; Liu, X.; Ma, S.; Zheng, X.; Du, Z.; Zhang, K.; Ouyang, D. Molecular Interactions for the Curcumin-Polymer Complex with Enhanced Anti-Inflammatory Effects. *Pharmaceutics* **2019**, *11*, 442. [CrossRef]
61. Teixeira, C.C.C.; Mendonça, L.M.; Bergamaschi, M.M.; Queiroz, R.H.C.; Souza, G.E.P.; Antunes, L.M.G.; Freitas, L. Microparticles Containing Curcumin Solid Dispersion: Stability, Bioavailability and Anti-Inflammatory Activity. *AAPS PharmSciTech* **2015**, *17*, 252–261. [CrossRef]
62. Chuah, A.M.; Jacob, B.; Jie, Z.; Ramesh, S.; Mandal, S.; Puthan, J.K.; Deshpande, P.; Vaidyanathan, V.V.; Gelling, R.W.; Patel, G.; et al. Enhanced bioavailability and bioefficacy of an amorphous solid dispersion of curcumin. *Food Chem.* **2014**, *156*, 227–233. [CrossRef] [PubMed]
63. Chakraborty, S.; Khandai, M.; Sharma, A.; Khanam, N.; Patra, C.; Dinda, S.; Sen, K. Preparation, in vitro and in vivo evaluation of algino-pectinate bioadhesive microspheres: An investigation of the effects of polymers using multiple comparison analysis. *Acta Pharm.* **2010**, *60*, 255–266. [CrossRef] [PubMed]
64. Fujii, M.; Okada, H.; Shibata, Y.; Teramachi, H.; Kondoh, M.; Watanabe, Y. Preparation, characterization and tableting of a solid dispersion of indomethacin with crospovidone. *Int. J. Pharm.* **2005**, *293*, 145–153. [CrossRef] [PubMed]
65. Jana, S.; Ali, S.A.; Nayak, A.K.; Sen, K.K.; Basu, S.K. Development of topical gel containing aceclofenac-crospovidone solid dispersion by "Quality by Design (QbD)" approach. *Chem. Eng. Res. Des.* **2014**, *92*, 2095–2105. [CrossRef]
66. Kenechukwu, F.C.; Ofokansi, K.C.; Ezugwu, R.O.; Attama, A.A. Improved dissolution and anti-inflammatory activity of ibuprofen-polyethylene glycol 8000 solid dispersion systems. *Int. J. Pharm. Investig.* **2016**, *6*, 139–147. [CrossRef]
67. Malipeddi, V.R.; Dua, K.; Awasthi, R. Development and characterization of solid dispersion-microsphere controlled release system for poorly water-soluble drug. *Drug Deliv. Transl. Res.* **2016**, *6*, 540–550. [CrossRef]
68. Al-Remawi, M.; Ali, A.M.A.; Khames, A.; Hamaidi, M. Meloxicam-Paracetamol Binary Solid Dispersion Systems with Enhanced Solubility and Dissolution Rate:Preparation, Characterization and In Vivo Evaluation. *J. Pharm. Innov.* **2017**, *12*, 206–215. [CrossRef]
69. Li, W.; Qing, S.; Zhi, W.; Yao, H.; Fu, C.; Niu, X. The pharmacokinetics and anti-inflammatory effects of chelerythrine solid dispersions in vivo. *J. Drug Deliv. Sci. Technol.* **2017**, *40*, 51–58. [CrossRef]

70. Onoue, S.; Nakamura, T.; Uchida, A.; Ogawa, K.; Yuminoki, K.; Hashimoto, N.; Hiza, A.; Tsukaguchi, Y.; Asakawa, T.; Kan, T.; et al. Physicochemical and biopharmaceutical characterization of amorphous solid dispersion of nobiletin, a citrus polymethoxylated flavone, with improved hepatoprotective effects. *Eur. J. Pharm. Sci.* **2013**, *49*, 453–460. [CrossRef]
71. Balata, G.; Shamrool, H. Spherical agglomeration versus solid dispersion as different trials to optimize dissolution and bioactivity of silymarin. *J. Drug Deliv. Sci. Technol.* **2014**, *24*, 478–485. [CrossRef]
72. Hwang, D.H.; Kim, Y.-I.; Cho, K.H.; Poudel, B.K.; Choi, J.Y.; Kim, D.-W.; Shin, Y.-J.; Bae, O.-N.; Yousaf, A.M.; Yong, C.S.; et al. A novel solid dispersion system for natural product-loaded medicine: Silymarin-loaded solid dispersion with enhanced oral bioavailability and hepatoprotective activity. *J. Microencapsul.* **2014**, *31*, 619–626. [CrossRef] [PubMed]
73. Heydari, I.; Radi, V.; Razmjou, S.; Amiri, A. Chronic complications of diabetes mellitus in newly diagnosed patients. *Int. J. Diabetes Mellit.* **2010**, *2*, 61–63. [CrossRef]
74. Zawar, L.R.; Bari, S.B. Preparation, characterization and in vivo evaluation of antihyperglycemic activity of microwave generated repaglinide solid dispersion. *Chem. Pharm. Bull.* **2012**, *60*, 482–487. [CrossRef] [PubMed]
75. Li, H.; Ma, L.; Li, X.; Cui, X.; Yang, W.; Shen, S.; Chen, M. A simple and effective method to improve bioavailability of glimepiride by utilizing hydrotropy technique. *Eur. J. Pharm. Sci.* **2015**, *77*, 154–160. [CrossRef]
76. Pokharkar, V.; Kutwal, M.; Mandpe, L. Pioglitazone Solid Dispersion System Prepared by Spray Drying Method: In Vitro and In Vivo Evaluation. *PDA J. Pharm. Sci. Technol.* **2013**, *67*, 23–34. [CrossRef]
77. Moreira, C.D.L.D.F.A.; Pinheiro, J.G.D.O.; Da Silva-Júnior, W.F.; Barbosa, E.G.; Lavra, Z.M.M.; Pereira, E.W.M.; Resende, M.M.; De Azevedo, E.P.; Quintans-Júnior, L.J.; Araújo, A.A.D.S.; et al. Amorphous solid dispersions of hecogenin acetate using different polymers for enhancement of solubility and improvement of anti-hyperalgesic effect in neuropathic pain model in mice. *Biomed. Pharmacother.* **2018**, *97*, 870–879. [CrossRef]
78. Gama, K.B.; Quintans, J.S.; Antoniolli, A.R.; Quintans-Júnior, L.J.; Santana, W.A.; Branco, A.; Soares, M.B.P.; Villarreal, C.F. Evidence for the Involvement of Descending Pain-Inhibitory Mechanisms in the Antinociceptive Effect of Hecogenin Acetate. *J. Nat. Prod.* **2013**, *76*, 559–563. [CrossRef]

MDPI
St. Alban-Anlage 66
4052 Basel
Switzerland
Tel. +41 61 683 77 34
Fax +41 61 302 89 18
www.mdpi.com

Pharmaceutics Editorial Office
E-mail: pharmaceutics@mdpi.com
www.mdpi.com/journal/pharmaceutics

www.ingramcontent.com/pod-product-compliance
Lightning Source LLC
Chambersburg PA
CBHW040819250926
51850CB00007BA/1215
9783036526393